Contents

Contents

Chapter 1
Overview

Rationale for this Book

This book gives a broad overview of technology and the many relevant technological structures that we use in today's world. Our goal is to provide students with an understanding of the roles of technology in a broader sense and introduce the science behind some of these technologies in each of the chapters. Technology has a direct relationship with all branches of science (physical, life, social, formal, or applied) because it is through the use of scientific principles, practices, and ideas that we design solutions, and those solutions come via machines, equipment, or methods created inside the broad scope of technology. This book also incorporates a number of assignments and tasks that can be used in a cafeteria-grading style course structure as well as in more traditional course/curriculum structures. Each chapter has clear objectives specified at the beginning and a variety of assessment options at the end.

Technology and its Management

Technology is pervasive in almost all areas of our lives and business, consequently the management of technology has become an increasingly important aspect in business and industry.

Technology Management

Technology management involves the application of management skills to the discovery, development, operation, and proper use of technology. Technology managers help create value for their organization by using technology and other resources to solve problems and improve efficiency and effectiveness. In short, twenty-first century technology managers help ensure that technology creates a better future for all. Technology management can have multiple areas of focus:

* Management with technological systems: Using technological systems to support management or decision-making of the organization
* Management of technological systems: Technological system management
* Technological system integration

Regardless of the three areas of emphasis mentioned earlier, technology almost always incorporates aspects of *project management*.

Technology management will be discussed in more detail in Chapter 3.

Project Management

Many technology management functions fall within the realm of projects. A project is a temporary endeavor undertaken to create a unique product, service, or result. According to Kathy Schwalbe, author of *An Introduction to Project Management*, "Many people and organizations today have a new or renewed interest in project management." Today's project management involves much more, and people in every industry and every country manage projects. New technologies have become a significant factor in many organizations, and the use of interdisciplinary and global work teams has radically changed the work environment. Throughout this book, you will see examples of project management in play.

Science, Technology, Engineering, and Math

The U.S. Department of Education has placed an emphasis on science, technology, engineering, and math, more commonly known as STEM. As a matter of fact, President Obama, back in 2010, set a priority for increasing the number of individuals who are proficient in these fields in a campaign called "Educate to Innovate." In this campaign, more than $250 million dollars was invested to help prepare over 10,000 new math and science teachers and train another 100,000 existing teachers since there has been a teacher shortage in these fields (Office of the Press Secretary, 2010). However, not only is there a limited number of teachers, there is also fewer American students pursuing these fields, even though they are vital and expected to grow. As a matter of fact, from 2010 to 2020 all occupations within STEM fields are expected to grow 14%, with math growing 16%, computer systems analysts growing 22%, systems software developers growing 32%, medical scientists growing 36%, and biomedical engineers growing a whopping 62% (U.S. Department of Education, 2015).

In 2013, the White House created a five-year strategic plan for STEM education. In this plan, STEM was made a priority in many of administration's education efforts. So, for example, "Department of Education's $4.3 billion Race to the Top competition offered states a competitive preference priority on developing comprehensive strategies to improve achievement and provide rigorous curricula in STEM subjects; partner with local STEM institutions, businesses, and museums; and broaden participation

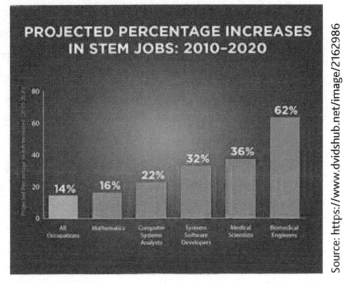

Source: https://www.dvidshub.net/image/2162986

of women and girls and other groups underrepresented in STEM fields" (Committee on STEM Education, 2013). Areas that play an increased role in STEM education include, but are not limited to, the Department of Education, the National Science Foundation, and the Smithsonian Institution. Investments are being made within each area as well as nationally overall to develop STEM skills, practices, and knowledge in both students and the general public. It is our hope that courses such as this will encourage students to consider careers these or related fields.

Cafeteria-Style Grading

This book makes use of a teaching and assessment methodology known as cafeteria-style grading. Cafeteria-style grading offers students the chance to tailor a course to their interests and learning styles by offering multiple modes of assessment for each curriculum and content area. It also permits them to explore new assessment methodologies without fear of penalty.

Cafeteria-Style Grading Defined

Cafeteria-style grading means students get to choose assignments, tasks, or tests that appeal to their own learning interests or styles and do not need to complete all the assignments to get an A grade. Rather, they complete those assignments desired in order to earn the applicable assessment points. In order for this to occur, the course must be structured to offer a significantly more points possible in the course than are required for an A grade, giving ample freedom to choose assignments. Our recommendation is to offer a minimum of twice as many points as available options.

Imagine this. You are a student starting a class on a topic you are not familiar with. You are a bit stressed because you want to maintain your grade point average. Plus, you have test anxiety, which tends to make your grades lower than you believe your actual comprehension and skill set to be. In this class, they have cafeteria-style grading. The instructor has a vast array of assignments to choose from (including tests but also hands-on activities, papers, interviews, and the like), but you do not need to do all of them. Actually, in most cases, you get to choose which ones you want to do. You earn points throughout the term to achieve the grade you want. If you don't do so well on a particular assessment, it doesn't kill your grade; instead, you try other assignments and continue to accumulate points. So, you start early and work hard. You try some things you have never done before, confident that your attempts at new things will not be grade killers. You also do some assignments that seem easy to you based on your current skill set. By the end of the term, you have a solid A grade, and you feel confident about the course materials.

The above scenario is just an example, but it is a realistic one. Some types of assignments available to students could include but are not limited to: construction of models or prototypes, interviews of community members, open book multiple-choice questions, assessment of materials, short academic papers, and recorded group discussions.

Success of Cafeteria-Style Grading

Three instructors on Utah Valley University's campus offered a total of 13 sections of a general education science class called "Understanding Technology" (covering the fundamentals of technology) from January 2012 to December 2013 using a cafeteria-style grading method. Over 400 students were involved in these courses. In each course, there were approximately 59 different assignment options. The total point value was a minimum of 781, while only 376 points were needed for an A grade. Point values for each assignment were established based on assignment difficulty and expected duration for completion.

The overall point values for grades, such as a 376 for an A, was based on requiring students to complete slightly less than half of all available points for an A. All three instructors used the same assignments and grading scales for the duration of this study. Even as assignments were refined, such as in above addition of points to interviews, the point scale remained the same. In some cases, instructors gave attendance points or offered some additional assignments such as attending a campus-wide lecture series event. In these situations, as well, the grade scale remained the same—the students simply had more options to choose from. Students were encouraged from the start of the term to complete assignments early. While assignments, quizzes, or tests closed after the due date, all were available as of the first day of the term. This made it plausible for some students to actually complete the course early if they desired, which about 5%–10% did.

Of 412 students who submitted assignments, 49.51% achieved an A grade, 25.24% scored 2% or more above an A (which could have been completed with one or two additional assignments). I think this information is cluttering the statistic making it difficult to understand. About 8.98% scored 5% or more above an A (which necessitated three or four additional assignments) it reads much more clearly without it, 3.88% scored 10% or more above an A (which necessitated five or six additional assignments), and about 1% actually completed 30% or more than was required for an A (which necessitated 16 or 17 additional assignments). This means that our instructors found out half of the students obtain an A grade and nearly one in 10 students go above and beyond the requirements of an A grade by at least 5%. Actually, about 4% of students complete more than is required by an additional 10% or more. You might assume this is because students did not know when they achieved an A, but this is not the case. Instructors notify the student as soon as they obtain an A grade. You might also assume students simply completed one additional assignment, but this is also not the case. The average assignment value is 1.77% of all the points possible, meaning they would need to complete over three additional assignments to achieve 5% beyond A level, which again 9% of all students did. This study shows that cafeteria-style grading actually provides a course structure that encourages students to go beyond expectations.

These study results coincide nicely with the literature on engaged learning techniques. Giving students the freedom to learn in modes where they are most comfortable encourages learning in any form. The result: an increased desire from the student to learn and an increased number of competencies learned—thus, an increase in learning objectives satisfied. When researching differentiated instruction that caters to the needs of different students with various learning styles (Huebner, 2010; Painter, 2009; Subban, 2006), it seems cafeteria-style assessment and grading is also a good fit. This study found that course components that take into account student differences and interests enhance student motivation to learn and encourage them to remain active in the course (Dotger & Causton-Theoharis, 2010; Subban, 2006).

More information on this study can be found in the article "Students reach beyond expectations with cafeteria style grading" as can be found in the *Journal of Higher Education* 2016, 8:1, pages 2 through 17.

Implementation of Cafeteria-Style Grading

Clear Communication

One important element for this course style to be successful is to clearly communicate to the students what a cafeteria-style course is and how it differs from traditional courses. In this regard, we recommend the following:

1. First, have a course outline that explains what a cafeteria-style course is and how grading works.
2. Second, have a required repeatable quiz at the start of class with questions addressing cafeteria-style grading and course expectations.
3. Third, send notification or communicate in class to the students during the first few weeks of class reiterating what a cafeteria-style course is and that not every assignment needs to be completed.
4. Fourth, at the midterm point of the course, remind students about the cafeteria-style grading method and how the course uses a points-based system instead of a percentage-based system.
5. Finally, once a student does have enough points for an A grade notify them of this fact.

Due Dates

We recommend having all assignments available at the start of the term. This way, students can peruse what is available and also potentially get a head start. We also recommend that once an assignment's due date has passed, that assignment is no longer an option to complete, meaning that students cannot simply wait until the end of the term and turn in a myriad of assignments in

hopes of earning enough points for a passing grade. We have found this helps keep students on track throughout the term.

Assignment Options

We recommend giving a variety of assignment options. This gives students the opportunity to use or develop multiple skills and experiment with alternative forms of assessment. We believe a cafeteria-style learning environment with multiple assessment option types encourages student engagement. When students are actively engaged in the subject material through discussing it or applying it, they learn the material better (Wehlburg, 2006). This may involve doing hands-on experiments, working with mentors or individual experience in the field, or other forms of student interaction or active learning. As Shulman (2002) states, "Learning begins with student engagement, which in turn leads to knowledge and understanding. Once someone understands, he or she becomes capable of performance or action" (p. 38).

One particular assignment we recommend helps students map a pathway to the grade they want. A few weeks into the term, you could have an assignment such as this:

What grade to you want?

By now, you should have a good feel for how our course works. For this assignment, please do the following:

1. Tell me what final grade you want for this course.
2. How many points do you need to achieve your desired grade? (See syllabus.)
3. Review all the assignments in the course and select those assignments that seem the most interesting to you based on their titles. (This is a planning list only. You can change your mind and see full assignment details as the course progresses.

The idea is that you have a plan in relation to how you might achieve the grade you stated in the first point in this list.)

4. Indicate how many total points the assignments you tentatively selected amount to. (Select enough assignments so that the point total will be 20% more than needed for your desired grade since it is likely you will not receive full points on all assignments.)

5. List by due date the assignments you plan to complete from #3 including their point values—all weeks, not just this week.

6. Submit this list in a table or spreadsheet with appropriate headings.

Remember, the purpose of this assignment is to ensure that you are planning ahead for the grade you want to achieve in this course; you are not locked into anything you state in this assignment.

Note: Generally, students who score higher on their final grade do a few assignments from each lesson.

Rationale for Cafeteria-Style Grading

Our practice of cafeteria-style grading is based on the premise that student choice is relevant and impactful. According to Bovill, Cook-Sather, and Felten (2011), student choice contributes to learners taking more responsibility for their individual learning. Not only does it help students take internal responsibility for their learning, it helps them identify learning and assessment methods that work best for them individually. "When students work with academic staff to develop pedagogical approaches, they gain a different angle on, and a deeper understanding of, learning" (Bovill et al., 2011, p. 138).

Cafeteria-style grading gives students more choice and also more ownership in their educational experiences. In his evaluation of student-centered learning environments, Cubukcu (2012) points out how student ownership of their goals and activities, including making decisions about their work and actions, encourages in depth understanding and more intrinsic motivational orientation. Additionally, such student-centered environments offer increased capability to incorporate learning opportunities that research shows are directly linked to student learning. As Adelman and Taylor (2005) describe, "Successful, engaged learners are responsible for their own learning. These students are self-regulated and able to define their own learning goals and evaluate their own achievement. They are also energized by their learning; their joy of learning leads to a lifelong passion for solving problems, understanding, and taking the next step in their thinking."

A final rationale for the student-centered assessment methods and options, which are employed within this textbook, is that such learning helps students understand themselves and their own learning styles better while offering them opportunities to try new things without fear of penalty in the form of lower grades. They can strengthen their existing talents as well as explore new areas for growth and development, potentially realizing they have valuable strengths and areas of interest.

Constructivist Approaches to Learning

The use of cafeteria-style grading lends itself to a more constructivist approach toward learning. In this approach, the learners actively construct their knowledge instead of passively receiving information that is transmitted to them from a teacher or textbook (Smart, Witt, & Scott, 2012). Schweitzer and Stephenson (2008) note that constructivism has informed both problem-based and learner-centered approaches. It can empower and enable the students if implemented successfully. In part, this means that instructors need not completely yield authority to mentor and guide but seek to dignify their students while also assuming responsibility for actively mentoring those learners. Dignifying our students' ability to learn while also actively mentoring them is exactly our goal in our cafeteria style of course structure and methods. As Schweitzer and Stephenson (2008) describe, the constructivist approach theorizes several key tenets: (a) knowledge is socially constructed; (b) learners physically construct knowledge—it must be embodied for the learner to acquire it; and (c) learners symbolically create knowledge, by fashioning their own representations for concepts and meanings. As they describe it, while not without limitations, "Constructivism requires that students play a large role in deciding what they want to learn and how they learn it, while also emphasizing higher order skills such as problem solving and self-evaluation."

Engaged Learning via Hands-on or Active Learning

Engaging students in their learning means more than permitting students to have a voice and role in their assessment choices and methods, it also relates to offering opportunities for hands-on or applied learning. When students actively engage in the material by discussing it or applying it, they learn the material better (Wehlburg, 2006). This may involve doing hands-on work, working with mentors or persons in the field, or other types of student interaction or active learning. As Gordon and Crabtree (2006) point out, "Though active learning and positive, appropriate feedback are essential components, the most powerful factor in building engagement is identifying and fully deploying talents in the classroom. And not only students' talents—the talents of everyone involved in education must be leveraged, from students, teachers, and principals to parents and community leaders" (p. 82).

When students make the transition from simply enacting what is required of them to learn, to analyzing consciously what constitutes and enhances that learning, they change not just what they know but who they are (Bovill et al., 2011). A component of this is active learning, which "implies not only a shift from passivity to agency but also from merely doing to developing a meta-cognitive awareness about what is being done" (Bovill et al., 2011, p. 2). In essence, active learning entails assessment and contemplation of the thoughts, perceptions, and emotions that come about during direct experiences (Cubukcu, 2012). When innovative educators are able to embrace and enhance students' previously existing talents, much progress can be made toward successful active and applied learning process.

Active Learning and Learner-Centered Teaching

Learner-centered teaching is more than just having active learning opportunities; it moves the focus in almost all ways from what the instructor does to what the students learn and offers formative assessment opportunities for students (Mostrom & Blumberg, 2012). As Mostrom and Blumberg summarize, "Students take greater responsibility for their own learning because the instructor's role changes from disseminator of information to facilitator of learning. Students actively engage with the course content and construct their own meaning of the topic. They move beyond memorizing information; instead, they understand the course material and can state the concepts in their own words and apply these to a novel situation" (p. 202).

Weimer (2002), in her book *Learner-Centered Teaching*, identifies five instructional practices that play a key role:

1. The balance of power shifted from the teacher to the students
2. The function of content as a means to qualitatively change a person's way of conceptualizing the world
3. The role of the teacher moving from content expert to promoter of learning
4. The responsibility for learning moved toward the student
5. Evaluation, purpose, and processes focused on assessing learning.

Our goal is to have educators incorporate each of these five practices via cafeteria-style grading.

Offering Assignment Options

Cafeteria style-based courses offer students a variety of assignment options for varying learning styles and interests. While there are other courses, instructors, and institutions who have implemented similar methods of student choice, we have found none that have offered the depth and variety we have elected to supply. Following are three examples of similar implementations:

Implementation 1: Nilson, of University of Pittsburgh, uses what she calls learning bundles. In this scenario each bundle is associated with one or more learning outcomes. "Some specs-graded courses could allow students to create their own bundles from a menu: To get a B, the student must complete four of 10 bundles, while an A requires six of 10 to be finished. In this setup, each student will achieve his or her own mix of outcomes" (Levine, 2014).

Implementation 2: In another example, although at a 12th grade level, the article *Learning Menus Tap into All Students' Special Skills* describes an Ohio high school Hilliard Darby that has begun experimenting with learning menus. With learning menus, students work within a choose-your-own assignment structure that's designed to tap into each individual student's strengths and weaknesses to help them reach their full learning capacity (2014).

Implementation 3: A third example is based on research of undergraduate exercise physiology content in which one strategy reported for addressing major-specific needs included development of assignments differentiated to each student group and use of multiple examples from the various disciplines during lecture. In this scenario, students may be asked to apply exercise physiology concepts to their discipline in assignments, projects, case studies, and entries. However, each assignment option is based on a specifically identified student group to which students have been categorized (Fisher, 2013).

Although the three earlier examples are limited in scope to the cafeteria style recommendations being offered here, each has been successful due, in large part, to their emphasis on student learning and success.

Effort and Knowledge

Not only do we want offer a strong variety of assignment options, we also want to assess student achievement on their effort as well as knowledge acquisition. Due to this, we encourage students to try new tasks, concepts, and topics. We note that any attempt made cannot hurt their overall points and according grade, but can only help. We have found limited research of other higher education academic institutions with similar goals. One example is Benedict College, which in 2004 started using a new grading policy they dubbed Success Equals Effort (SE2) for its freshman and sophomore students. In this situation, students were assessed not just on content learning but also based on effort. The way it worked was that students were given two grades—one for effort and one for content learning. Each was then weighted based on if the student was a freshman or sophomore in college. They concluded, "The effort grade affects the knowledge grade positively and significantly across all specifications. This is strong evidence that more student effort does lead to increase learning" (Swinton, 2010).

While we do not directly distinguish between content learning and effort as primary categories, our assessment rubrics entail both means of assessment. Perhaps of even more importance, similar to methods used at the University of Denver, the assignment descriptions and associated rubrics clearly define what is expected for each assignment. These expectations are based on broad course objectives (DeLyser et al., 2003). This way, students have a clear idea of criteria, expectations, and outcomes. According to Fishman (2014), students who perceived the capability to achieve academic outcomes were more likely to feel internally obligated to produce such outcomes. This is based on what Perry, Hladkyj, Pekrun, Clifton, and Chipperfield (2005) refer to as the student's perceived academic control.

Student Academic Control

Raymond Perry has done extensive research on perceived academic control and failure in college students. After initial research and then studying a large group of students over three years, he concluded that students who feel they have high perceived academic control do better scholastically. Perceived academic control is a person's belief that they can intentionally influence and predict events and outcomes in their academic environment. In judging one's capacity to influence an event, a student assesses whether a certain attribute will produce a specified achievement outcome and whether he/she possesses that attribute (Perry et al., 2005; Stupnisky, Perry, Renaud, & Hladkyj, 2013). Further research by Stupnisky et al. have found matching results. Self-esteem is a relatively weak and unreliable predictor of college student's academic achievement, whereas college students with high perceived academic control over academic outcomes are more successful academically (2013). Specifically, "perceived academic control was found to have a stronger influence on students' achievement-based emotions, whereas self-esteem more strongly predicts students' stress and health" (Stupnisky et al., 2013).

However, Perry also noted that it was only students with high academic control who also have high failure preoccupation. "Preoccupation with failure refers to the amount of attention students devote to monitoring the successes and failures of their goal-striving efforts" (p. 538, 2005). Based on his three-year research, "high-academic-control students did better scholastically over 3 years than their low-control counterparts in terms of higher grade point averages, lower voluntary withdrawals, and fewer departures—but only if they were high in failure preoccupation" (p. 552, 2005). While student autonomy and self-reliance are important, they are not enough on their own. A student must also have a desire to succeed. No matter how good the teacher or curriculum, without the latter they will not perform as well overall. In essence, only when high-control students are preoccupied with failure to they outperform low-control students (Perry et al., 2005; Fishman, 2014).

We are able to impact student's perception of academic control by offering a variety of assignments, scoring methods, difficulty levels, and assessment tools; however, we cannot control the internal failure preoccupation or concern entrenched in individual students. The closest we can come to altering their failure preoccupation is reminding students of the relevance of the educational experience and encouraging them to find methods to follow to success.

Competency-Based Learning

Student's progress upon demonstration of learning by applying specific skills and content in competency-based education. For some, this is seen as entirely positive, regardless of age, since students advance to higher level work with demonstration of mastery, not age (Sturgis & Patrick, 2010). As described by the U.S. Department of Education (n.d.), "Transitioning away from seat time, in favor of a structure that creates flexibility, allows students to progress as they demonstrate mastery of academic content, regardless of time, place, or pace of learning. Competency-based strategies provide flexibility in the way that credit can be earned or awarded, and provide students with personalized learning opportunities. These strategies include online and blended learning, dual enrollment and early college high schools, project-based and community-based learning, and credit recovery, among others. This type of learning leads to better student engagement because the content is relevant to each student and tailored to their unique needs. It also leads to better student outcomes because the pace of learning is customized to each student."

Competency-based approaches to education have been criticized in the past, however. It has been suggested that attempts to create unambiguous outcomes for learning is essentially a futile attempt to make simple what is complex. It has been argued that training needs to allow the unexpected and mysterious outcomes, which occur with the meeting of minds; which cannot be replaced nor predetermined without placing severe limits on the outcomes that could be achieved. As attempts are made to increasingly specify outcomes, competency-based approaches end up in a never ending spiral of specifications (Gonczi, 1999). Gonczi (1999) lists a number of arguments against competency-based approaches that have emerged in the literature over the last few years. These can be classified into four categories:

1. Arguments against behaviorist approaches to competency. Such arguments are based on the assumption that a behavioral approach is the only approach to competency-based standards.
2. A more sophisticated version of this anti-behaviorist argument is the claim that all competency approaches, irrespective of their intention, are necessarily behaviorist. The proponents of these arguments, unlike those in the first category, are aware that there are other approaches, but they claim that any attempt to set predetermined standards or outcomes for learning cannot avoid being behaviorist.

3. Arguments against the generic approach to competence. These suggest that competence is only able to be understood within particular contexts or fields of practice.
4. Arguments against any competency approach on the basis of its normative assumptions about the nature of "the good." These arguments are often also about the aims of vocational education—the extent to which it should meet the needs of industry/professions (as opposed to the individual).

One question is if there are particular starting points that are better or worse for introducing competency-based models. However, there remains inadequate research to determine if any one starting point is more valuable than another (Sturgis & Patrick, 2010). Here are two examples of such competency-based plans with different starting points as offered by the U.S. Department of Education (n.d):

❖ Michigan Seat Time Waiver: Michigan passed legislation in 2010 providing a seat time waiver to districts that want to offer pupils access to online learning options and the opportunity to continue working on a high school diploma or grade progression without actually attending a school facility.
❖ Ohio's Credit Flexibility Plan: This plan, adopted by the State Board of Education in 2009, allows students to earn high school credit by demonstrating subject area competency, completing classroom instruction, or a combination of the two. Under this plan, subject area competency can be demonstrated by participation in alternative experiences including internships, community service, online learning, educational travel, and independent study.

Even without the identification of clear starting and exit points, competency-based learning is still seen as a potential strong point for not only higher education but in high school as well. As the National High School Center points out, through outcome-focused approaches, competency-based instruction in high school enables and supports the following: flexibility, multiple assessments, and responsiveness to individual student needs (n.d.). And as Sturgis and Patrick (2010) state, "demonstrating proficiency on learning objectives is strikingly similar to earning merit badges in camp or after school." (p. 22)

There are some key elements though that may be necessary for the success of competency-based learning. Software and Information Industry Association (SIIA) Symposium participants in 2010 jointly identified the following top 10 essential elements and policy enablers of personalized learning:

Essential elements:

1. Flexible, Anytime/Everywhere Learning
2. Redefine Teacher Role and Expand "Teacher"

3. Project-Based, Authentic Learning
4. Student-Driven Learning Path
5. Mastery/Competency-Based Progression/Pace

Policy enablers:

1. Redefine Use of Time (Carnegie Unit/Calendar)
2. Performance-Based, Time-Flexible Assessment
3. Equity in Access to Technology Infrastructure
4. Funding Models that Incentivize Completion
5. P-20 Continuum and Non-Age/Grade Band System (Association for Supervision and Curriculum Development, 2010; Wolf, 2010).

While there remain questions as to the ultimate success and implementation of competency-based learning, it is certainly considered a viable option. As noted by Wolf, "Today's industrial-age, assembly-line educational model—based on fixed time, place, curriculum and pace—is insufficient in today's society and knowledge-based economy. Our education system must be fundamentally reengineered from a mass production, teaching model to a student-centered, customized learning model to address both the diversity of students' backgrounds and needs as well as our higher expectations for all students." (Wolf, 2010)

Differentiated Instruction

Cafeteria-style grading falls under the instructional theory of differentiated instruction. Willis and Mann (2000) define it as, "a teaching philosophy based on the premise that teachers should adapt their instruction to student differences" (p. 1). Differentiated instruction can be achieved by allowing students to learn in different ways through diverse content, process, or products of the curriculum (Willis & Mann, 2000). Differentiated instruction caters to the needs of different students with various learning styles (Huebner, 2010; Painter, 2009; Subban, 2006) "by providing entry points, learning tasks, and outcomes tailored to students' learning needs" (Watts-Taffe et al., 2012, p. 304).

Differentiated instruction is also a type of pedagogical instrument used to facilitate the learning process by adapting instruction and assessment methods to address student differences via diverse content, processes, or products of the curriculum (Subban, 2006; Willis and Mann, 2000). It caters to the needs of varying learning styles among students by providing differing entry points, learning tasks, and even outcomes based on the students' learning needs. Course components that take into account student differences and interests enhance student motivation to learn and encourage them to remain active in the course (Dotger & Causton-Theoharis, 2010; Huebner, 2010; Painter, 2009; Subban, 2006; Watts-Taffe et al., 2012, p. 304).

Research conducted by Goodwin and Gilbert (2001) shows that students who choose optional course components demonstrate an increased desire to select course components that involve ongoing active participation in the course (p. 491) and equally reports that by having a choice in their course components they have more of an incentive to take the time to complete the course work. However, it is also noted that students who lack intellectual maturity might select the assignments that require the least perceived work and not rise to the challenge of taking full responsibility for their learning and growth, which this type of instruction offers to students (Goodwin & Gilbert, 2001). That said though, even in these cases the students are still performing, just in ways they perceive as the least challenging and rigorous for them.

Conclusion

We believe the types of assessment methods and options presented here help students understand themselves and their learning styles better, while offering them opportunities to try new things without fear of penalty in the form of lower grades. "For many students, the most important discoveries may be the realization of where their greatest talents lie and where the flow of experiences occur" (Gordon & Crabtree, 2006, p. 83). In this environment, students are encouraged to experiment, learn, and explore.

References

Association for Supervision and Curriculum Development. (2010). *Education leaders identify top 10 components of personalized learning.* Retrieved from http://www.ascd.org/news-media/Press-Room/News-Releases/Education-Leaders-Identify-Top-10-Components-of-Personalized-Learning.aspxy

Bovill, C., Cook-Sather, A., & Felten, P. (2011). Students as co-creators of teaching approaches, course design, and curricula: Implications for academic developers. *International Journal for Academic Development, 16*(2), 133–145. doi:10.1080/1360144X.2011.568690

Committee on STEM Education. (2013). *Federal science, technology, engineering, and mathematics (STEM) education five-year strategic plan.* A report from the committee on STEM education national science technology council. Retrieved from https://www.whitehouse.gov/sites/default/files/microsites/ostp/stem_stratplan_2013.pdf

Cubukcu, Z. (2012). Teachers' evaluation of student-centered learning environments. *Education, 133*(1), 49–66. Retrieved from http://goo.gl/HDN8mZ

DeLyser, R. R., Thompson, S. S., Edelstein, G., Lengsfeld, C., Rosa, A. J., Rullkoetter, P., ... Whitt, M. (2003). Creating a student centered learning environment at the University of Denver. *Journal of Engineering Education, 92*(3), 269–273.

Fisher, M. M. (2013). Current practices in the delivery of undergraduate exercise physiology content. *Physical Educator, 70*(1), 32–51.

Fishman, E. J. (2014). With great control comes great responsibility: The relationship between perceived academic control, student responsibility, and self-regulation. *British Journal of Educational Psychology, 84*(4), 685–702.

Gonczi, A. (1999). Competency-based learning a dubious past—An assured future; New dimensions in the dynamics of learning and knowledge. In *Understanding learning and work.* (pp. 180–195). New York, NY: Routledge. Retrieved from http://www.researchgate.net/publication/27467878_New_dimensions_in_the_dynamics_of_learning_and_knowledge/file/60b7d518a492068afb.pdf

Goodwin, J. A., & Gilbert, B. D. (2001). Cafeteria-style grading in general chemistry. *Journal of Chemical Education, 78*(4), 490–493. Retrieved from http://cpltl.iupui.edu/media/cfaac0df-062a-4322-81e1-f470cd828810/-1748818363/cPLTLContent/2013/PLTL%20Literature%20PDFs/Goodwin%20Gilbert_2001.pdf

Gordon, G., & Crabtree, S. (2006). *Building engaged schools: Getting the most out of America's classrooms* (1st ed.). New York, NY: Gallup Press.

Huebner, T. A. (2010). What research says about... differential learning. *Educational Leadership, 67*(5), 79–81. Retrieved from http://www.ascd.org/publications/educational-leadership/feb10/vol67/num05/Differentiated-Learning.aspx

Levine, M. (2014, May 15). Advocating a new way of grading. *University Times, 46*(18). Retrieved from http://www.utimes.pitt.edu/?p=30598

Mostrom, A., & Blumberg, P. (2012). Does learning-centered teaching promote grade improvement? *Innovative Higher Education, 37*(5), 397–405. doi:10.1007/s10755-012-9216-1

Office of the Press Secretary. (2010). *White House.* President Obama Expands "Educate to Innovate" Campaign for Excellence in Science, Technology, Engineering, and Mathematics (STEM) Education Retrieved from https://www.whitehouse.gov/the-press-office/president-obama-expands-educate-innovate-campaign-excellence-science-technology-eng

Painter, D. D. (2009). Providing differentiated learning experiences through multigenre projects. *Intervention in School and Clinic, 44*(5), 288–293. Retrieved from http://isc.sagepub.com/content/44/5/288.full.pdf+html

Perry, R. P., Hladkyj, S., Pekrun, R. H., Clifton, R. A., & Chipperfield, J. G. (2005). Perceived academic control and failure in college students: A three-year study of scholastic attainment. *Research in Higher Education, 46*(5), 535–569.

Schweitzer, L., & Stephenson, M. (2008). Charting the challenges and paradoxes of constructivism: A view from professional education. *Teaching in Higher Education, 13*(5), 583–593. doi:10.1080/13562510802334947

Shulman, L. S. (2002). Making differences: A table of learning. *Change, 34*(6), 36–44. doi:10.1080/00091380209605567

Smart, K. L., Witt, C., and Scott, J. P. 2012. Toward Learner-Centered Teaching: An Inductive Approach. *Business Communication Quarterly.* 75(4): 392–403.

Stupnisky, R. H., Perry, R. P., Renaud, R. D., & Hladkyj, S. (2013). Looking beyond grades: Comparing self-esteem and perceived academic control as predictors of first-year college students' well-being. *Learning and Individual Differences, 23*, 151–157. Retrieved from https://scholar.google.com/citations?view_op=view_citation&hl=en&user=xcxWioIAAAAJ&citation_for_view=xcxWioIAAAAJ:Se3iqnhoufwC

Sturgis, C., & Patrick, S. (2010). *When success is the only option: Designing competency-based pathways for next generation learning.* Quincy, MA: Nellie Mae Education Foundation. Retrieved from http://www.inacol.org/wp-content/uploads/2015/03/iNACOL_SuccessOnlyOptn.pdf

Subban, P. (2006). Differentiated instruction: A research basis. *International Education Journal, 7*(7), pp. 935–947. Retrieved from http://files.eric.ed.gov/fulltext/EJ854351.pdf

Swinton, O. H. (2010). The effect of effort grading on learning. *Economics of Education Review, 29*(6), 1176–1182. doi:10.1016/j.econedurev.2010.06.014

U.S. Department of Education. (n.d.). *Competency-based learning or personalized learning.* Retrieved from https://www.ed.gov/oii-news/competency-based-learning-or-personalized-learning

U.S. Department of Education. (2015). *Science, technology, engineering and math: Education for global leadership.* Retrieved from http://www.ed.gov/stem

Watts-Taffe, S., Laster, B. P., Broach, L., Marinak, B., McDonald Connor, C., & Walker- Dalhouse, D. (2012). Differentiated instruction: Making informed teacher decisions. *The Reading Teacher, 66*(4), pp. 303–314. Retrieved from http://professionallearning.typepad.com/files/differentiated-instruction-making-informed-teacher-decisions.pdf

Wehlburg, C. M. (2006). *Meaningful course revision: Enhancing academic engagement using student learning data* (1st ed.). Bolton, MA: Anker Publishing Company.

Weimer, M. (2002). *Learner-centered teaching: Five key changes to practice.* San Francisco, CA: Jossey-Bass.

Willis, S., & Mann, L. (winter 2000). Differentiating instruction: Finding manageable ways to meet individual needs. *Curriculum update.* pp. 1–3. Retrieved from http://www.ascd.org/publications/curriculum-update/winter2000/Differentiating-Instruction.aspx

Wolf, M. A. (2010). *Innovate to educate: System [re]design for personalized learning a report from the 2010 Symposium.* Software and Information Industry Association, Washington, DC. Retrieved from http://www.ccsso.org/Documents/2010%20Symposium%20on%20Personalized%20Learning.pdf

Chapter 2
Curriculum Options

Outline

- ❖ Objectives and outcomes
  - ➢ Course learning objectives
  - ➢ Learning outcomes
  - ➢ Chapter learning objectives
- ❖ Methods for handling assessment and assignments
  - ➢ Types of assessment methods
    - ◼ Formative assessments
    - ◼ Formal methods of assessment
    - ◼ Informal methods of assessment
    - ◼ Summative assessments
  - ➢ Handling assessment and assignments
  - ➢ Grading methods
  - ➢ Full cafeteria
  - ➢ Base assignments
  - ➢ Traditional
- ❖ Course structure
  - ➢ Modules/sections
  - ➢ Chronological/progressive
- ❖ Inquiry-based learning and undergraduate research
- ❖ Assignment types
  - ➢ Discussions
  - ➢ Tests
  - ➢ Research
    - ◼ Papers and reports
    - ◼ Annotated bibliography
    - ◼ Article or literature review

(Continued)

(*Continued*)

- Develop a research question
- Lab or experiment write ups
- Posters
- ➤ Design and/or Build projects
  - Technical reports
- ➤ Assessment tasks
  - Case studies
  - Argument papers
- ❖ Rubrics and other grading and assessment tools
  - ➤ Terms of agreement quiz
- ❖ Quiz instructions
- ❖ Quiz questions
  - ➤ Rubrics
- ❖ References

Objectives and Outcomes

All courses and their content, whether in modules or categories or any other format, should clearly state course objectives and expected outcomes. Two terms are commonly used in this area: (a) course objectives and (b) learning outcomes. We will address each here and explain their importance. In essence, these are used to identify key concepts or skills that students are expected to learn.

Course Learning Objectives

Course learning objectives are clear and concise statements that express the goals and intentions for the course. They drive how the course is structured and implemented as well as express what you intend students to learn by the end of the course. In almost all cases, objectives are specified at the overall course level, in some cases they are also specified or related to distinct areas of the course, such as for each module. The course learning objectives should tie directly to the learning outcomes.

Learning Outcomes

Learning outcomes are a formal and clearly communicated statement of specifically what students are expected to learn in the course (similar to objectives) and are used to assess what students actually learned per observation (different from objectives). They refer to knowledge, skills, or professional development that instructors expect the students to learn or master during the course and should be stated in a specific and measurable form. They are also referred to as expected learning outcomes, expected outcomes, or student learning outcomes (SLOs).

Major characteristics of learning outcomes as described by Suskie (2004) include:

1. They specify an action by the students/learners that is observable
2. They specify an action by the students/learners that is measurable
3. They specify an action that is done by the students/learners (rather than the faculty members)

If this is done, learning outcomes can then be successfully assessed.

When specifying how students will be assessed or graded, faculty need to ensure the methods used to tie in to desired learning outcomes. As Levine (2014) notes in describing the experience of a faculty at the University of Pittsburgh, "the lowest passing grade must ensure that students demonstrate they have learned the course's fundamentals. This is particularly crucial when a course is not an elective and is part of a sequence of courses, each dependent on the student attaining a certain level of knowledge successfully."

At California State University in Fresno's Lyles College of Engineering, six faculty members from six different programs experimented with course grading based on SLOs in 2011. Here is what they found, "SLO-based grading changes the way the instructor looks at students' work. It calls for intentional, objective grade assigning that is far beyond grading based only on the mechanics of the work." Not only that, "Compared to traditional grading, SLO-based grading seems to produce higher grades on average for students who have strength in certain traits because it allows significant partial credits, though the final product may be incomplete. The grade is usually assigned based on attaining the learning outcome across the whole paper regardless of the number of questions or problems. For example, for certain outcomes, such as critical thinking, the student may demonstrate attainment of the outcome by responding correctly to some but not all of the questions that require critical thinking. It becomes clear that the incorrect answers to some of the questions are due to a lack in other outcomes, such as grasp of the subject matter. When the work is evaluated based on the two outcomes equally weighed, the final grade becomes higher than if each question were assigned a certain percentage" (Bengiamin & Leimer, 2012).

Chapter Learning Objectives

We recommend supplying learning objectives to each chapter as well as to the overall course. In this manner each chapter can, in some ways, stand alone in a module-type format. Module-level objectives help clarify and delineate what it is you want students to learn from that module. They are useful from both the curricular perspective but also can serve as an outline for students.

Syllabus sample

The following is an example of how course objectives and learning outcomes could be incorporated in to your syllabus.

Course objectives (goals of course)

1. The course will illustrate the role of various technologies and how they operate.
2. The course will illustrate how various technologies are applied in a practical sense into jobs and products.
3. The course will comprehensively address how technology can affect society and how societal demands can drive new technology developments.
4. The course will demonstrate how developments in one technology can affect many other technologies.
5. The course will analyze the relationship between science, engineering, and technology.
6. The course will demonstrate how new technology comes about, inclusive of invention and creativity.
7. The course will analyze some of the ethical dilemmas inherent in existing and emerging technologies.

Learning outcomes (expectation of student mastery)

1. The student will demonstrate how various technologies work by designing or creating simple products.
2. The student will describe how various technologies are applied in a practical sense into jobs and products, which relate directly to their own lives.
3. The student will articulate how technology can affect society and how societal demands can drive new technology developments.
4. The student will describe how developments in one technology can affect many other technologies via discussions, reports, or similar.
5. The student will compare and contrast relationships between science, engineering, and technology in multiple fields or areas of emphasis.
6. The student will summarize how new technology comes about via invention and creativity among other prompts.
7. The student will share and discuss some of the ethical dilemmas inherent in existing and emerging technologies as they relate to your own work environment.

Methods for Handling Assessment and Assignments

In assessment of learning outcomes, the aim is to understand what students are learning and how well they are learning in relation to the stated expected learning outcomes for the course. Methods for assessing learning outcomes are any activities used to understand what students are learning and how well they are learning. The end of each chapter will have a variety of alternative assessment options.

Types of Assessment Methods

Based on your situation, you may opt to use formal and/or informal methods of assessment as well as formative or summative types of assessments.

Formative Assessments

Formative assessment methods assess and monitor student learning during the learning process itself to identify areas of comprehension and lack of understanding. They commonly include both formal and informal methods of assessment.

Formal Methods of Assessment

Formal methods of assessment include techniques such as graded quizzes, papers, in-class activities, homework assignments, and other class deliverables that occur throughout the term.

Informal Methods of Assessment

Informal methods of assessment include classroom assessment techniques such as ungraded class discussion, written reflection, polls or surveys, nongraded quizzes, and the like that occur throughout the term.

Both formal and informal assessment methods can be used to investigate how well students are accomplishing the expected learning outcomes for the course.

Summative Assessments

Summative assessments are used to evaluate student learning after completion of the course or modules. These measure learning outcomes specifically and determine if students have achieved the desired levels of understanding and competency. Summative assessments include techniques such as graded quizzes, exams, papers, and portfolios. The difference from the formal formative methods that have similar items listed is that these occur at the end of the term and are intended for SLO assessment only and not for course objectives fulfillment and course guidance.

Handling Assessment and Assignments

Each situation is different based on the course objectives, teaching styles incorporated, institutional norms and expectations, and more. Therefore, we will address here some grading and assessment options that you might potentially choose from based on your own circumstances. At the core of our recommended methods, though, is the use of a points-based grading method that makes use of rubrics for consistency. While there are a number of other assessment practices that exist, for the purposes of this book, we believe points-based methods to be the best fit and rubrics a logical addition to that method. Within this realm, we will address a number of alternative pathways and methods that could be used.

Grading Methods

Linda B. Nilson of University of Pittsburgh uses a grading system where students earn tokens that they can trade for missed homework or classes. She also has her classwork grouped into grade-based bundles where students self-select a grade level they attain. Added to that, she operates her grading system around specifications for a specific grade instead of by ranking how well or poorly the student does on a test or assignment (Levine, 2014). A number of similarities to Nilson's work can be found in our methods, which also do not penalize for missed homework or classes, has grade-based bundles of assignments where students self-select a grade level they desire to attain, and place little emphasis on the results of a particular test or assignment in isolation.

That said, our methods do use grading with points. As Feldman, Alibrandi, and Kropf (1998) found in their research, point systems provide a mechanism for keeping detailed accounting of student work that can demonstrate unambiguously why the teacher assigned a specific final grade, and secondly helps keep students on task. Our point system does not differentiate between the sources of points. Once the points are awarded to the students, they are indistinguishable from one another, this also matches point system methodology used in the Feldman and Alibrandi research (1998) and differs from other grading methods such as modified point systems, weighted averages, and true averages.

We believe, as does Mostrom and Blumberg (2012), that improved instructional design and more appropriate, authentic assessment techniques may lead to increased student learning. We also want to ensure that our assessment methods have constructive alignment. As Bell, Mladenovic, and Price (2013) describe, "Constructive alignment exists where learning outcomes, learning activities and assessment are clear, related and integrated. In an aligned learning environment, it should be clear to students where they are going (learning outcomes), how they will get there (learning activities, feedback and reflection) and what is expected of them (assessments, marking guides and grade descriptors)."

To the degree possible, we want to address the needs and expectations of both students who simply need an idea of assessment standards and those who are seeking precise guidance. As Bell et al. (2013) point out, the two categories reflect differing student perceptions about their roles as well as the role of the instructor. As an aspect of this desire to satisfy the needs of both groups, we have implemented grading rubrics on nearly all assignments. This ensures consistency of assessment between instructors and offers students more clear assessment and evaluation expectations. In some cases, we also offer references to examples, which could be used as models or templates by students, assuming we do not see it as limiting their perceived creativity, problem-solving practices, or methods for completion.

Full Cafeteria

In full cafeteria courses, the students are not required to complete any specific assignments or tasks; not even the final exam. Instead, students are truly given full rights to determine the tasks they wish to complete and can emphasize areas of most interest to them both in course content and in methods of assessment. One potential drawback to this method is that it is possible for students to skip course curriculum topics entirely and focus only on those areas in which they have the most interest. While we have found this not to be the case in our own implementation of such systems, the risk exists.

As noted in the article *Students Reach beyond Expectations with Cafeteria Style Grading* found in the *Journal of Applied Research in Higher Education* and written by two of the authors of this book in 2016,

> While some students completed a greater variety of assignments than others, one overall tendency was for students to complete all of the online quizzes and exams. These exams were open book and unlimited time. A benefit of students pursuing this route was that the full curriculum was covered since the quizzes were allocated across all topic areas. One concern with this style of grading had been that students could complete all their assignments in the first segment of the course, thus not gaining actual experience with the full curriculum content or learning objectives. However, even for students who completed the course early, their strong inclination instead was to complete assignments from throughout the curriculum instead of doing each assignment sequentially until the desired grade was achieved (Arendt, Trego, & Allred, 2016, p. 7)

Base Assignments

We have what are called base assignments. This is an assignment or assignments that all students must complete and earn a prespecified score before other assignment options become available to them. In order to accomplish this, these particular assignments can be repeated by students if necessary. They also presume fundamental core curriculum topics are assessed and sufficiently understood before students can proceed. A student, however, needs to complete assignments above and beyond these limited base assignments if they desire

© Maksim Kabakou/Shutterstock.com

a passing grade in the course. Once the base assignment is successfully completed, a variety of assignment options become available that the student can choose from if they desire. These latter assignments can be attempted one time each. By completing these additional assignments, students are able to earn additional points toward the grade they desire to attain.

Here is a syllabus sample:

In this course, assignments are handled differently. Assignments are served up cafeteria-style. This means you get to choose to do those assignments that appeal to your own learning interests and you do not need to complete all the assignments to get an A grade. Rather, you complete those assignments desired in order to earn the applicable points. There are over twice as many points possible in this course than are required for an A. So, for example, there are 1130 points possible in the class but you only need 452 points to attain an A grade—once you get to 452 points (40% of all available points) you have an A (see grading scale below for details).

For each module, you need to either obtain a minimum of 80% on the module quiz (unlimited attempts permitted on quizzes) or complete and submit the highest point value assignment (one if more than one is available). At that point, the proceeding module will become available. You do NOT need to do all of the assignments in each module, refer to paragraph earlier. Once an assignment's due date has passed, though, that assignment is no longer an option to complete. Only quizzes and the midterm will remain open throughout the term, so you can always progress from module to module.

We recommend offering a variety of assignment options for the students. In this book, you will find, for each chapter, assignments of the following types that you could choose from or work from to create your own customized assessments.

Traditional

In some instances, a traditional course format works best. This often includes a curriculum that is followed chronologically, a single set assessments for all students, and more commonly consists of primarily oral recitation by an instructor. We feel our book will also function well in these environments. While it is unlikely you will use all of the assessment methods offered in each chapter, you can make use of the ones that are most fitting in ways that work for your particular circumstances.

Course Structure

Modules/Sections

We recommend the course content from this book be offered via modules or sections. We have written it in a manner conductive to this and have worked to ensure each chapter can stand independently of the others. We have also added assessments at the end of each chapter that are chapter specific. While the content can certainly work in a chronological or even completely unstructured format, the author intention was modularly. Each chapter of the book, for example, has its own module-based learning objectives.

Chronological/Progressive

A chronological or progressive course structure moves in sequence from one topic to the next, often with each progressing lesson building upon the ones before it. This book was not written in a chronological format overall but could be used for these purposes based on the structure of the course and curriculum.

Inquiry-Based Learning and Undergraduate Research

Critical thinking and creativity are essential skills for success. Students can explore creativity by addressing problems and investigating appropriate solutions. Research, scholarly and creative activities are student-centered activities that enable them to use creativity and improve critical thinking skills. Through undergraduate research, students learn how knowledge is constructed, how to collect and analyze data, and how to solve problems. Students also learn to be independent learners and gain considerable self-confidence that prepares them for future careers in research and other disciplines (Lopatto, 2010). In addition, students learn how to deliver their findings to multispectral audience. These experiences should be available for students early in their careers to provide those students the opportunities to explore new career paths. As Adedokun et al. (2010) note, "Through guided participation and extensive collaboration, long-term observation and practice, the novice researcher gradually acquires the skills and expertise needed for effective performance in the profession" (p. 3).

Inquiry-based learning (IBL), inside and outside the classroom, includes research, scholarship, and creative activity. IBL is imperative to efforts to retain and engage students in higher education (Kuh, 2008), particularly at teaching institutions, because they are robust and effective interventions. Students who receive the greatest benefit from IBL are those who are at-risk of dropout or failure, minority background, or first-generation status (Ishiyama, 2002).

IBL significantly contributes to the development of highly skilled professionals and productive adults. When combined with other traditional educational opportunities, these activities help prepare students for successful career paths after graduation.

* IBL improves SLOs. Empirical studies of IBL yield robust effects that demonstrate long-term impacts on students' thinking, writing, problem-solving, networking, and later employment opportunities. "The results also indicate that students involved in research projects demonstrate deeper learning through higher order thinking, integrative learning, and reflective learning" (Hoffman, 2009, p. 23).

❖ IBL is a high-impact practice that increases retention and completion. "Two significant outcomes of this program are the long-term retention and higher GPAs of students compared with those in the matched control group throughout the student's college experience." (Schneider, Bickel, & Morrison-Shetlar, 2015, p. 42).

❖ IBL prepares students for successful careers. "Student's research work has been helpful in improving skills related to networking, improving professional credentials, and contributing to a body of knowledge … skills that will help prepare undergraduates for their future careers" (Salsman, Dulaney, Chinta, Zascavage, & Joshi, 2013, p. 8).

❖ IBL creates a positive experience for students in the present as well as in the future. "A majority of respondents indicated that undergraduate research was a significant positive factor in their actual admission to graduate school, employment, or both" (Schmitz & Havholm, 2015, p. 1).

Nationally, institutions of higher education report improvements in retention and completion by emphasizing IBL (Malachowski et al., 2015).

When combined with other traditional educational opportunities, undergraduate research helps in developing and preparing students for successful career paths after graduation. "Undergraduate research is like role-playing. I mean no disparagement of the research—role-playing is a critical part of life. Children learn how to be adults in part by trying on grown-up clothes and imitating a parent who is, say, driving a car or vacuuming a rug. Similarly, undergraduates can learn the conventions of research through imitation and practice" (Chapman, 2003, para. 7). It is often said that "practice makes perfect," but it is probably better said that practice makes prepared. "Through guided participation and extensive collaboration, long-term observation and practice, the novice researcher gradually acquires the skills and expertise needed for effective performance in the profession" (Adedokun et al., 2010, p. 3).

In order to impact retention, students need to be involved in research early and often. This is particularly true in science, technology, engineering, and math (STEM). "Students, who participate in research early, during the first year and second year, are more likely to succeed and graduate with college degrees in STEM disciplines. Such students are also likely to advance to graduate school in STEM areas or proceed to professional schools" (Fakayode et al., 2014, p. 663). Much of the current research in the open literature has focused on increased retention rates as one of the main benefits of undergraduate research. "Despite the difficulty of isolating the main effects of research opportunities from other activities the student is exposed to on campus, the descriptive evidence indicates that undergraduate research programs are a vital component of an integrated retention strategy for STEM students" (Hoffman, 2009, p. 22).

Central to the concept of IBL as a form of engaged pedagogy is the role of the instructor in the design of students' scholarly and creative experiences, the integration and preparation of student skill development leading toward competence and autonomy in selecting their own questions or goals for their work, and the active use of best practices and SLOs to shape the student experience.

Assignment Types

Some types of assignments available to students include but are not limited to: construction of models or prototypes, interviews of community members, open book multiple-choice questions, the creation of teaching modules, short academic papers, and recorded group discussion. We have categorized them generally as:

- ❖ Discussions
- ❖ Tests
- ❖ Research
  - ➢ Papers and reports
  - ➢ Annotated bibliography
  - ➢ Article or literature review
  - ➢ Develop a research question
  - ➢ Lab or experiment write ups
  - ➢ Posters
- ❖ Design and/or build projects
  - ➢ Technical reports
- ❖ Assessment tasks
  - ➢ Case studies
  - ➢ Argument papers

Next we will discuss best practices for a number of these assignment types.

Discussions

Online (or in class) discussion topic ideas can be found at the end of each chapter. We recommend you use these as potential assignments. Discussion forums/boards are a means for students to engage with one another and with the instructor as well, fostering group cohesiveness and camaraderie. They provide discussion space for exploring understanding of key content in collaboration with others. It should be noted that while the instructor should have a presence in the discussions, the students should lead the conversations. Substantive student dialogue is an essential part of the learning process.

Faculty should provide guidelines regarding the method and quality of the student postings or responses to others, particularly when used for assessment, in which criteria for grading should be established.

Discussion instructions could read something like this:

Post an original discussion in the online discussion board on the following topic: Discuss how society drives technological changes and vice versa. What are your thoughts and why? (200 word minimum). Next, comment on the post(s) of a minimum of one other student in a thoughtful and academic way that enhances the conversation. See rubric for grading and assessment measures.

Sample discussion rubric

Quality of original post: 3 points
Level of participation: 2 points
Quality of feedback to others: 3 points
Timing of posting: 2 points
Grammar and mechanics: 2 points
Total points: 12

Tests

Tests might consist of quizzes to assess student absorption of the course material throughout the course to determine if course objectives are being met along with measuring SLOs, or exams, which are given at the end of the course specifically to measure SLOs. Some test formats might include objective methods such as multiple-choice questions, matching, true or false questions, or fill in the blank. Or subjective methods such as short answer, essays, or problem-solving.

One potential option is to offer collaborative tests where students work together to find the correct answer. This works best with small groups. We recommend you first have each student complete the test independently. Then, have the students meet as a group to discuss their answers. Have the group turn in a single test, which is then assessed. Give the students the average score of their own test plus the group test.

Research

Research assignment ideas can be found at the end of each chapter. Research assignments, for the purposes of this book, are any assignments that encourage or implement undergraduate research. This might include research essays, annotated bibliographies, literature reviews, posters, lab or experiment write ups, or similar.

Note: you have to determine for yourself and your institution what you consider to be scholarly/academic work that can be referenced (such as peer-reviewed journals) and in what contexts. Generally speaking, though, scholarly journals are peer reviewed. EBSCO summarized well:

Academic Journals: EBSCO defines academic journals as journals that publish articles which carry footnotes and bibliographies, and whose intended audience is comprised of some kind of research community. It is a broad classification that includes both "peer-reviewed" journals as well as journals that are not "peer-reviewed" but intended for an academic audience.

Scholarly (Peer Reviewed) Journals: Scholarly (Peer Reviewed) Journals are journals that are intended for an academic audience and are peer-reviewed. EBSCO has established specific guidelines for what we consider a peer-reviewed journal. For more information, please see: *What are (Scholarly) Peer Reviewed publications?*, which can be found at http://support.ebsco.com/knowledge_base/detail.php?id=976

Papers and reports

For research-related papers and reports, we recommend you include writing guidelines similar to the following:

❖ Use American Psychological Association (APA) or Modern Language Association (MLA) format as your style guide. You can learn more about each at http://www.apa.org/ or https://www.mla.org/

❖ 1" margins 12 point, Times New Roman font
   ➢ Two double-spaced pages (not including reference pages or cover page/name and course information)
   ➢ You are not required to include an abstract
   ➢ Reference page with at least two references that are used within the paper (i.e., both on reference list and in-text citation).

❖ All direct quotes and paraphrasing must be referenced. Otherwise, your paper will be deemed plagiarism. Papers without references will receive a zero
   ➢ Proofread carefully (deductions will be made for spelling, grammatical, or style errors)

We also recommend a rubric for assessment with the following general categories at a minimum:

Format/layout/length: 2 points
Content/information: 6 points
Quality of writing: 2 points
References and use of references in text: 2 points
Total Points: 12

Annotated bibliography

An annotated bibliography is a list of citations to books, articles, and documents. Each citation is followed by a brief descriptive and evaluative summary (a few paragraphs). The annotation informs the reader of the relevance, accuracy, and quality of the sources cited, and provides a foundation for further research.

Students can be requested to make a short bibliographic list with a minimum of (x) reliable sources. This would be a good time to describe what you consider to be scholarly/academic work (such as peer-reviewed journals). Then, have the students completed the following for each citation:

❖ Citation in APA or MLA style
❖ Short summary (main points, perspective)
❖ Evaluation/critique
❖ Notes

Article or literature review

With an article review, the student is given an article to read. They then evaluate or critique the data, research methods, and results. A literature review, on the other hand, identifies key areas across literature. It helps the researcher understand current thinking and find gaps in the research that could be potentially filled. In most cases, a review of the literature is directly related to a topic or problem under study, followed by an explanation of how the research question grows out of that review. The general purpose of the literature review is to know what others have written before you begin your own investigation; so your study is grounded in a particular context of what is known about a subject in order to establish a foundation for the topic (or question) being researched.

Develop a research question

How do you develop a usable research question? Choose an appropriate topic or issue for your research, one that actually can be researched. This might not be as easy as it sounds. You have to choose a question that is neither too broad nor too narrow. Plus, it has to be something that can actually be answered which hasn't been answered already by someone else. After all, the question guides and centers your research.

How do you formulate a good research question? Duke University Thompson Writing Program (n.d.) suggests the following: Choose a general topic of interest, and conduct preliminary research on this topic in current periodicals and journals to see what research has already been done. This will help determine what kinds of questions the topic generates. Once you have conducted preliminary research, consider: Who is the audience? Is it an academic essay, or will it be read by a more general public? Once you have conducted preliminary research, start asking open-ended "How?" "What?" and "Why?" questions. Then evaluate possible responses to those questions.

You want to ensure your question is doable in your given time frame and with your available resources; so consider what information will be needed. You want to choose something that not only solves a new, previously unsolved question, but is also something others will care about. One you have accomplished the above, the answer to your research question should be your thesis statement.

Lab or experiment write ups

The idea of a lab write up is for the student to show evidence of his/her ability to interpret evidence or data and relate the interpretation to a theory of the academic discipline. These could take a variety of forms based on the circumstances. Overall, though, a lab format includes the following: purpose, introduction, and background; materials and equipment; procedures/methods; results/data; data analysis and observation; conclusion and discussion; and sources used.

Posters

Posters are widely used in the academic community, and most conferences include poster presentations in their program. Research posters summarize information or research concisely and attractively to help publicize it and generate discussion. A poster is a quick way of visually conveying information about an area of work (research) and should communicate its point effectively without someone needing to be there to explain it. A poster is usually a mixture of a brief text mixed with tables, graphs, pictures, and other presentation formats.

Design and/or Build Projects

Problem-based learning (PBL) and project-based learning promote metacognitive strategies, critical thinking, and problem-solving capabilities. These activities promote student engagement and motivation for learning (Stolk & Harari, 2014). PBL allows instructors to use dialogic spaces differently creating spaces for reflection and shared vision with students (Colvin, 2005). Based on this, students are asked to complete authentic tasks with design and build projects. This might involve a paper-and-pencil concept drawing, basic mechanical drawings, building prototypes or building physical goods, and describing items used via bill of materials or otherwise. In some cases, it might include having the student describe to others how to build it or even writing technical instructions. Design and/or build project assignment ideas can be found at the end of each chapter.

Technical reports

Technical reports are used in industry to communicate technical information. A technical report (also known as a scientific report) is a document that describes the process, progress, or results of technical or scientific research or the state of a technical or scientific research problem. They essentially take technical and scientific language and concepts and translate them in to explanations that nonexperts can understand and act on as necessary. Doing this usually necessitates explaining complex concepts with examples, graphics, metaphors, or the like to make the concepts clear to a nonscientific/technical audience.

One way to handle these types of assignments is to tell students that the audience they are writing for is either (a) junior high school students, (b) high school students, or (c) college students not in this area of study. This gives them a specific target audience to shoot for which is not above and beyond their own personal scope.

Assessment Tasks

In assessment tasks, the student is asked to evaluate or make a judgement about something. Often this is in the form of an opinion piece or via addressing perceived benefits and drawbacks. These types of assignments offer the student an opportunity to become more self-aware as well as express their individual thoughts and ideas.

Case studies

A case study provides a description of a particular situation or scenario, which is then critically analyzed by the students. Case studies unite theory and practice. In a case study, the student examines a situation, identified the positives and negatives of the situation, and makes recommendations. In writing a case study, the tone should be factual, authoritative, concise, and easy to follow. There is often in executive summary if the case study is longer.

When assessing a case study, we recommend the following rubric:

Identification of main issues	20%
Connections to theoretical or empirical research	20%
Analysis and evaluation of issues	25%
Assessment of the facts in light of relevant research	
Evaluation of consequences	
Recommendations on effective solutions or strategies	20%
Mechanics and formatting	15%

Argument papers

In an argument paper (also known as an argumentative paper, argument essay, persuasive essay, or persuasive paper), the writer is tasked with investigating an issue, taking a logical stand on the issue, and finding and incorporating a multitude of evidence in a logical manner to support the overall claim. Usually argument papers include: an introduction, a refutation segment that gives the opposing view and then refutes it (i.e., the author understands and has considered opposing viewpoints but rejects them for sound reason), presentation of the author's argument (usually with multiple points on justifying the viewpoint offered), the conclusion, and sources used.

When assessing an argument paper, we recommend the following rubric:

Introduction and background	15%
Presentation of counter argument	15%
Presentation of affirmative argument (evidence)	25%
Logic and validity of argument	15%
Grammar/Spelling/Proofreading/Style	15%
Format/Referencing	15%

Rubrics and Other Grading and Assessment tools

Instruction is most effective if the teacher includes specific requirements, which align with learning outcomes that must be met for the assignment to be complete (Dotger & Causton-Theoharis, 2010). The teacher should also track the progress of each student to verify that true learning is reflected in the students' assignments (Willis & Mann, 2000). Each of these practices can be seen in the methods used below.

Terms of Agreement Quiz

No matter what course structure and assessments you choose, we recommend you have students complete a terms of agreement quiz, also known informally as a syllabus quiz, at the start of the term. This ensures the students have read and understand the course structure, goals, and expectations. The following is a sample terms of agreement quiz, which students must complete with a score of 100%

© enterlinedesign/Shutterstock.com

before additional course content becomes available to them. The quiz can be repeated as many times as necessary.

Quiz Instructions

This quiz will ensure that course expectations are clear. This is a multiple attempt quiz and must be passed with a perfect score, which releases the full course materials.

Quiz Questions

I understand that this course is served up cafeteria style, meaning that there are more assignment options than need to be done for an A grade, as explained in the syllabus. I also understand that I need to complete either the quiz with an 80% or more or submit the highest point-value assignment for each module in order to unlock the next module.

True
False

I understand that the due dates are firm and assignments cannot be turned in late without prior approval from the instructor. I also understand that technological problems are not a justifiable reason for late work.

True
False

I understand that for some assignments, I may be required to purchase some basic supplies such as aluminum foil, block of wood, a mouse trap, or a syringe. Otherwise, I can opt out of these assignments.

True
False

I understand that some assignments will require the uploading of video or still images. However, I can opt out of these assignments.

True
False

I understand that this is an unlimited attempt quiz, and that when I have answered all of the questions correctly acknowledging my agreement to abide by the course policies, the full course materials will automatically become available.

True
False

Rubrics

A rubric is used to communicate expectations and scoring/grading criteria used. The rubric ensures that students and instructors alike are clear on expectations and measures to be used. It also ensures consistency in grading and clarity of overall feedback. As Rosenow (2014) points out, "Task-specific rubrics are particularly beneficial because they increase discussion about the different components of a specific assignment" (p. 32). Rubrics also offer more clarity and transparency about assignment criteria. We recommend, at a minimum, rubric use to list criteria and components of a specific assignment. Taken further, it could also incorporate descriptors that express how a decision is made if the criteria have been met.

References

Adedokun, O. A., Dyehouse, M. Bessenbacher, A., & Burgess, W. D. (2010, April). *Exploring faculty perceptions of the benefits*. Paper presented at the Annual Meeting of the American Educational Research Association, Denver, CO. Retrieved from http://files.eric.ed.gov/fulltext/ED509729.pdf

Arendt, A., Trego, A., & Allred, J. (2016). Students reach beyond expectations with cafeteria style grading. *Journal of Applied Research in Higher Education, 8*(1), 2–17. Retrieved from http://dx.doi.org/10.1108/JARHE-03-2014-0048

Bell, A., Mladenovic, R., & Price, M. (2013). Students' perceptions of the usefulness of marking guides, grade descriptors and annotated exemplars. *Assessment & Evaluation in Higher Education, 38*(7), 769–788. doi:10.1080/02602938.2012.714738

Bengiamin, N. N., & Leimer, C. (2012). SLO-based grading makes assessment an integral part of teaching. *Assessment Update, 24*(5), 1–16.

Chapman, D. W. (2003). Undergraduate research: Showcasing young scholars. *The Chronicle Of Higher Education*. Retrieved from http://chronicle.com/article/Undergraduate-Research-/9284

Colvin, J. C. (2005, January). *Peer tutoring and its impact on problem-based learning*. Paper presented at the Institute for Transforming Undergraduate Education, Newark, DE.

Dotger, S., & Causton-Theoharis, J. (2010), "Differentiation through choice: using a think-tac-toe for science content", in *Science Scope*, Vol. 33 No. 6, pp. 18–23.

Duke University Thompson Writing Program. (n.d.). What makes a good research question? Retrieved from http://twp.duke.edu/uploads/media_items/research-questions.original.pdf

Fakayode, S. O., Yakubu, M., Adeyeye, O. M., Pollard, D. A., & Mohammed, A. K. (2014). Promoting undergraduate STEM education at a historically black college and university through research experience. *Journal of Chemical Education, 91*(5), 662–665. doi: 10.1021/ed400482b.

Feldman, A., Alibrandi, M., & Kropf, A. (1998). Grading with points: The determination of report card grades by high school science teachers. *School Science and Mathematics, 98*(3), 140.

Hoffman, J. R. (2009). Applying a cost-benefit analysis to undergraduate research at a small comprehensive university. *Council on Undergraduate Research Quarterly, 30*(1), 20-24.

Ishiyama, J. (2002). Does early participation in undergraduate research benefit social science and humanities students?. *College Student Journal, 36*(3), 380–386. Retrieved from http://goo.gl/VHBpQ9

Kuh, G. D. (2008). Excerpt from high-impact educational practices: What they are, who has access to them, and why they matter. Retrieved from https://www.aacu.org/leap/hips

Levine, M. (2014, May 15). Advocating a new way of grading. *University Times*, 46(18). Retrieved from http://www.utimes.pitt.edu/?p=30598

Lopatto, D. (2010). Undergraduate research as a high-impact student experience. *Peer Review*, *12(2)*, 27–31.

Malachowski, M., Osborn, J. M., Karukstis, K. K. and Ambos, E. L. (2015). *Realizing Student, Faculty, and Institutional Outcomes at Scale: Institutionalizing Undergraduate Research, Scholarship, and Creative Activity Within Systems and Consortia*. New Directions for Higher Education, 169, 3–13. doi:10.1002/he.20118

Mostrom, A., & Blumberg, P. (2012). Does learning-centered teaching promote grade improvement?. *Innovative Higher Education*, 37(5), 397–405. doi:10.1007/s10755-012-9216-1

Rosenow, C. R. (2014). Collaborative design: Building task-specific rubrics in the honors classroom. *Journal of the National Collegiate Honors Council*, 15(2), 31–34.

Salsman, N., Dulaney, C. L., Chinta, R., Zascavage, V., & Joshi, H. (2013). Student effort in and perceived benefits from undergraduate research. *College Student Journal*, 47(1), 202–211.

Schmitz, H. J., & Havholm, K. (2015). Undergraduate research and alumni: Perspectives on learning gains and post-graduation benefits. *Council On Undergraduate Research Quarterly*, 35(3), 15–22.

Schneider, K. R., Bickel, A., & Morrison-Shetlar, A. (2015). Planning and implementing a comprehensive student-centered research program for first-year STEM undergraduates. *Journal Of College Science Teaching*, 44(3), 37–43.

Stolk, J., & Harari J. (2014). Student motivation as predictors of high-level cognitions in project-based classrooms. *Active Learning in Higher Education* 15(3): 231–247. DOI: 10.1177/1469787414554873

Suskie, L. (2004). *Assessing student learning: A common sense guide*. Bolton, MA: Anker Publishing Company.

Willis, S., & Mann, L. (2000). Differentiating instruction. Finding manageable ways to meet individual needs. *Curriculum Update*, winter 2000, 1–3. Retrieved from http://www.ascd.org/publications/curriculum-update/winter2000/Differentiating-Instruction.aspx

Chapter 3

Managing Technological Systems

(Continued)

(*Continued*)

> ➤ Design and/or Build projects
> ➤ Assessment tasks
> ❖ Terms
> ❖ References
> ❖ Further reading

Chapter Learning Objectives

❖ Define the Internet of Things (IoT)
❖ Identify components of smart sensors
❖ Explain how the IoT works
❖ Examine interrelationships of different technological systems
❖ Describe the concept of convergence
❖ Summarize how globalization is changing the way we live and work

Overview

In the United States, we live in a society of interconnected technological systems. These technological systems are interconnected, interrelated, interacting, and interdependent structures, which are organized to accomplish specific goals. A system can consist of processes, people, materials, machines, organizations, parts, plans, structures, rules, and arrangement of items that work together at some level. Systems can be independent of other systems but today we are seeing a convergence of systems. We are also seeing an increasingly interconnected world of technological systems at the global level.

Our homes are connected to many systems including electrical, gas, cable, satellite, internet, phone (though many households no longer have landlines and rely entirely on the cellular network), and in some cases security systems. We have complex communication systems that allow many of us to work remotely from home with colleagues around the world on a variety of projects. We have plethora of goods that are manufactured and produced globally. Many of the goods that are in your home and workplace are made of components that were built thousands of miles away. Just take a look in your kitchen for example and see how many countries are represented by the products you use and the purchases of food that you made.

Systems

Each of the chapters in this textbook discusses systems. Even the structure of this book can be considered a system that is embedded within the structure of the higher educational system. You as a student are part of this system, the online course management system is also part of the system. The interactive assignments and links are part of this system, the constructivist approach to learning and the offering of assignment options and student control are all parts of an interrelated interdependent structure. When you graduate from college and look for a position in the work force, again you will be part of a larger economic system that was dependent upon the higher education system of which this textbook and class are part of.

In this chapter, we will look at the systems involved in each chapter and the integration and convergence between systems will be discussed. We will also be looking at technological enterprises and the various systems involved in different technological enterprises. Management of technology (MoT) and technological systems will be included in this chapter. The concept of the "Internet of Things" will also be addressed. Globalization and the realization that we live in an exceedingly interconnected interdependent world will be explored.

©kentoh/Shutterstock.com

Design Thinking and the Scientific Method

A common theme throughout the text is the integration and convergence of systems. The concepts of the scientific method and design thinking are integral components of each of the following areas including: information technologies, aerospace and transportation, location and tracking technologies, autonomous and semi-autonomous technologies, mediated and virtual reality, medical and biotechnologies, nanotechnology, advanced manufacturing and production technologies, technologies used in business intelligence and analytics, robotics and artificial intelligence and energy and environmental technologies.

The chapter on design thinking discusses the importance of the development of the scientific method. This is a system in which one observes what is occurring, then they ask a question, they do research, they make a hypothesis, do an experiment, make conclusions, and then report their findings. The scientific method is used in some part of each and every chapter that follows, as are the elements of design thinking. The steps involved in design thinking start with empathizing with the person who will be using the design or system, then defining the problem, followed by ideating or developing a large variety and quantity of ideas, from there one goes on to the prototype stage where the initial development takes place, then you have to test the prototype to see if it works or perhaps the design can be improved upon. This is a cyclical process and can be considered a system that is used widely until a favored design or process is developed. Design thinking can also be used to improve products and processes and is a very important component of innovation. Failure is also actually part of this design process, and out of failure, many good things can result, including improved design and new knowledge. There are even journals devoted to failure and negative results, including the *Journal of Negative Results*, the *Journal of Negative Results in BioMedicine*, and the *Journal of Pharmaceutical Negative Results*. (Goodchild van Hilten, 2015).

We can see the principals of design thinking in information technologies that we use every day. Consider the design of your smart phone or the tablet you are using. It is probably small, lightweight, internet connected, Wi-Fi enabled, has global positioning system (GPS) and there are at least 20 apps or more on it. We click on icons to launch applications. At one time, professors in computer science thought that the concept of a graphical user interface as used in many of the systems we use today would never take off because too much memory was used; now this design is ubiquitous in computing systems.

In the aerospace industry examples of the use of the scientific method abound, from the first flight in hot air balloons to putting the first man on the moon, a series of experiments were involved and design thinking was used to develop the airplanes and rockets. Location and tracking methods also use design principals and our GPS use design principals and information discovered through the use of the scientific method for putting satellites in space. These complex systems are all interdependent and interconnected. The use of virtual reality can be applied to training that astronauts and pilots received. This then carries us into the development of autonomous and semiautonomous vehicle. The design and infrastructure for self-driving cars is still being worked out and experiments are taking place daily in this area. Soon, there will be standards and infrastructures developed that support the use of autonomous and semiautonomous vehicles. They will use GPS in navigating, which can be integrated into collision avoidance systems as well as real-time mapping that notifies the user that traffic is being rerouted due to construction.

The medical and biotechnology areas are based on the use of the scientific method. There are experiments occurring daily in these fields also and new methods of treating disease are continually being found. Processes and technologies used in the design of medical devices and procedures are under constant development and refinement. Regulatory agencies, such as the US Food and Drug Administration (FDA) also have input into the development process of medical devices as they give final approval for their use in the United States. For example, the FDA (2016) just finalized their recommendation on the "Use of International Standard ISO 10993-1, Biological evaluation of medical devices Part 1: Evaluation and testing within a risk management process." This document provides guidance on risk assessment, such as biocompatibility (are there chemical of physical characteristics of the device that could cause unwanted tissue responses), to medical device manufacturers on the complex process of developing devices that are or will be used in patients. A human-centered approach to design is especially important in the medical field and you will see that the application of design thinking principals is readily apparent in medical technologies used in hospitals and other health care facilities. Manufactured components such as imaging equipment are developed and designed to provide the necessary images while being relatively comfortable for the patients and for the clinicians that use the equipment.

Completing the Human Genome Project required the use of high degrees of computational power and applications along with extensive use of robotics. The integration of these systems required extensive use of design processes. Laboratories had to be set up that integrated a large array of different types of equipment, such as robotic equipment, computer stations, areas where samples are kept in temperature controlled environments and even spaces where mice are kept for analysis.

Nanotechnology is another area where design principals and the scientific method are used. Just think about what you need to design at the nano level, a nanometer is small, very small, in fact it is only one billionth of a meter or 1/1,000,000,000. Nanotechnology crosses many aspects in science including chemistry, biology, computer science, material science, and energy science and integration of these scientific disciplines is necessary in order to capitalize on the benefits of nanotechnology. The "U.S. Department of Energy's Argonne National Laboratory has created a new collaborative center known as *Nano Design Works* (NDW)" (http://nanoworks.anl.gov).

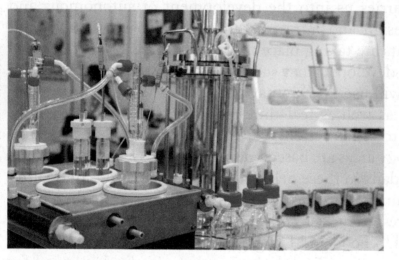

©Chekunov Aleksandr/Shutterstock.com

They help with the fabrication and synthesis of nanomaterials. They have the electron and scanning probe microscopy capabilities used in characterizing the nanomaterials and in the development of predictive modeling at the nano level, which assists in translating fundamental research into actual products. You will learn more about nanotechnologies in Chapter 12.

Information Technologies

The material covered in Chapter 5 is based on large and complex systems with several important components that interact with many other systems. It is a system that includes computers that process and store information, as well as retrieving and sending information. Layered upon that are the systems that ensure the secure transmission of data and information, which is critical in many areas especially when dealing with financial transactions and personal health-related information.

Information technology applications are an integral part of every area discussed in this textbook. In addition, they were critical in the writing of this textbook. Documents were housed in the cloud where the multiple authors could access and see the work that the other authors were doing. Research was able be to done online, library databases were accessed online, and word processing programs were used.

Information technologies are used in the design of many items. Consider the use of a computer-aided design (CAD) program that is used in conjunction with concurrent engineering to design and develop machining processes used to manufacture the mobile phone you have. For example, someone designs the back of the case of the mobile phone in Milan Italy, while someone else designs the lens used in the embedded camera in Japan, and another person or team of people are working on the software applications in Oregon, which will enable the person using the phone in Utah to take and store pictures, as well as edit them. People in all of these countries are part of an integrated team and they will be able to send and receive files or the files will be housed in the "cloud" retrievable at any time by the different people working on the project. Once the design of the back of the case is finalized then another team designs the facility and machinery, which will build the components necessary for the manufacture of the phone. Thousands of people can be involved in all of the processes that are required to design, build, and distribute the phone to the consumer.

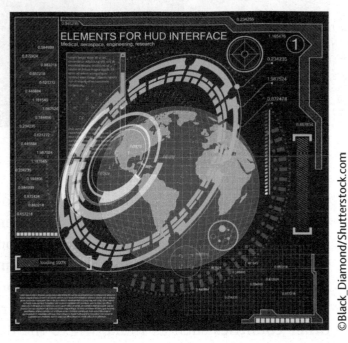

©Black_Diamond/Shutterstock.com

Information systems are critical to the aerospace and transportation industries. For example, photos taken with the Hubble telescope from space are sent back to earth with the use of information technologies. Large amounts of data are produced from these images and the telescope was a key component in the scientific discovery of dark energy (Hubble space telescope, n.d.). Navigation systems of aircraft are also moving to the sky with the use of accelerometers and laser ring gyros and GPS as well as ground-based Distance Measuring Equipment. You will learn more about these technologies in Chapter 6.

Location and tracking technologies also require extensive utilization of information-based technologies. The use of GPS and geographical information systems (GIS) generate large amounts of data that can be collected by many different organizations. Government offices can use GIS for a variety of information-based tasks, including keeping track of road work projects, mapping out critical infrastructure elements such as gas, water and electric lines and knowing where police and fire service personnel are. GIS are also very important in planning of city developments. Industry also uses GIS systems in a variety of manners, distribution companies can track shipments in real time, companies using GIS technologies will frequently ask if it is OK to track you via your cell phone or tablet when you download an app.

They want to know where you are and what you may be doing. If you go to a store, that information can be tracked and stored for future use. These technologies and their relationship to information technologies are discussed in detail in Chapter 7.

In addition to information technology use in the aerospace and transportation industry, and use in location and tracking, information systems are also integral components of autonomous and semiautonomous technological systems. Unmanned aerial vehicles relay on the ability to communicate with ground-based technologies.

©Sergey Nivens/Shutterstock.com

In military applications communication systems, interoperability and modularity, security systems for research and intelligence gathering as well as weaponry use all rely on secure transmission of information.

Mediated and virtual reality systems in addition rely heavily upon information-based technological systems. Training can take place remotely using information-based technical systems coupled with virtual reality systems. People can play computer games such as Madden with other people that live on the other side of the country with this technology, which some of you may do when not studying.

We see the integration of information technologies, life sciences, and physical sciences in the diagnosis and treatment of disease and illness. In the medical arena, the advent of electronic health records is based on information technologies and has changed the way doctors, nurses, and clinicians work. A growing amount of information is contained within these records. People can see test results online and schedule appointments online. Sometimes radiology reports are sent electronically to other countries such as India, where the data is interpreted and the results are sent back electronically. People can also be informed patients and look up information regarding a disease or diagnosis easily online. It is important to note that this does not take the place of a visit to your doctor for diagnosis and treatment. Researchers can cull data found in the records. Biotechnology in medicine is creating vast amounts of data that require analysis resulting in a growing need for informationalist programs and computational bioinformatics because of the growing amount of information in the biomedical and biotechnology area.

Nanotechnology applications require computational programs in order to produce to materials at the nano level. Efforts to produce tiny computer chips lead to many of the early applications in nanotechnology fabrication and most of the techniques used in that industry include some type of information-based technology.

In advanced manufacturing and production settings, information technologies are used every day. Forecasting buyer behavior along with production needs assists plant managers with purchasing and inventory decisions. Concurrent engineering for development of products occurs with the assistance of information technologies. Without information technologies, we would not have additive manufacturing

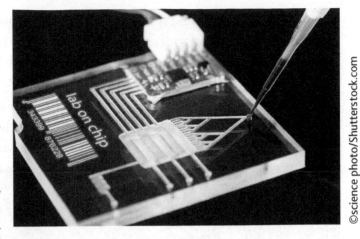

using 3D printers that are now being used to print parts for satellites while in space. We have come a longway literally!

The use of computing power is the underlying technological component used in modern day business analytics and intelligence research. Data are gathered online and stored in large databases where it is "mined" and interpreted for useful information. So for example, if you go to Amazon and look for a book on humming birds, you will also be directed to purchases that other customers made who had looked for books on humming birds. You may end up purchasing a humming bird feeder to go along with the humming bird book even though that was not your original intention. All of the reward cards that you own are used to collect data on you and your purchases, this information can also be bought and sold to other businesses, which is another method of revenue generation for firms. If you enable the ability of apps to track your location that is even better for some companies. They know when you are getting near a store and your driving habits. It is truly amazing the amount of information that is accumulated on you using business intelligence and analytics.

Of course, robotics and artificial intelligence are also dependent on computing power and information technologies. Robots are being used in many areas today, including industrial applications such as welding the exterior panels of car bodies and other components. There are also educational and recreational robots. We even have surgical robots, and surgery can be done across thousands of miles with the use of telemedicine and robots. These systems are highly integrated and interdependent. We are on the way to artificial intelligence and more information about these topics can be found in Chapter 15.

Energy and the Environment

All of the technological systems discussed in the text require the use of energy and they have an impact on our environment. Just think of information systems and all of the components within that system that require energy. First of all, you need to consider all of the energy that went into the manufacturing of the computers, laptops, phones, servers, cables, routers, tablets, cell towers and switching devices, and all of the other peripheral components; then you need to think about how all of that equipment got to its final destination; boats, trucks, and planes are used to transport these goods (these also had to be built and that required energy). Once the equipment is in place you have to have power to use it, you also need power to keep servers and massive storage device used for cloud computing applications cool. It is estimated that the Internet is responsible for approximately 2% of global energy consumption today (Giles, 2011).

Think how much energy is saved though with the use of the information technologies and the Internet. You can do your research right from home instead of driving to the library, and finding the information there. You will most likely have limited amounts of information housed in an old school physical library unless they have Internet access to large databases, which most of them do today. Then you have copy the papers that you find and transport them back home. Also, consider that you can have meetings across the miles with video and audio conference calls. You can have someone on the line from China, another person in Michigan, and someone from Utah all on the same call using Skype (and it can be at a very low cost or free if you have free Wi-Fi access). To have all three people travel to the same location for a meeting would be very costly in terms of energy use.

The aerospace and transportation systems are constantly looking for ways to reduce energy consumption that leads to a reduction in costs. There is a continual quest by automobile manufacturers to reduce energy consumption. Cars are being made that are much lighter with the use of new materials, such as aluminum frames that are being welded together with the use of robotics, and the vehicles are more aerodynamic (don't forget the design systems and their tie-in here). Alternative fuels and battery-powered cars are being built today. The infrastructure to charge these vehicles or to fuel them in the case of natural gas vehicles is still being developed. Location and tracking technologies combined with information technologies can aid in the development of more efficient delivery routes. Let's not forget that drones and the use of semiautonomous and autonomous vehicles are also being explored for delivery of items which could potentially lead to a reduction in energy usage.

Biotechnologies and nanotechnologies are used in the cleaning up the environment. Bioremediation uses biologically based material to improve the conditions of soil and water that have been polluted from industrial processes or things such as oil spills. The "confluence of environmental biotechnology and nanotechnology will lead to the most exciting progress in the development of nano-devices having bio-capabilities in novel metal remediation strategies" (Rajendran & Gunasekaran, 2007). See Chapter 11.

Training done on simulators using virtual or mediated reality systems can also lead to a reduction in energy costs. Aspiring pilots frequently use simulation instead of actual airplane time to do some of their training for licensing. Manufacturing and production systems can also be made more energy efficient and steps are continually being made in this area due to the reduced costs associated with using less energy not to mention the positive effects on the environment when we use less energy.

Business intelligence and analytics are also used to help in the reduction of energy use. We have smart meters installed on our homes and hotels are starting to use on demand energy for heating, cooling, and lighting applications. For example, FirstFuel Software, using remote assessment and analytics has helped the Department of Defense identify areas where energy use could be reduced in over 100 buildings, which could result in approximately $4 billion in savings (Clancy, 2014).

©Khakimullin Aleksandr/Shutterstock.com

Convergence and Integration of Systems

Throughout the text and in the world today, you are seeing a convergence and integration of the various technological systems that are in use. These systems are highly interdependent. Convergence crosses traditional disciplinary boundaries and integrates tools

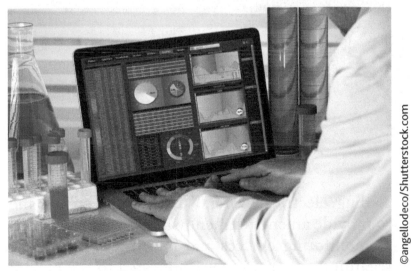

©angellodeco/Shutterstock.com

and knowledge from different disciplines including physical science, life sciences, engineering, physics, and mathematics (National Academy of Science [NAS], 2014). The development of a convergent-based approach to solve societal problems is becoming increasingly important. Economic, social, and behavioral sciences along with humanities research is needed to inform and support convergence-based research. Remaining in disciplinary silos will no longer be adequate to address problems facing society such as the need for clean fuels and energy storage systems, maintenance of secure food supplies and treatments for chronic diseases (NAS, 2014). We have seen some convergence in some of the newer disciplines such as nanotechnology and biotechnology, but more support is needed for convergent research to take off.

Biomedicine is one of the key areas that is now using convergence to solve problems. Some of the pressing areas where a convergent approach to research is needed are, brain disorders, infectious diseases and immunology, and cancer treatments. Research and progress is already being done with advanced imaging technologies, drug and therapy delivery using nanotechnology, engineering of body parts, and data and health information. There are many more potential applications though. For example:

"Convergence techniques could enable rewiring the genes of mosquitoes to eliminate Zika, dengue, and malaria. They could help solve the emerging threat of drug-resistant bacterial strains, which infect over two million people in the U.S. every year. Convergence-based immunotherapy could activate a person's immune system to fight cancer, reprogramming a person's T-cells or antibodies to find and attack tumor cells" (Massachusetts Institute of Technology, 2016).

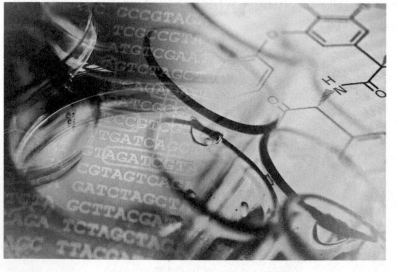

©isak55/Shutterstock.com

When considering biotechnology in medicine, you will note that there are biological systems, chemical systems, mechanical systems, information systems, and energy systems involved and within each of these systems, there are subsystems. Integration of these systems is critical to their optimal functioning. In the energy system used in hospital setting, you have lighting, heating, cooling, electrical, water, Wi-Fi, smart sensors, and gas systems all working together to provide an environment where patients are comfortable, where technologies such as monitoring devices and imaging devices can run, and where vital information regarding patients is communicated to the nursing staff, and where the pathologist in the lab can send results from genetic testing. These systems are or will soon be, part of the smart grid and the IoT.

We also see a convergence of different systems in many other areas including autonomous and semiautonomous vehicles, aerospace and transportation, mediated and virtual reality, advanced manufacturing, nanotechnology and energy, and environmental applications. Consider the integration of GPS with GIS and smart sensors and informatics and robotic technologies in the transportation, aerospace and semi-autonomous vehicles industries. Do not forget that the vehicles have been produced using advanced manufacturing techniques and nanomaterials, and that bioremediation may have been used to assist in the disposal of the waste generated from the manufacturing processes or the materials could have been obtained from the recycling of electronic waste.

We will continue to see an increased integration and convergence of systems and disciplinary knowledge. Items will be interconnected to each other in manners yet to be thought of. The amount of information in the world will continue to accelerate as well. Investments in knowledge capital and innovation are occurring as a result and have even exceeded investment in physical structures in some countries. (OECD, 2013)

Internet of Things

The IoT will eventually change the way we live, work, and communicate with each other. It is all about connectivity and convergence of systems. "The Internet of Things (IoT) is a system of interrelated computing devices, mechanical and digital machines, objects, animals or people that are provided with unique identifiers and the ability to transfer data over a network without requiring human-to-human or human-to-computer interaction" (Techtarget.com, n.d.). Objects will be interconnected with each other through the Internet. These objects can be anything from the refrigerator in your home, to the drug delivery device in a hospital, to your car, your wearable health and fitness tracker, or your electricity meter. Basically, anything that can be connected to the Internet with an Internet Protocol (IP) address and that can send and receive data will be part of this. We already have electronics embedded in many of our devices and connecting them together in useful ways is the next step.

The IoT has the potential to change business models and improve processes in business as well as reducing costs and creating value. Decision-making, including resource allocation and use will be improved as a result of the improved ability to obtain data and process and store it with fewer input errors. Businesses need to think about how the IoT will impact their processes. They will need to have leadership that will assist them in development and implementation of a business strategy that optimizes the IoT. Their production processes and retail operations will change, as will the ways that they structure and use data, all as a result of the technological integration of the various systems in the IoT (Rudman & Sexton, 2016).

What is Needed for the IoT to Work?

There are three important elements that are required for the IoT to function: (1) the communication and sensing technologies, (2) the ability to store data and analyze it, and (3) the applications developed to optimize or make decisions based on the data that have been received. These three elements work in conjunction with each other to enhance lives and improve business performance. This network of interconnected objects will provide real-time information needed for analysis and responses tailored to the situation.

The communication and sensing devices can be an Radio frequency identification (RFID) tag or smart sensor and a wireless network can be the communication device. Smart sensors are critical, as they collect the input from the environment and perform functions that have been predefined. Minimally smart sensors have a microprocessor and some type of communication technology along with the sensor itself. The ability for the smart sensor to compute is critical to the physical design of the mechanism; the sensor incorporates software that does the predefined functions such converting data, communicating with other external devices, and digital processing. The smart sensor is an essential and primary element in the IoT.

For example, your fitness tracker notes that you have walked 10,000 steps for five days in a row, this information could automatically be sent to your local sports store. The "Sports R Us" store works with the fitness tracker manufacturer and they then send you a congratulatory note on your achievement and a discount coupon on your smart phone for new walking shoes (as the shoes have to be wearing out if you are walking 10,000 steps a day). If you are in the area where the store is located (remember your GPS and the ability to track your whereabouts) you may even get a notification that hey, walking shoes are on sale now and look at how close you are to the store. It could even be possible that the sensor is part of the shoe and it will notify you when you need new ones.

The use of wearable sensors is already occurring and many applications are being developed for wearable smart sensors in the health field. For example, there are devices one can wear when sitting at a computer that monitor for posture, and will encourage you do to exercises to relieve lower back pain. There are also wearable monitors for people who have asthma and ones that will transmit vital health information to your health care provider. (Wearable Technologies: n.d.).

Vast amounts of data will be generated from these devices and this data will most likely be stored in the cloud. The cloud is basically a network of remote servers that are used to store and retrieve data from. Instead of storing the information on laptops, desktop computer or servers located in a business organization, data files are now being stored in the cloud. This makes information and files accessible from anywhere. Programs and applications can also be stored and retrieved from the cloud. You will no longer have to carry a laptop with you to get to the file you need. Companies can and do have people from all of the world working on the same files from their various locations in real time.

Challenges in Development and Use of IoT

Several challenges are associated with the design and implementation of the IoT. First of all, standards have not been decided upon yet. As with many emerging technologies, it may take a while for standards to be developed and this will make implementing the technology into existing business is somewhat difficult and expensive at first. The development of standards is especially important for data integrity and security. You will want sensors that can be used on different platforms, which means that they could be used on Windows-, Android-, iOS-, and Linux-based devices as well as working with different industry applications.

©a-image/Shutterstock.com

Companies need to have technical expertise and skills to develop the physical infrastructure and to perform the data analysis. They will have networks of sensors, computers, and actuators that move or control something. Investments will be required to build the necessary infrastructure both at the network and software levels. Companies will also need to have control of access, because if the objects have sensors that are connected to the Internet that poses a security risk. In addition, companies need to ensure that the object has a unique identifier that can be verified to make sure that the information that is being received is in fact from the object intended versus being from another object—otherwise the correct information will not be sent.

Security and integrity of data is critical to the functioning of IoT. You may have heard about Chrysler recalling more than 1.4 million vehicles because of security risks in the onboard computer system in the vehicles. Hackers were able to remotely take over control of the Uconnect dashboard computer on a Jeep, this computer-controlled steering, transmission, and brake functions (Greenberg, 2015). It would be terrifying if someone was able to hack into your car and disable your brakes or steering. Chrysler collaborating with Sprint was able to fix the software vulnerability in their vehicles by sending a USB drive to owners and having them update the onboard computer system. Preventing future attacks on vehicles' computerized systems will be of great importance. Most of today's cars have computers that assist in the control of different functions and that are connected. In many ways we benefit from the connectivity of our vehicles, traffic jams can be seen on the display screen of new vehicles and you can be rerouted to avoid them. There are even integrated apps that will notify you via text message that your automobile needs an oil change or other service.

Hospitals and health care is another area where security is extremely important. Personal data could be leaked from your fitness and health-tracking devices (Mansfield-Devine, 2016). There are millions of pieces of medical equipment installed in hospitals today, many of which are interconnected to networks. If a hospital's network or a health insurance company's network is breached by hackers, the hospital or insurer could incur fines and their brand name could be damaged. It becomes even scarier though if the medical device in the hospital or doctor's office is hacked. For example, the medication infusion pump (a system that delivers drugs to patients usually through an intravenous [IV] line) that was produced by Hospira Symbiq was able to be accessed remotely through a hospital's network. The FDA issued a warning to hospitals, nursing homes, and other facilities that used this infusion pump about the cybersecurity vulnerability and they "strongly encourage that health care facilities transition to alternative infusion systems, and discontinue use of these pumps" (FDA, 2015). Not having a secure device cost Hospire Symbiq a great deal of money and there was damage to their brand name.

Technology Management

All of the technologies and technological systems discussed in this chapter and within the textbook need someone to manage them. There is rapid technological change occurring in society, coupled with increasing complexities that are involved in the convergence of different industries. Businesses have to be effective, create value, and do things quickly as technologies are constantly changing. There is also constant pressure to allocate limited resources effectively.

Technology management, also known as MoT, is a broad and growing discipline that crosses many industry sectors including but not limited to: the information technology industry, aerospace and transportation industries, supply chain management and distribution, the medical and health care industry, biotechnology, nanotechnology, manufacturing and production, analytics and business intelligence, robotics and artificial intelligence, and energy and environmental industries.

There are several challenges associated with managing today's high-tech industries. We are moving to an interconnected knowledge-based economy. There is rapid technological change occurring across multiple industries in the global economy. The increased connectivity requires collaboration among varied partners, including academics, industry, and governments to optimize the benefits. Innovation in product lines, services, and processes requires keeping employees up-to-date and proficient in the use of the new technologies and this poses challenges to managers in today's businesses. Industry needs creative highly skilled employees to provide the knowledge intensive products and services that people desire. A high level of information is required, and communication capabilities are necessary to provide access to all of the information that is being generated today. You cannot apply the same management practices that are applied to manual, service, or data work to creative unstructured work that is driven by knowledge and the expertise of the employees (Maier, 2010).

©Rawpixel.com/Shutterstock.com

Technology managers plan, organize, and coordinate many areas; they cultivate operational capabilities and provide guidance and leadership in the development of business strategy. There are several core competencies associated with the MoT: management of technological change, management of organizational change, project management, assessment and evaluation of technology, quality MoT, information and knowledge management, innovation and product development, and strategic MoT (Becker, 2008).

The management of technological change, such as the increasing use and incorporation of smart sensors, often brings about a corresponding change in organization structure. For example, data that once had to be gathered and inputted by someone now are automatically generated from a smart sensor, so the person who had previously collected the data is no longer needed for data collection, but now the vast amount of data generated needs to be analyzed by someone in order to turn the data into useful information.

There is also an increasing need to manage projects that is in many ways brought about the increased levels of technological change. New systems are constantly being brought online at organizations. The deployment of the new technology is a project with start and end dates. A variety of tasks are necessary to keep the project moving forward. In addition, there are a variety of constraints including budget, time, suppliers, and workforce that must be managed.

The assessment and evaluation of technology is an important part of technology management. Managers need to be able to assess risks associated with bringing new technologies onboard and they need to be able to evaluate existing technologies. Managing quality is a critical component of technology management. All of the systems mentioned in this chapter and within the textbook have to meet certain quality standards in which to function. Business processes are put in place which attempt to ensure that certain quality standards are met. Continuous improvement processes are cyclical and employed. ISO 9000 standards are used internationally to measure quality practices within organizations.

Managing information and knowledge is becoming increasingly important in today's highly interconnected world. People can obtain information about products easily on the Internet, and social media plays a central role in many consumers decisions on whether or not to make a purchase. Access to knowledge is critical and having informed knowledgeable employees has become essential.

Innovation, product development, and design are integral to technology management. There is constant innovation in products and processes today. The manner in which innovation and product development is approached and supported is a key strategic areas that business leaders must address. If organizations fail to adopt innovative practices they may be out of business.

Strategic planning is required by technology managers today. They must have the ability to constantly scan the environment in which their organization lies and be able to make decisions that will move the organization forward. They need to decide what the important directions in which to steer to the organization are. They have to determine what types of resources will be needed for moving in the desired direction.

Strategic MoT

Strategic MoT entails having a long-range perspective on the wide-ranging effects that technologies will have on all levels and functions of an organization. It involves establishing business strategy. Often strategic management has to do not just with the technologies used but also with innovation. It commonly has a focus on value of the enterprise or company as a whole as related to current profits and cash flow and, perhaps more importantly, how it is sustained in the future. Strategic MoT also emphasizes systematic MoT through its expected life cycle from birth to decline. After a technology is created, assuming it is accepted and utilized, it will at some point hit maturity and ultimately decline and be replaced.

Management Information Systems

Management information systems (MIS), as compared to technology management, are instead formal discipline within business education that bridges the gap between computer science and the well-known business disciplines of finance, marketing, and management. Its core function is to determine business requirements for information systems as well as to work to improve efficiency and effectiveness of strategic decision-making based in system-based information. A typical job title in this area might be something like Business Systems Analyst. Actually, the system that an individual works with in this field is often known as the companies' Management Information System (singular). In this context, an MIS is generally a computerized database of financial and related information that is organized and used to produce reports on operations and the like. It gives managers feedback about company performance.

Globalization

The world we live and work in has become interconnected and information is readily accessible with modern day technologies. Information, people, ideas, technology, products, culture, and capital investment all flow easily between nations today (Kushwaha, Chaurasiya, & Chohan, 2015). We can have

©sdecoret/Shutterstock.com

conference calls inexpensively with people around the world. Data are transmitted in seconds to the other side of the globe. Breaking news can be heard everywhere. We travel in a matter of hours to other countries. Goods are produced and sent throughout the world. Business is no longer confined to one state, one country, or one continent, the world is now where business is conducted.

Many firms are multinational today; they perform different aspects of business in different countries. Science and technology transfer are integral aspects of globalization (Sidhu et al., 2015). A firm may have engineering staff in the United States, have production facilities in China, have buyers throughout Asia and the Americas, and the controlling interest in the firm is held in Japan. The ability to communicate easily and inexpensively with people from other nations has made doing business in multiple countries feasible for a growing number of firms.

The globalization of the economy and industry has also lead to the exchange of culture and ideas. We now have people from different countries working together to solve complex problems. The synergy created by science and technology applications and the collaborative efforts of scientists and technologists is changing and will continue to change the way we live and work in the world.

Career Connections

Information Assurance Analysts

Analysts who work in information assurance plan and design software systems that are deployed to prevent unwanted attacks on data and systems. They focus on security issues and risk assessment of software and hardware. They make sure that both internal employees and external customer's data and information are not accessible nor can it be manipulated by unauthorized users. Most positions require a bachelor's degree. Growth in this area is high with 18% growth projected between 2014 and 2024 (Bureau of Labor Statistics, 2015a).

Engineering Technicians

These technicians Help engineers design and develop a wide variety of equipment such as communication devices, computers, medical equipment, electronics, and navigational devices. They can do product evaluation and testing, measure, adjust, and repair diagnostic equipment, they also can work with manufactures to develop and deploy automation equipment. At least an associate degree is required to enter this field.

Energy Technicians

An engineering technician can specialize in solar, wind, nuclear, gas, and electrical areas. Some college is required for entrance into this field and there may be on-the-job training opportunities. Solar and wind technicians generally work outdoors. Growth for wind turbine technicians is expected to grow by a staggering 108% in the next eight years according to the Bureau of Labor Statistics (2015b).

Technology Managers

A Technology manager works in a variety of industries including manufacturing, information technology, medicine, education, transportation, biotechnology, and aerospace to name a few. They plan, organize, and coordinate many areas, as well as cultivate operational capabilities and provide guidance and leadership in the development of business strategy. Most technology managers have a bachelor's or master's degree. This field is expected to grow.

Computer and Information Systems Managers

Computer and information systems managers plan, coordinate, and direct computer-related activities in an organization. They often are responsible for implementing computer systems which meet the information technology goals of the company. Entry-level education for this field is commonly a bachelor's degree and the job outlook is faster than average. Added to that, the median pay is $131,600 per year. Note, however, that you can expect to work more than 40 hours a week (Bureau of Labor Statistics, 2015c).

Modular Activities

Discussions

* Post an original discussion in the online discussion board on the following topic: Discuss how society drives technological changes and vice versa. What are your thoughts and why? (200 word minimum). Next, comment on the post(s) of a minimum of one other student in a thoughtful and academic way that enhances the conversation. See rubric for grading and assessment measures.
* Post an original discussion in the online discussion board on the following topic: Discuss, with examples, challenges and risks in managing technological systems. What are your thoughts and why? (200 word minimum). Next, comment on the post(s) of a minimum of one other student in a thoughtful and academic way that enhances the conversation. See rubric for grading and assessment measures.

Tests

* Online graded quiz on overall chapter content, written in multiple choice format. When submitted, we recommend giving the correct answer along with the page number in which it is found for questions students did not answer correctly.
* Take the self-assessment quiz "How Good Are Your Management Skills?" found at https://www.mindtools.com/pages/article/newTMM_28.htm and submit your final score along with a list of the items you chose "often" or "very often" to.

Research

* Often management of technological/information systems is associated with databases or server rooms. However, it is more than that. Research an area of technological/information systems management and write up a two page informative paper. You must use a minimum of two sources that must be properly cited.

❖ Research a company that is using strategic MoT as described earlier in this chapter. Submit to your instructor the following: (1) company name and description, (2) description of how it uses strategic MoT (200 word minimum), (3) description of the impact or expected impact of strategic MoT (200 word minimum). Use a minimum of two external sources for your research and cite them appropriately.

Design and/or Build Projects

❖ Develop a schematic that shows all of the people and positions necessary to design, build, and distribute a microwave oven. You will be assessed on the inclusiveness of the schematic.

❖ Imagine the role you see the IoT playing in the next 10 to 20 years. Create a poster, collage, or diagram showing the role you predict it to have in the future. You will be assessed on the quality, thoughtfulness, and professionalism of your work.

Assessment Tasks

❖ Compare and contrast technology management with information systems management and write a two page summary of your views on how they are similar and different. You will be assessed on your thoughtfulness, depth, accuracy, and relevance.

❖ Technology management is concerned with development, planning, implementation, and assessment of technological capabilities within an organization or entity. A simple way to think of it is management of the use of technology for human advantage. What, in your opinion, are the most important aspects of technology management and why? Write a 200-word minimum summary of your thoughts with rationale for your assertions. Use a minimum of two external sources for your assessment.

Terms

Convergence—Convergence, generally, is to move toward one point and join together. In this context, it crosses traditional disciplinary boundaries and integrates tools and knowledge from different disciplines including physical science, life sciences, engineering, physics, and mathematics

Internet of Things—a system of interrelated computing devices, mechanical and digital machines, objects, animals, or people that are provided with unique identifiers and the ability to transfer data over a network without requiring human-to-human or human-to-computer interaction.

References

Becker, P. R., & Eastern Michigan University. Department of Leadership and Counseling. (2008). Core curricular elements of effective undergraduate technology management academic programs.

Bureau of Labor Statistics. (2015a). U.S. Department of Labor. *Occupational Outlook Handbook, 2016–17 Edition,* Information Security Analysts. Retrieved from http://www.bls.gov/ooh/computer-and-information-technology/information-security-analysts.htm

Bureau of Labor Statistics. (2015b). U.S. Department of Labor. *Occupational Outlook Handbook, 2016–17 Edition,* Information. Wind turbine technicians. Retrieved from http://www.bls.gov/ooh/installation-maintenance-and-repair/wind-turbine-technicians.htm

Bureau of Labor Statistics. (2015c). U.S. Department of Labor. *Occupational Outlook Handbook, 2016–17 Edition.* Computer and Information Systems Managers. Retrieved from http://www.bls.gov/ooh/management/computer-and-information-systems-managers.htm

Clancy, H. (2014). 10 Companies to watch in energy analytics. *Forbes.* Retrieved from http://www.forbes.com/sites/heatherclancy/2014/12/31/10-companies-to-watch-in-energy-analytics/#5123add821c7

Giles, J. (2011). Internet responsible for 2 percent of global energy usage. *New Scientist.* Retrieved from https://www.newscientist.com/blogs/onepercent/2011/10/307-gw-the-maximum-energy-the.html

Goodchild van Hilten, L. (2015). Why it's time to publish research "failurs": Publishing bisas favors positive results; now there's a movement to change that. *Elsevier Connect.* Retrieved from https://www.elsevier.com/connect/scientists-we-want-your-negative-results-too

Greenberg, A. (2015). After Jeep hack, Chrysler recalls 1.4M vehicles for bug fix. *Wired.* Retrieved from https://www.wired.com/2015/07/jeep-hack-chrysler-recalls-1-4m-vehicles-bug-fix/

Hubble Space Telescope. (n.d.). *National Geographic.* Retrieved from http://science.nationalgeographic.com/science/space/space-exploration/hubble/

Kushwaha, P. K., Chaurasiya, P., & Chohan, N. S. (2015). Globalization and technology—an economic dimension. *Golden Research Thoughts, 5*(3), 1–7.

Maier, R. (2010). Forward. In G. Passiante (Ed.), *Evolving towards the internetworked enterprises: Technological and organizational perspectives.* New York, NY: Springer. Retrieved from http://download.springer.com.ezproxy.emich.edu/static/pdf/734/bok%253A978-1-4419-7279-8.pdf?originUrl=http%3A%2F%2Flink.springer.com%2Fbook%2F10.1007%2F978-1-4419-7279-8&token2=exp=1466802796~acl=%2Fstatic%2Fpdf%2F734%2Fbok%25253A978-1-4419-7279-8.pdf%3ForiginUrl%3Dhttp%253A%252F%252Flink.springer.com%252Fbook%252F10.1007%252F978-1-4419-7279-8*~hmac=1d5ba1fb8c876edd0e950aba780ef13f77c2b26106a473a17884d27f743b2f51

Mansfield-Devine, S. (2016). Securing the internet of things. *Computer Fraud & Security, 2016*(4), 15–20. doi:10.1016/S1361-3723(16)30038-0

Massachusetts Institute of Technology. (2016, June 23). A strategy for 'convergence' research to transform biomedicine: Report calls for more integration of physical, life sciences for revolutionary advances in biomedical research. *ScienceDaily.* Retrieved from www.sciencedaily.com/releases/2016/06/160623145936.htm

National Academy of Sciences. (2014, May 7). National coordination needed to advance convergent research, report finds. *ScienceDaily.* Retrieved from www.sciencedaily.com/releases/2014/05/140507114803.htm

OECD, (2013). *"Introduction and overview," Supporting in investment in knowledge capital, growth and innovation, OECD Publishing.* Retrieved from http://www.keepeek.com/Digital-Asset-Management/oecd/industry-and-services/supporting-investment-in-knowledge-capital-growth-and-innovation/introduction-and-overview_9789264193307-4-en#page1

Rajendran, P., & Gunasekaran, P. (2007). *Nanotechnology for bioremediation of heavy metals.* In S. N. Singh & R. D. Tripathi (Eds.), *Environmental bioremediation technologies* (1. Aufl.; 1st ed.). New York, NY; Berlin: Springer. doi:10.1007/978-3-540-34793-4

Rudman, R., & Sexton, N. (2016). The internet of things. *Accountancy SA, 22*–23. Retrieved from http://ezproxy.emich.edu/login?url=http://search.proquest.com/docview/1794510817?accountid=10650

Sidhu, L. S., Sharma, J., Shiny, & Shivani. (2015). Information technology, globalization through research and social development.*Compusoft, 4*(5), 1760.

Techtarget.com. (n.d.). *The internet of things.* Retrieved from http://internetofthingsagenda.techtarget.com/definition/Internet-of-Things-IoT

U.S. Department of Energy's Argonne National Laboratory. (2015). *Nano design works (NDW).* Retrieved from http://nanoworks.anl.gov

U.S. Food and Drug Administration. (2016). *Use of International Standard ISO 10993–1, Biological evaluation of medical devices Part 1: Evaluation and testing within a risk management process.* Retrieved from http://www.fda.gov/downloads/medicaldevices/deviceregulationandguidance/guidancedocuments/ucm348890.pdf?source=govdelivery&utm_medium=email&utm_source=govdelivery

U.S. Food and Drug Administration. (2015). *Cybersecurity vulnerabilities of Hospira Symbiq Infusion System: FDA Safety Communication.* Retrieved from http://www.fda.gov/MedicalDevices/Safety/AlertsandNotices/ucm456815.htm

Watts, S., & Noh, J. (2014). Going green with management-management technology comparison within green companies: China, USA, and Korea.*International Journal of e-Education, e-Business, e-Management and e-Learning, 4*(3), 160. doi:10.7763/IJEEEE.2014.V4.323

Wearable Technologies. (n.d.) Retrieved from https://www.wearable-technologies.com/2015/04/wearables-in-healthcare/

Further Reading

Albulescu, V. L., Neagu, C., & Doicin, C. (2015). *Critical issues concerning the technological resources management in innovative enterprises. Managerial Challenges of the Contemporary Society. Proceedings, 8*(2) 22.

O'Donnell, J. (n.d.). *How smart sensors are transforming the Internet of Things.* Retrieved from http://internetofthingsagenda.techtarget.com/definition/smart-sensor

Oh, E., Chen, K., Wang, L., & Liu, R. (2015). Value creation in regional innovation systems: The case of Taiwan's machine tool enterprises. *Technological Forecasting and Social Change, 100,* 118. doi:10.1016/j.techfore.2015.09.026

Pelser, T. G. (2014). The enigma of technology management in strategy deployment. *International Business & Economics Research Journal (Online), 13*(5), 915.

Rankin, J. A., Grefsheim, S. F., & Canto, C. C. (2008). The emerging informationist specialty: A systematic review of the literature. *Journal of the Medical Library Association: JMLA, 96*(3), 194–206. http://doi.org/10.3163/1536-5050.96.3.005

Chapter 4

Design Thinking

(Continued)

(Continued)

- ➤ Entrepreneurship and design
- ➤ Designer researchers
- ➤ Business designers
- ➤ Social innovator
- ❖ Modular activities
  - ➤ Discussions
  - ➤ Test
  - ➤ Research
  - ➤ Design and/or Build projects
  - ➤ Assessment tasks
- ❖ Terms
- ❖ References

Chapter Learning Objectives

- ❖ Define design thinking
- ❖ Identify key components commonly found in design thinking models
- ❖ Examine the relevance and impact of design thinking methods
- ❖ Describe steps involved in design thinking methods
- ❖ Distinguish design thinking from scientific method

Overview

In this chapter, we will explore the design thinking process beginning with what it is and why it's important in today's world. Design thinking is essentially a creative problem-solving method, which uses both divergent and convergent thinking to come up with unique innovative solutions. Traditionally, many engineering and science programs teach variations of the design thinking process in connection with the scientific method as applicable in their fields of study. However, it is only in recent years that this process of creative problem-solving has become more popular across multiple industries, particularly in the areas of health care, information technology, computer science, industrial design, and organizational management.

While not the first to define the design thinking process, Stanford University's Hasso Plattner Institute of Design (also known as d.school) and IDEO (a design company) together have created the most widely accepted standard process and are well known for their combined efforts to promote design thinking in business, education,

Sergey Nivens/Shutterstock.com

Figure 1.   Innovative solutions and design thinking.

and scientific research.[i] The broad overview of the design thinking process included in this chapter will be drawn from the Stanford's d.school standard, as well as other sources. Once the process is understood, the applications become nearly limitless. Students of the design thinking process use design thinking to solve problems in their education, work, social relationships, and community.

Introduction

Why do we keep creating new technologies? The simple answer is because the world has no shortage of problems that need solving. As long as people are sick with cancer, scientists and researchers will look for ways to find a cure. In doing so, they will create new drugs, surgical tools, and imaging, and diagnostic equipment. As long as people on the planet struggle to find clean drinking water, we will search for better ways to use, preserve, purify, and transport water. As long as people have a desire for instant easy communication, we will continue to develop new and smarter devices to speak to each other. The smart phone is a technology that is constantly adapting to solve new problems. For every challenge we face, there are many solutions and different ways to reach those solutions.

In research and development, the scientific method has been the standard problem-solving method for centuries. The scientific method is used to prove or improve on previous knowledge, investigate new phenomena, or discover new knowledge. The steps are: observe, ask a question, research, hypothesize, experiment, draw

---

i   Ideo and d.school standards for Design Thinking. Retrieved April 2016 from http://designthinking.ideo.com and dschool.standford.edu/dgift/

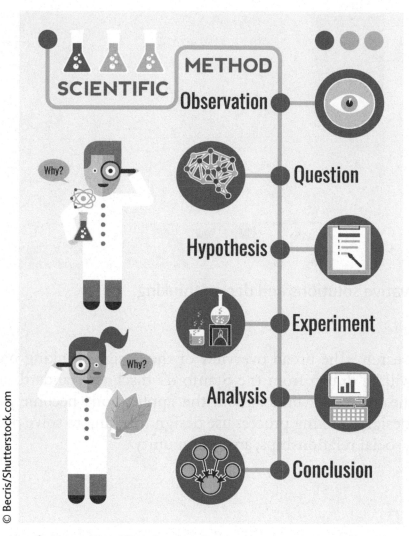

Figure 2.   Steps of the scientific method.

conclusions, and report. This method of problem-solving is very effective in working systematically through a problem and measuring results. However, the scientific method can sometimes be too systematic and time consuming for modern innovation. Also, there is an empirical nature to the scientific method that does not always take into account the user experience or human reaction to the discovery.

Scientific Method

In the United States, most middle-school science students learn the basics of the scientific method. In an elementary experiment, the steps can be completed in a short time. Let's say a seventh grade science class is told to boil water. Teams observe water in a container, ask the question: "How hot does water have to get to boil?" Perhaps the teacher poses this same question. Then student teams research and discover that others have determined fresh water typically boils at 100°C. The class hypothesizes that the water in their containers will boil if heated to 100°C. The experiments begin as teams heat the water using various sources, and the water boils at 100°C each time. They document all of the results and report them back to the teacher. This is a simple test and can be completed in minutes. For more complicated questions, the research and experimentation stages can take years. Often in research, science experiments are repeated with small variations to determine the validity of the results and to build new information upon the previous findings. For our boiling water experiment, one variation might be to add salt to the water and measure the change in boiling point temperature or heat different liquids and measure the rates and temperatures at which they boil.

In educational settings such as science classes, experiments like our boiled water may be repeated by millions of students a year, always getting the same results. If the outcome is different, then the student must determine what changed (went wrong) to get the variation. It is easy to assign grades to projects that use proven experiments because the outcomes are well documented and easy to measure and analyze. In experiments like these, the user experience is not a factor for the outcome. Competence and following instructions are keys in achieving the results. Depending on the problem that is being solved and the research being done, following the scientific method can be the best way to reach discovery of new data. The outcome or result of the experiment is supported empirically (by the numbers) but it might not be the "best" result for the user or effected group.

Chemotherapy is one example of an innovation in science with limited regard for the user experience. Yes, the chemicals used in chemotherapy have been proven to reduce the number of cancer cells and to sometime prevent further growth of those cells.

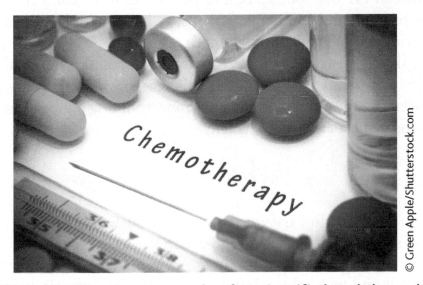

Figure 3. Chemotherapy is an example of a scientific breakthrough with a bad user experience.

© VLADGRIN/Shutterstock.com

Figure 4. New methods of thinking allow for better designs.

The numbers don't lie. However, the chemical treatments are also poisonous to the human body, can take the patient to the brink of death, and occasionally cause death from side effects of the treatment. It may be the best option we have to combat many forms of cancer, but it is a bad experience for the patients. A lot of people—scientists, doctors, and others—are working on creative ways to both minimize the effects of this poisonous medicine and find alternatives to using it. With each new breakthrough in this field, there are new challenges uncovered or created. For this reason, whether using the scientific method for discovery or the human-centered design thinking innovation method, the process must be repeated for each new challenge that arises.

Design Thinking

A problem-solving method that has emerged in response to the rapid innovation of the last few decades and the information age where the user experience is very important is the design thinking method. The design thinking method is very effective in product development and technology, but can be applied to many industries and societal challenges as well.

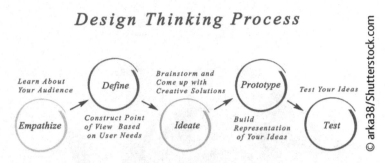

Figure 5. Steps of design thinking process.

In design thinking, the steps begin with looking at how a problem affects people and how people are affecting the problem. This focus on people is called "human-centered" problem-solving, and design thinking is a process that attempts to create human-centered solutions. The design thinking steps as explained by the d.school are:

❖ Empathize
❖ Define
❖ Ideate
❖ Prototype
❖ Test

Principles of Design Thinking

While the steps of design thinking are very straightforward, there is more to the process than simply going through the steps. The principles of design thinking that accompany the process are important to ensure that the process is properly implemented. The principles of design thinking include collaboration, creative confidence, brainstorming, respect for others' ideas, divergent thinking, convergent thinking, and embracing "failure" as a learning tool. We will discuss each of these areas.

In order for design thinking to be an effective problem-solving tool, it is necessary to approach the process with an open mind and a desire to find a human-centered solution. These principles will be explained in the context of the process description. However, it will be beneficial to discuss what *creative confidence* and *divergent and convergent* thinking are in a little more detail before we proceed.

Creative Confidence

Creative Confidence is the ability to come up with new ideas and the courage to share them. Artist Pablo Picasso said, "Every child is an artist. The problem is how

© DeeMPhotography/Shutterstock.com

Figure 6.   Holding onto creative confidence as we grow.

to remain an artist once [we] grow up." [ii]There are multiple studies that show people begin life with a strong sense of creative confidence. Do you remember being an artist? As children, particularly young children three to five years old, we created and shared artwork, ideas, and our view of the world with an eagerness almost nonexistent in adults. Children are excited about doing something new and putting their own twist on things. If given a picture of a boy and his dog to color, children might use any number of colors for the boy's skin and might make the dog purple. It isn't about what is real or what is accepted, it's about what colors look good to the child for any number of reasons. It doesn't matter that the child has never seen a purple dog. Tapping into

---

ii   Retrieved April 2016 from http://www.forbes.com/quotes/7314/

the creative confidence of our youth is important for the discovery of new innovative ideas. In order to open the imagination, the design thinking process must include a safe environment for the sharing of ideas to encourage creative confidence. Innovation sometimes requires creating something that has never been seen before.

Divergent and Convergent Thinking

Divergent and convergent thinking are contrasting ways of approaching a problem. Albert Einstein may have captured the difference when he said, "Logic will get you from A to B. Imagination will take you anywhere." In this example, convergent thinking is logical. It leads from point A to point B. Divergent thinking, on the other hand, is imagination that leads you anywhere. Divergent thinking is considering not just one path but all possible paths to any destination.

The business definition of divergent thinking is:

Idea generation technique (such as brainstorming) in which an idea is followed in several directions to lead to one or more new ideas, which in turn lead to still more ideas. In contrast to convergent thinking, (which aims at solving a specific problem) divergent thinking is creative, open-ended thinking aimed at generating fresh views and novel solutions.[iii]

The key difference between convergent thinking and divergent thinking is the outcome. For example, imagine two groups of people are given the task to solve the problem of bikes being stolen on campus. Group A will use convergent thinking and Group B will use divergent thinking. Group A would work toward a solution of keeping the bikes safe from theft, perhaps installing better bike racks and distributing locks or setting up better surveillance to catch the thief.

Group B, the divergent thinkers, would begin by wondering why students ride bikes to school and whether or not there is a better solution to riding a bike. All ideas

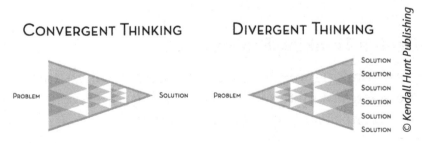

Figure 7.   Convergent and Divergent Thinking Perspectives. Image provided by Tanner Wheadon.

iii   Retrieved March 2016 from http://www.businessdictionary.com/definition/divergent-thinking.html#ixzz41Zqha3yJ

© Rawpixel.com/Shutterstock.com

Figure 8. Convergent and divergent thinking.

would be explored with no predetermined solutions being presented. This group might discover solutions like the use of golf carts, which are harder to steal and could create an all-weather shuttle service from transit hubs on campus to get students to class on time, or they may come up with a policy that allows students to store their bikes in the back of the classroom for increased security. It is nearly impossible to guess Group B's solutions, because the solution will be a product of their combined creativity. They may find it difficult to select just one solution to focus on and may not be as productive in the end as Group A, who is working directly to a goal. However, Group A will probably not come up with as many creative solutions.

It is a combination of divergent and convergent thinking that makes design thinking an effective process. If Group B only uses divergent thinking, they might never come to a conclusion about the best solution to test and implement. The creativity of divergent thinking and the discipline of convergent thinking when applied at the appropriate times in the problem-solving process increase the power of design thinking. Group A, using convergent thinking more closely related to the scientific method would come up with solutions, even good solutions. However, Group B, using divergent thinking and design thinking, might come up with revolutionary solutions.

Steps of the Design Thinking Process

The steps of design thinking are not rocket science, although they do have a lot in common with the scientific method and engineering processes traditionally used in rocket science. In recent years, even rocket scientists are recognizing the power of design thinking.

As a creative strategist and product designer at NASA's Jet Propulsion Lab, Jessie Kawata[iv] is using design thinking in space exploration.

---

iv   Kawata (2015).

*The Jet Propulsion Laboratory, NASA's lead robotic center of solar system exploration, is a metropolis of astrophysicists, astronomers, climate scientists and planetary geologists— to name a few—not to mention the most brilliant aerospace engineers on Earth. One of the newest areas of interest in this city is The Studio, the first in-house art and design consultancy that NASA has ever had—*Jessie Kawata, October 2015.

Whether design thinking is used in creating better interstellar vehicles or to produce a better trash can, the steps to solving the problem are the same. As we discussed earlier, according to Stanford's d.school, the five steps of design thinking are: Empathize, Define, Ideate, Prototype, and Test.

In this next section, we explore the individual steps of the design thinking process in greater detail. Let's use two fictional people and see how they go through the steps of the problem-solving process. Mary and Hugo are nothing alike. Mary is a biologist who loves to categorize things and use the scientific method to solve problems. She is analytical and likes to measure outcomes. Hugo, on the other hand, is an architect who embraces human-centered design and learned early in his studies that the most beautiful building in the world is not complete without bathrooms. He likes to balance the creative and the practical in his work.

Mary is on a team using the scientific method. She has assembled expert scientists to assist her. Hugo is on a multidisciplinary team with a variety of people from different fields and will be using design thinking to solve the problem. They are given an example of a fictional small town where more than half the people are turning orange. In fact, their skin is bright orange!

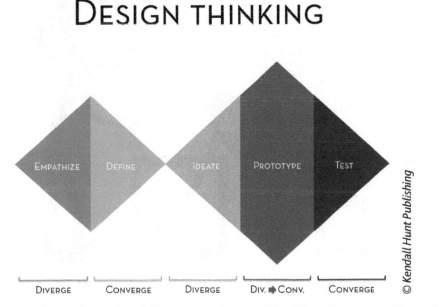
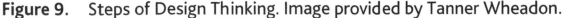

Figure 9. Steps of Design Thinking. Image provided by Tanner Wheadon.

Empathize

It may seem strange at first glance to think of starting a design process with empathy, which means to be able to put oneself in another's position or to understand how someone feels by putting yourself in their shoes. However, this is the very heart of design thinking. It is about people and how they interact with and use the technology or solutions being produced. David M Kelley, founder and chairman of IDEO said, "The main tenet of design thinking is empathy for the people you're trying to design for. Leadership is exactly the same thing, building empathy for the people that you're entrusted to help."

How do we empathize? It begins much like the scientific method with observation. In this step, the designer or problem solver observes people and how they behave in the context of their lives. Often people say or do things differently when faced with questions or when they know they are being observed. How they feel can often be more easily observed than it can be described. For this reason in the empathize step, it is important to collect information about what people in the situation are saying, doing, feeling, and thinking.

The next part of observation is engaging and interviewing those you are trying to design for—the user. Finally, there is an immersive element to empathizing—when you put yourself in the user's position, as much as possible. The idea in this step is to learn as much as you can about the people who might use the solution. This portion of the process is divergent thinking in that the information being gathered for observation could lead in many directions. No assumptions are made, ideas are all

© Toby Bridson/Shutterstock.com

Figure 10. Understanding who you are designing for and how they will use the technology.

accepted during the interviews, and all impressions are considered in understanding the situation.

Mary and Hugo would approach this first step of problem-solving with observation. Mary and her team of scientists would observe and probably take some samples of the environment, the town, perhaps blood tests for those who have turned orange. The scientists might observe whether everyone was the same shade of orange, or if there were shades of orange. They would find out if the orange skin people all turned at the same time, their ages, where they work, if everyone in a household turned orange or just some of them, and what they ate. They would be looking for observable patterns in the outbreak and a cure for those turning orange.

Hugo and his team would most likely start by watching people in the town. He has innovative thinkers from across many industries on his team. They begin by simply sitting in the local restaurant and observing how people talk about turning orange. Then after seeing how people are acting toward each other and listening to them, the team would begin interviewing people from the town, asking similar questions as the scientists. They would probably find out if there is any advantage to being orange and if it is painful or causing any health issues. Then some of the team members might even paint their skin orange and spend the day working in the town to see how people with orange skin are treated and how it feels to turn orange. They would suspend any of their bias against orange skin and be open to any information they could find.

© ALMAGAMI/Shutterstock.com

Figure 11.   Empathy and observation begin in the same way.

Define

Defining the problem is not enough in this step. It is about looking at all the information from the empathy step and focusing on a specific scope and a meaningful challenge. It is a convergent thinking process of sifting through information and experience with the user and narrowing to an actionable statement of the challenge from your point of view.

With our example of the town of orange people, this is where Mary starts to formulate and ask questions and research causes for people turning orange. Are there any other instances of towns in history turning orange? Her team cannot discover any physical cause or illness that is causing orange skin. Age doesn't seem to be a factor either. The only physical trait of the orange skin people other than the color change is that they tend to burn more easily in the sun. She forms a hypothesis based on her observations and research that the orange people produce more of a certain enzyme and she believes that with the right treatment they will no longer be orange. She prepares to experiment so as to come up with a cure.

On the other hand, Hugo and his team take all of the information from the empathy phase and begin to distill it. Examining what the social outcomes of being orange are, what the physical benefits or problems come from being orange, and what the most meaningful challenge is for the town as a whole. What he discovers is that people in the town think it is cool to be orange. The local high school football team captain is orange. The mayor is orange. Those who aren't orange, try to eat more carrots and tangerines because they heard that eating a lot of orange food will turn your skin orange. The team also discovers that people who have been orange for many years have no more physical side effects than those who turned orange more recently. The only physical problem seems to be sensitivity to sunlight. They burn more easily, but no other illnesses have been attributed to the orange skin. They did not find anyone

© Andrey Burmakin/Shutterstock.com

Figure 12.   Distilling information to define the heart of the challenge.

in the town who wanted to be cured, only people who wished they were orange. They define the challenge to help people become orange and open a sunscreen shop on Main Street.

Ideate

The ideation step is all about variety and quantity of ideas. The designers again exercise divergent thinking to open the imagination for creative solutions to the defined challenges. During this part of the process, all ideas are welcome and recorded—the more imaginative the better—as they apply to the challenge. It is vital during this stage that the environment be safe for sharing new concepts and connections that might not currently exist. Many teams in this phase will set a rule outlawing the word "no". All responses are "Yes and …" This is the step of creative confidence and respect for others. Some of the most outrageous ideas may uncover new avenues of thought on the subject.

Mary and the scientists in this step are forming a hypothesis and planning the experiments that will test their theories. They are beginning to identify the best test subject, namely those few in town who are more interested in answering the mystery of why people are orange than they are of gaining any prestige from being orange.

During the ideate step, Hugo and his team gather in a room with a table full of sticky notes and white boards and start throwing ideas out about how to duplicate the orange skin phenomenon. His team was especially selected to offer a variety of opinions and points of view. They soon have a room full of seemingly disconnected ideas from orange facial tattoos to tinted spray tan. They have a color wheel on the board to explore complimentary colors people might want their skin painted for special occasions. They are ready to start prototyping a few ideas.

Figure 13.   All ideas are welcome during ideation.

Prototype

In the prototype step, the ideas and concepts of the ideate step begin to take physical form in rough representations of the idea. It is important to keep early prototypes, rough and easy to change. Many times modeling materials like clay or play dough are used for this stage along with cardboard and paperclips. The point is to make something more solid and the idea can become a reality and change it until there is a good chance it might be a solution. This is where the principle of embracing failure comes in more than any other. Prototypes may be built and destroyed and rebuilt many times. The beginning of the prototyping phase of the process continues to some degree with divergent thinking. Expanding the variety of prototypes and deepening the understanding of the user experience by getting initial user reactions to the prototypes. However, the results of the feedback and the feasibility of creating the solution begins to move designers back to convergent thinking as the field of choices is again narrowed to the most realistic and best solutions emerging for the problem. In the end, only a few prototypes, perhaps only one, will move forward to the test phase.

For Mary and her team, this is the experimentation phase, and it is critically different from design thinking in that for her experiment, it is important to have a control group who will not change. She needs orange skin subjects who will not receive the antidote she is testing. This is often a blind study where the subjects don't know if they are in the control group or in the active testing group. As she sees changes in the people who receive the antidote, she can measure them against the control group to see how much change is happening. If she were to try and operate like a prototyping stage, it would be very difficult to pinpoint reasons for the changes if everyone is given a different antidote and there is no control group.

© Viktor88/Shutterstock.com

Figure 14.   Prototyping and iteration.

Alternatively, Hugo and his team can show prototypes of different skin painting techniques on different users and see what color they like most. He has a spectrum of orange tint spray tan that he sprays on hands and arms. He even has a few people who are trying the blue and purple prototypes just for fun. In the meantime, they are measuring how long it takes for the spray tan to wash off, so they can see how often people would have to reapply. They finally have a formula they feel is popular and easy to produce. It is tangerine orange and lasts a month at a time. It is time for them to test the rough formula on more people.

Test

The test step is crucial to the process because it takes the rough prototype and further refines it. Ideas are clarified as they are built. At this time, the user experience is everything. If you have created a solution that no one wants to implement it becomes clear and may require going back to the drawing board to see if something has been missed or overlooked. Then again, if you have a solid solution that is popular this is the phase where the designer begins to understand the feasibility of duplicating the solution and distributing it.

This is when the scientists using the scientific method start to collect the results of experimentation, tweak the dosages, and examine successes in order to ensure the results are repeatable. Conclusions will be drawn and formerly orange people begin to lose the unnatural coloring. The scientist team duplicates the antidote and makes it available to anyone who wants to reverse the orange skin syndrome. After six months, only 2% of orange skin people are using the antidote. All is not lost, however, because the antidote and Mary's research actually cures a skin disease in another country that was making people scaly and itchy.

Figure 15.   Tangerine dream body pigment test campaign ad (fictional product).

Hugo and his team, using the design thinking process, take the tangerine dream spray pigment and test it on a wider group of people in many different situations. The first iterations of the tanning spray ran in streams if people showered within the first 48 hours of application, but the team changed the formula and were able to get it to stay on six weeks with no streaking. The final week of fading is still a problem as people tend to look yellow for a few days before reapplying the solution. Yellow is an unpopular color in the town and one city council member hid in his house to avoid being photographed during the fading phase. The team is now working on expanding the line of spray pigments to other colors. They have found that green is a very popular color in jungle regions especially among the youth.

Rinse, Repeat, Reiterate

The process of design thinking is a cycle that repeats over and over. Once an idea has been tested it is then time to take the knowledge gained and begin observing how the prototype is used and empathize, define, ideate, prototype, and test again. Each new solution creates opportunities or challenges that then go through the process. In the case of the orange people, things such as the culture or climate may affect the attitudes of the people and new innovations may need to be created to assist for instance with the fading stages of the spray pigment. There are many things that can be investigated and taken through the design thinking process.

Storytelling

One final concept in both design thinking and scientific research is the sharing of the solution. It does not benefit the world if solutions are not shared, if products sit on the shelf because no one knows that they solve a problem, or if a doctor has the cure for measles but doesn't find a way to get the message out to the people suffering from measles. Once a solution has been validated, it should be communicated broadly.

Differences between Scientific Method and Design Thinking

As can be seen from the hypothetical example illustrated earlier, each approach can result in vastly different solutions—but each method found viable solutions.

As we have discussed, new technologies and innovations come primarily from the desire to solve a problem. Both the scientific method and design thinking have their strengths and limitations. They are both processes used to solve societal, environmental, business, and global problems. The scientific method is especially useful for development of medications and other products where a control group and measurement of changes can be systematically studied. Astrophysics and the studies built on previous theoretical research are also natural applications for the scientific method.

Design thinking is frequently utilized in product development, industrial design, computer technologies, and the user experience portion of most problem-solving. If a new miracle medication is discovered by science and distributed to those who could benefit from it, but people are still suffering because they won't take the medicine, it is a great time for design thinking to take over and solve the problem of why people won't take the miracle cure.

Types of Prototypes

Earlier we discussed prototypes generally. Again, a prototype is first, typical or preliminary model of something, especially a machine, from which other forms are developed or copied. Here we will look at some different types and discuss their relevance. In some cases, prototypes are used in an iterative process. This could include a concept/experimental prototype, feasibility/technical prototype, representational nonworking model, alpha prototype, beta prototype, and preproduction prototype. In other cases, you might have a horizontal/use interface prototype or vertical prototype as well. Let's discuss each in a bit more detail:

❖ *Concept prototypes* have the overall purpose of analyzing your approach. It is characterized by high level, overall vision, and is often completed at the concept definition stage.

❖ *Feasibility/technical prototypes* determine the feasibility of various solutions. These are characterized by proof of concept for specific ideas or issues. This is also often completed at the concept definition stage. These prototypes generally verify that critical components are available, would be able to interoperate and are capable of meeting expected needs.

❖ *Representational nonworking models* are just like they sound. These models are built to show the appearance, design, and general features but cannot actually function. In some cases, this might be known as an appearance prototype, a miniature, or a scale model.

❖ A *horizontal prototype* clarifies the scope and requirements, particularly in relation to a user interface. As an example from programming, this type of prototype might have a model of the system including dummy menus, screens, and processes, and the like but without the actual coding and core functions written.

❖ A *vertical prototype* generally contains a simplified version of the core functions of a system. As an example from programming, this type of prototype would offer a minimally running system that can demonstrate how the full version would operate, or at least how the core functions would operate.

❖ *Alpha prototypes* are often the first functional version and are intended to assess whether a product might work as intended.

❖ *Beta prototypes* are often the second functional versions and are used to assess reliability and stability. This version is what is tested with identified consumer types to ensure they operate and are used as intended.

❖ A *preproduction prototype* is generally a factory model. It demonstrates what is to be produced before full production begins.

Each of these types of prototypes can be useful for learning, communication, and management of design ideas. They reduce uncertainty as well as clarify purpose.

The Science behind Design Thinking

The science utilized in the design thinking process varies depending on the type of problems being identified and solved. From civil engineering to anthropology, the solutions must match the challenges. It is difficult to pinpoint any particular discipline of science that would not benefit from the application of design thinking. However, there are few scientific areas of study that are natural fits for the design thinking process.

© Andrey_Kuzmin/Shutterstock.com

Figure 16.   Science of design thinking.

Anthropology

Anthropology is the study of human beings, cultures, societal relationships, traditions, ancestry, and distribution. One principle in anthropology is that every person believes their traditions and viewpoint is natural and that other's traditions, beliefs, and viewpoints are strange. This makes true empathy and understanding of people who do not share your heritage and culture difficult. In design thinking, it is important to try to overcome this principle and remove bias during the empathize and define steps. If a designer brings their cultural bias into the process it can limit the solutions and outcomes may be less than optimal.

Anthropologists try to remove cultural bias using careful observation, engagement with the people and immersion in the culture. A scientist studying another culture will eat the same foods, live in the same conditions, often learning the language, working alongside members of the society, and interviewing people. This immersion takes time and information. True understanding is realized when the cultural norms of a group begin to seem normal to the observer regardless of their native and inherited norms.

Organizational Science

Organizational science is the study of human beings in organizational settings and groups. This discipline of science researches how people work in groups, interactions within organizations, how groups are formed, and management of organizations. It is a field of study that has taken a close look not only at individuals operating within a group but also at the organizations themselves. Communication, structure, and activities of organizations are researched. This is important in applying design thinking both in the team of designers and how they are composed, but also in the users or groups the designers are trying to help.

Psychology

Psychology is the study of behavior and the mind. In psychology, many aspects of human thought and emotion are researched to find relationships and causes for both behaviors and perceptions. As human-centered design is the focus of design thinking, the more designers understand humans the more effective they can be with all the steps of the process.

Neuroscience

Neuroscience is the study of the nervous system. Professionals in this branch of science research the brain from molecular and cellular research to thought process and reaction to environment. Neuroscience can help designers understand on a biological level why people react a certain way to or interact with products and situations.

Career Connections

With the increasing acceptance of design thinking as an innovation tool in multiple industries, there are also an increasing number of career opportunities in human-centered design and problem-solving, such as positions in industrial design, social innovation, user experience (UX) professionals, and business designers. In 2014, Tim Brown, of IDEO, identified several career opportunities in design thinking that emerged in recent years.[v] The following examples are adapted from Brown's list. It is by no means a comprehensive list of opportunities, as they are continually increasing in this field.

Designer Programmer

One emerging prototyping medium is computer code. Programmers who can innovate and come up with new ideas and then create strong user experience to take ideas successfully to market are in high demand. In a digital age where apps and online information are the muscles of the economy, designers who can also create code have a great advantage.

System Designer

There are systems that have been reluctant to change and now find disrupting technologies are forcing innovation. Health care and education system designers are increasingly finding positions in companies like GE Healthcare, Johnson & Johnson, and others. Necessity for innovation in these systems is creating opportunities for design thinkers to find their niche within the industry and create meaningful change.

Entrepreneurship and Design

"Simplicity is the ultimate form of sophistication," said Leonardo da Vinci. In the start-up world, entrepreneurs are striving to capture simplicity and sophistication of design. From founding board to venture capital, designers are finding themselves more valuable in the communities of entrepreneurs and innovators. "Many of the fastest growing companies are succeeding because they've designed a highly appealing product or service," notes Tim Brown of IDEO.

Designer Researchers

More data are being collected than ever before in history. In the past, design researchers were anthropologists or psychologists, but the needs of the market are shifting

---

v   Brown (2014).

from general behavioral study to unlocking the secrets of real-time data processing to find user behaviors and strategy. For those who love to work with people and have a passion for data, a career in design research is a possible fit.

Business Designers

Business is about more than just a product or service. Designers looking at business model, strategy, marketing, supply chain, and operational innovations are important new professionals. The spread of the *Internet of Things* that is connecting the physical world to the digital world has raised many challenges in the business model of the past. It is the perfect time for designers to apply their skills to innovate almost every aspect of doing business from manufacturing to customer interaction.

Social Innovator

Those who have the heart of philanthropist and the skills of a designer are finding more opportunity than ever to apply design solutions to social challenges and impact the world in a positive way. Design thinking is being utilized in global impact organizations and nonprofit organizations. Applying the principles of design thinking to clean energy solutions, clean water, poverty alleviation, and microenterprise can bring a great personal reward and a good career path.

Modular Activities

Discussions

* Post an original discussion in the online discussion board on the following topic: Discuss how you could use the design thinking process in your own life either now or in the future (200 word minimum). Next, comment on the post(s) of a minimum of one other student in a thoughtful and academic way that enhances the conversation. See rubric for grading and assessment measures.
* Read the case "Think designing scissors is simple? Think again! Q&A with Colin Roberts, Industrial Designer at Fiskars Americas" found at http://www.core77.com/posts/18114/think-designing-scissors-is-simple-think-again-qa-with-colin-roberts-industrial-designer-at-fiskars-americas-18114 and post an original discussion in the online discussion board on the following topic: What are your thoughts on the design process described? What was surprising to you? (200 word minimum). Next, comment on the post(s) of a minimum of one other student in a thoughtful and academic way that enhances the conversation. See rubric for grading and assessment measures.

Tests

❖ Online graded quiz on overall chapter content, written in multiple choice format. When submitted, we recommend giving the correct answer along with the page number in which it is found for questions students did not answer correctly.

❖ Use the Web site Socrative found at http://www.socrative.com/ (free of charge) to set up a live interactive quiz where students can instantly see the overall results for the class. Ask the following questions:

➤ What do you feel is the most innovative company in the world? (short answer)
➤ What industries do you feel design thinking would work best in? (short answer)
➤ I believe children are better at design thinking than adults (true/false)
➤ I believe design thinking and scientific method are essentially the same things (true/false)
➤ I believe industries are more likely to succeed if they work to make people want things instead of trying to make things people want (true/false)
➤ I agree with the statement from Steve Jobs, "Design is not just what it looks like and feels like. Design is how it works" (true/false)

Research

❖ Research report on divergent/convergent thinking: Research divergent and convergent thinking write a three page informative report about it with a minimum of three sources.

❖ Create an annotated bibliography of design thinking resources in APA or MLA format. Include a minimum of five (5) items on your list. You must include the following: (1) information about the article or source (author, title, publication date, etc.); a summary of the article or source written in your own words (150 words minimum); and a short assessment of the source (100 word minimum). You can learn more about annotated bibliographies at https://owl.english. purdue.edu/owl/resource/614/03/

❖ Interview a person who works in a design field. As the following questions at a minimum and then come up with four (4) questions of your own:
➤ What role does design thinking play in your job or field?
➤ What are challenging aspects of your job?
➤ What are your favorite parts of your job?
➤ What would you recommend to someone going in to this field?

❖ Submit both the questions and the answers.

❖ Research one of the following higher education academic areas:
➤ Pre-engineering
➤ Engineering graphics and design technologies

➤ Engineering
➤ Digital media
➤ Entrepreneurship

This might include review of offerings or degrees at your institution or at another location, contacting a professional/faculty in the field, assessing higher education level academic events or educational resources, and so on. Submit a 200 word minimum summary of your findings. Make sure to include your source(s).

Design and/or Build Projects

❖ Try your hand at design thinking. Design and prototype a solution for the following problem: You frequently go on out-of-town trips that last a week to two weeks. While you love house plants, you simply cannot seem to keep them alive because of the lack of watering while you are gone. Having someone come water them is not an option because you live in a high security compound and guests are only permitted in the home if you are present with them. You want to create a solution that has never been thought of before. Note: You will be graded on your originality, thoughtfulness, effectiveness, and professionalism.

❖ Original invention idea: Describe your invention idea completely using the following outline.

➤ General Purpose—State in general terms the purpose and the object of the invention (100 words minimum)

➤ Background—Describe the prior art (patents already exist that affect your own invention idea; identify by patent number or journal citation) and indicate how the invention differs and is more advantageous than prior art. Note: if sufficient research is not completed then points will be lost. Almost all invention ideas are affected by some previously existing patent. Some recommended locations to look are http://patft.uspto.gov/ or http://www.uspto.gov/patents/process/search/ or https://www.google.com/?tbm=pts

➤ Description and operation—Describe completely (sufficient to permit the preparation of a patent application) the construction of the invention using reference characters to identify components of attached illustrations. Give a description of one complete operational cycle. If the invention relates to the synthesis or identification of a new composition of matter, describe the product in a structured form, if possible, and the process for making it. Include all the available information regarding its physical characteristics and all test data evidencing its utility.

➤ Nontechnical Description—Describe (in 300 words or less) the invention in terms understandable to non-scientists.

➤ Sketches, prints, photos, and any pertinent manuscripts—Include sketches, prints, photos, and any pertinent manuscripts

Note: The abovementioned questions are actually a slice of those found in www .research.va.gov/programs/tech_transfer/report-invention-cert.doc, you only need to complete the above mentioned questions, but I wanted to point out these are used in an actual governmental setting.

❖ You decide to build a tree house for an eight-year-old child you know and adore. Their property has strong, solid, and large maple and oak trees so finding a tree sufficient for your project will not be the challenge. Instead, your primary challenge is that you only have a grand total of $500 for the project as a whole and the most important aspect must be the safety of the child and others. At the same time, you want it to be enjoyable to children. Design a tree house that could fulfill these needs and explain your rationale for components you added. Note: You will be graded on your feasibility, thoughtfulness, potential safety of design, potential appeal to children, and professionalism.

Assessment Tasks

❖ Visit the IDEO web site at https://www.ideo.com/ and locate some of the projects they are working on (a category they call "work"). Find one that looks interesting to you and complete the following:
  ➤ Title of project
  ➤ Short description in your own words (cut and paste is not acceptable)—200 word minimum
  ➤ Your assessment of the relevance, impact, and value of the project—200 word minimum
❖ Complete the e-study found at https://www.oercommons.org/courses/adapting-evaluation-for-local-contexts-in-a-globalized-world which looks in to adapting evaluation to local contexts. General information within this case is publically accessible online. Upon review of the case, assess your findings. Which aspects of the e-study seemed most and least useful to you and why? Write a 400-word minimum summary of your assessment.
❖ How might design thinking be used in physical sciences (astronomy, physics, chemistry, or earth sciences) or math fields (algebra, trigonometry, calculus, geometry, etc.)? Write a 200 word summary of your thoughts.

Terms

Brainstorm—to suggest a lot of ideas for a future activity very quickly before considering some of them more carefully.

Convergent Thinking—is a term coined by Joy Paul Guilford as the opposite of divergent thinking. It generally means the ability to give the "correct" answer to standard questions that do not require significant creativity, for instance in most tasks in school and on standardized multiple-choice tests for intelligence.

Divergent Thinking—is a thought process or method used to generate creative ideas by exploring many possible solutions.

Empathy—the ability to understand and share the feelings of another.

Evaluate—to judge or determine the significance, worth, or quality of; to assess.

Human-centered design—a process that starts with the people you're designing for and ends with new solutions that are tailor made to suit their needs.

Prototype—a first, typical, or preliminary model of something, especially a machine, from which other forms are developed or copied.

Scientific Method—a method of procedure that has characterized natural science since the seventienth century, consisting in systematic observation, measurement, and experiment, and the formulation, testing, and modification of hypotheses.

References

Brown, T. (2014, July 22). *5 New design careers for the 21st century (Thoughts from Tim Brown).* Retrieved from http://designthinking.ideo.com/?p=1386

Kawata, J. (2015, October 19). *Is-design-thinking-rocket-science.* Retrieved from http://www.core77. com/posts/41855/

Read more: Divergent thinking. BusinessDictionary.com. Retrieved from http://www.business dictionary.com/definition/divergent-thinking.html.

Chapter 5

Information Technology and Communications

(Continued)

(Continued)

- ❖ Freedom of speech/expression
  - ➤ First amendment to the US constitution (Religion and Expression)
  - ➤ USA patriot act
- ❖ Destruction of information
  - ➤ Right to be forgotten
  - ➤ Self-destruction
- ❖ Role of informatics
- ❖ The science behind the technology
  - ➤ Network protocol standards
    - ■ Open systems interconnection basic reference model
  - ➤ Radio waves and signals
- ❖ Career connections
- ❖ Modular activities
  - ➤ Discussions
  - ➤ Tests
  - ➤ Research
  - ➤ Design and/or Build projects
  - ➤ Assessment tasks
- ❖ Terms
- ❖ References

Chapter Learning Objectives

- ❖ Define information technology and information systems
- ❖ Identify expert systems and social media
- ❖ Explain communication technology
- ❖ Examine the need for information security
- ❖ Describe mobile technology
- ❖ Distinguish open source from private resources and materials
- ❖ Summarize the role of informatics
- ❖ Address issues relating to right to privacy and destruction of information

Overview

Our dependence on information technology (IT), from our desktops to our cell phones, is evident. As of 2015, in the United States, cell phone ownership among adults is above 90% and 68% have a smartphone, up from 35% in 2011 If the household income is $50,000 or more, then cell phone ownership jumps to 98% for adults. Regardless of income, if a person is 18–29 years old, they also have a 98% chance of having a cell phone and an 86% chance of having a smartphone (Anderson, 2015; Pew Research Center, 2014; Rainie, 2013). So, it would appear cell phones (or perhaps more accurately mobile devices) have become critical resources for a majority of the US population. Actually, the same can be said worldwide. Globally, there are 97 mobile phones in use for every 100 citizens. In some countries, such as the Russian Federation, Argentina, Libya, Panama, and Saudi Arabia, there are well over 1.5 phones for every citizen in the country (World Bank, 2015).

When it comes to computers, the numbers are also quite high. In 2013, 83.8% of the US households reported computer ownership, with 78.5% of all households having a desktop or laptop computer, and 63.6% having a handheld computer. In that same year, 74.4% of all households reported Internet use, with 73.4% reporting a high-speed connection (File & Ryan, 2014). Again, the United States is not alone. Russia, Chile, Poland, Lebanon, and Venezuela are a few other locations that have at least 60% computer ownership (adults who have a working computer in the household) (Pew Global, 2015). However, not everyone sees the increasing level of devices and corresponding Internet access as a good thing overall. Pew Research Center asked 32 emerging and developing nations about their views on Internet usage and they found that while it as seen as a generally positive thing for education (64% said it is a positive influence compared to 18% who said it is a negative influence), personal relationships (53% said it is a positive influence compared to 25% who said it is a negative influence), and economy (52% said it is a positive influence compared to 19% who said it is a negative influence), it is not seen as nearly as positive for politics (36% said it is a positive influence compared to 30% who said it is a negative influence) and is seen as not positive generally for morality (29% said it is a positive influence compared to 42% who said it is a negative influence) (Pew Research Center, 2015).

It should be noted as well that not all technology devices have been on the rise. E-reader device ownership has fallen, so as of 2015 only 19% reported ownership, down from 32% in 2014 in the United States. Ownership of Mp3 players also has not risen. Instead, it has hovered around 40% since 2008 (Anderson, 2015).

Information Technology

First, what constitutes IT? It is a broad spectrum. In essence, it is the study or use of systems (in particular, computers or telecommunications) for storing, processing, securing, retrieving, exchanging, and sending information. It includes the development, maintenance, and use of computer systems, software, and networks that process and/or distributes data in any way.

In the not so distant past, IT might have been seen as a powerful calculator that could process and manipulate numbers at an astounding speed. Later, it might have been seen as a large mainframe machine that could rapidly store and process massive amounts of data. As of the 1970s and 1980s, it incorporated what we now know as the

Internet and the mobile phone. Now, it entails interoperability, interdependence, and intertwined resources. While some of the fundamental technologies remain intact—such as with telephone, radio, or television—the complexity and scale of today's systems are never before seen. Throughout it all, IT as a whole has been changing our lives. This is not to say that technologies that came before such as electric power or the transistor did not have a profound impact. They certainly did. Actually, IT has been build off the advancements and developments of predecessors such as those just mentioned. In working with these new developments, the challenges are not just technical but also managerial. As we become increasingly dependent on IT and as it becomes increasingly complex, keeping the systems themselves running is just one of the issues. Ensuring they are accurately, timely, secure, sustainable, and use properly and for good intent are just some of the many added issues.

At its core, IT and the related systems are generally based on models, protocols and standards. It is in these ways the systems can work together. This does not mean that the varying technologies necessarily work in cooperation; in some cases, it is the opposite. However, as a whole, at a grand scale, for IT to work most effectively the many players and systems must work with one another.

The Internet and all the technologies that rest upon it are, after all, now our go-to means of obtaining information quickly on virtually any and all topics. It is our go-to for communication with others. As a society as a whole, we are increasingly dependent on it for our daily functioning. If we are to offer equal access to individuals within our communities, it would seem, therefore, that this would entail open access to all of these resources. This, however, is not necessarily the case. According to the United States Census bureau:

❖ In 2013, 83.8% of US households reported computer ownership, with 78.5% of all households having a desktop or laptop computer, and 63.6% having a handheld computer

❖ In 2013, 74.4% of all households reported Internet use, with 73.4% reporting a high-speed connection

If we consider the global environment, access is far lower. According to the International Telecommunication Union (ICU) (2015), while we had 4000 million internet uses worldwide in 2000, as of 2015 there were 3.2 billion internet users. It notes, however, "4 billion people from developing countries remain offline, representing 2/3 of the population residing in developing countries" and continues, "Of the 940 million people living in the least developed countries (LDCs), only 89 million use the Internet, corresponding to a 9.5% penetration rate" for 2015. All in all, this means that global internet penetration grew from 6.5% in 2000 to 43% in 2015. A relevant segment is based on mobile broadband penetration, which actually reached 47% in 2015 globally.

Information Systems

Information systems are data- and communication-based processes and devices. They are used to gather, manage, store, and use information. Actually, data itself is a part of an information system, along with the hardware, software, and users. That said, when the term information systems is used in a standard business setting, the speaker is usually referring to solutions to meet organizational and management needs for information and decision support in some fashion.

Cloud-Based Technology

In its simplest definition, one might think of cloud-based technology as anything that involves hosted services over the Internet. This refers to applications, infrastructure, services, or resources made available to others on demand (subscription or other payment) via the Internet from another entities' servers. Some other terms used include cloud computing, cloud services, internet-based computing, on-the-line computing, or simply the cloud. According to the U.S. Department of Commerce National Institute of Standards and Technology, "Cloud computing is a model for enabling ubiquitous, convenient, on-demand network access to a shared pool of configurable computing resources (e.g., networks, servers, storage, applications, and services) that can be rapidly provisioned and released with minimal management effort or service provider interaction. This cloud model is composed of five essential characteristics, three service models, and four deployment models." (Mell & Grance, 2011)

One could think of its origins as reaching all the way back to late 1960s, when ARPANET (Advanced Research Projects Agency Network) was created by J.C.R Licklider with the idea of being an intergalactic computer network where everyone in the globe could be interconnected and accessing programs and information from anywhere. Others, though, give credit for the cloud concept to John McCarthy, who originated the idea of computation being delivered as a public utility (Mohamed, 2009).

To get a better understanding of cloud-based technology, you might envision how your social media sites work. When you want to use your social media (Facebook, Twitter, etc.), you log in on an internet-enabled device. All of your files, communications, and contacts are on this site's servers, not on your own machine. This way, you can get to it from anywhere and on any system that has internet capability.

Cloud-based systems may be public, private, or somewhere in between. Your social media accounts likely have some stuff that is totally private, some stuff that is visible to only some people, and some stuff that is visible to anyone. This is a hybrid cloud. A public cloud, on the other hand, has information and resources that are available to anyone with access to that cloud space. A private cloud is the opposite, and only you or a very specific group has access to a segment of the cloud space.

Some of the capabilities offered on the cloud include:

❖ Software-as-a-Service (SaaS): Gives subscriber's remote access to resources, software, and applications without the need for a physical copy of the software on your local machine. A subset of this segment is application service provider.
  ➤ Application-service-provider (ASP): Gives subscribers access to particular software applications.
❖ Platform-as-a-Service (PaaS): Gives subscribers the ability to develop and operate applications over the Internet via programming resources and tools provided and controlled by the cloud vendor.
❖ Infrastructure-as-a-Service (IaaS): Gives subscribers access to infrastructure such as storage space, processing power, network-related services, or specific hardware

As things stand now, "37% of United States small businesses are adapted to the cloud, but an anticipated 78% will be fully cloud operational by 2020. 65% are conducting back office work including bookkeeping and accounting with cloud-based apps today. Nearly half (43%) of small business owners use a smartphone as the primary device to run their operations" (Columbus, 2015). It should be noted that not only we are increasingly users of cloud resources, we are also increasingly developers of the same. The U.S. Department of Commerce International Trade Commission also notes that the United States also has strong export markets for offering cloud-based services, particularly in Canada, Japan, United Kingdom, Brazil, and South Korea (Pardo, Flavin, & Rose, 2016).

When using cloud-based resources, users depend on the external company for consistency, stability, and security. The user no longer has direct control of the data or the system. However, this also means that much of the responsibility and risk are also transferred to the external entity as well, which may be a good thing if they have more resources and expertise.

Communication Technology

IT plays a key role in communication regardless of distance. It might involve analog or digital audio, text-based systems, multimedia, or systems and applications. In each case, signals are passed between devices send and receive information. Usually, physical wires are used to transmit signals, but they can also travel in the air via radio frequency (RF) waves. In many cases, signals in communication technology are specified by amplitude/voltage and frequency/time.

Commonly, data are transferred in one of two ways using electronic signals: analog or digital. Analog signals are commonly limited to some type of range with minimum and maximum values. Digital signals (square waves), on the other hand, are not continuous. Instead, they have a limited set of possible values (even if that set is large). Electronics that use these signals can be analog, digital, or a combination of the two. That said though, commonly older technological devices are analog, such as clocks, record players, and traditional telephones, while more recent devices like CD players, DVD players, or computers are digital.

Computers as Binary Machines

Computers (binary machines) understand everything as a combination of two digits, zero and one. A single unit of information is known as a bit (or binary digit), which is either a zero or one. These bits are combined to create a code the computer can interpret. A combination of eight bits is known as a byte and represents a character (such as a letter in the alphabet or a number or a symbol) along with including a means for the computer to error check. Here is a basic byte:

It takes lots of bytes to create coding that the computer can interpret and act on. Thus, we have other common sizes we reference. A kilobyte (KB), for example, is 1024 bytes. A megabyte (MB) is 1024 KB. A gigabyte (GB) is 1024 megabytes, and so on. When communicating on a network, these bits and bytes take the form of network packets. Although the packets are simply a sequence of bits, their format has a specific meaning based on the protocol the packet is using. Information or communication technology systems, functions, and duties have vast and varied roles. Packets play a role in construction/production, collection/compilation, access/restriction, interpretation/translation, organization/depth, preservation/retrieval, storage/backup, maintenance, dissemination/distribution, transformation/manipulation, and usage/handling.

Telecommunications

Telecommunication is any communication or exchange of information over a distance in the form of electromagnetic signals via cable and satellite, telegraph, telephone, radar, broadcasting, or the Internet. The communications occur between two or more stations, each equipped with a transmitter to transmit information and a receiver to receive information. In this field, the focus is on characterization, representation, transmission, storage, design, install, support, and securing of telecommunications and related network infrastructures. Electrical engineering plays a strong role in digital signal processing, controls and optics or similar.

Wireless and Mobile Technology

Wireless and mobile technology is playing an ever increasing role in our society whether local, national, or global. Technically speaking, mobile technology just means something that is portable. However, in current society, we tend to use the terms wireless and mobile interchangeably. In reality, wireless technology uses radio, micro-wave, or similar to transmit signals for broadcasting, computer networking, or other communications. In simple terms, it uses electromagnetic waves rather than some type of wire to carry a signal across some communication path. This includes items such as our garage door openers or remote controls although usually the first thing that comes to mind is our connection to the Internet via devices we carry around with us (i.e., mobile devices with wireless access). Wireless technology, while it can be in a fixed location such as a printer connected in our home to a wireless modem, is often used in portable/mobile devices.

Increasingly it is the means for individuals to access Internet-based information. Yet a number of libraries currently restrict the use of cell phones or other mobile devices within their physical buildings. Perhaps, this came to be during a time when users actually used their cell phones to talk to others, thus disrupting other patrons. However, this is no longer the case. Nowadays people increasingly rarely use their cell phones to actually use their vocal cords to speak to another. Instead, they are tex-ting, e-mailing, browsing, running applications, or even recording. This is particularly true of the younger population. Consider this statistic from Pew Research regard-ing cell phone use by young adults in 2011, "cell owners between the ages of 18 and 24 exchange an average of 109.5 messages on a normal day—that works out to more than 3,200 texts per month—and the typical or median cell owner in this age group sends or receives 50 messages per day (or 1500 messages per month)." However, this study notes, if we ask how often cell owners use their devices for voice communication we find "cell owners make or receive an average of 12 calls on their cells per day, which is unchanged from 2010."

Consider some of these statistics presented by Cisco (2016):

* Global mobile data traffic grew 74% in 2015. Global mobile data traffic reached 3.7 exabytes per month at the end of 2015, up from 2.1 exabytes per month at the end of 2014.
* Mobile data traffic has grown 4000-fold over the past 10 years and almost 400-million-fold over the past 15 years.
* Globally, smart devices represented 36% of the total mobile devices and connec-tions in 2015; they accounted for 89% of the mobile data traffic.
* Mobile video traffic accounted for 55% of total mobile data traffic in 2015.
* Smartphones (including phablets) represented only 43% of total global handsets in use in 2015, but represented 97% of total global handset traffic. In 2015, the typical smartphone generated 41 times more mobile data traffic (929 MB per month) than the typical basic-feature cell phone (which generated only 23 MB per month of mobile data traffic).

Wireless Networks

In a wireless network, just as in their copper- or fiber-wired predecessors, data are shared with packets except over radio or similar waves. They use a protocol to communicate called 802.11 for which there are standards such as 802.11b which uses the 2.4-gigahertz (GHz) spectrum or the 802.11a which operates in the 5-GHz spectrum.

Radio Frequency Identification

RF identification (RFID) is a process in which items are identified and/or tracked using radio waves (RF electromagnetic fields). The items to be tracked or identified have tags with electronically stored information on them and can be read without making physical contact or requiring a direct line of sight. As the U.S. Food and Drug Administration describes them:

> Radio Frequency Identification (RFID) refers to a wireless system comprised of two components: tags and readers. The reader is a device that has one or more antennas that emit radio waves and receive signals back from the RFID tag. Tags, which use radio waves to communicate their identity and other information to nearby readers, can be passive or active. Passive RFID tags are powered by the reader and do not have a battery. Active RFID tags are powered by batteries. RFID tags can store a range of information from one serial number to several pages of data. Readers can be mobile so that they can be carried by hand, or they can be mounted on a post or overhead. Reader systems can also be built into the architecture of a cabinet, room, or building (2014).

The technology has been around since the 1970s and is now commonplace and found in items such as car keys, identification cards (such as patient identification in health care), inventory (for inventory control), toll tags, and access cards to name a few. An example of an RFID tag is an identification and security tag placed on higher cost items in a department store. In some cases, an alarm will sound if the RFID tag is not removed. In other cases, you might leave with the RFID tag still intact on the item you bought but the store will know exactly when it left the facility (and left their range of tracking) and can use the tag to instigate a financial transaction for the purchase of the item. It should be noted that no personal information is usually stored on an RFID tag, instead it is a number, which points to information stored elsewhere on databases.

As noted earlier, RFID uses radio waves to communicate information. The reader consists of a transceiver and an antenna, so it can both transmit and receive communications. RFID tags contain an integrated circuit (chip) that transmits data to the RFID reader with the help of an antenna. While the tags could have batteries, this is not necessary. Some RFID tags simply supply information hard-coded on the chip. These are known as passive RFID tags. One nice thing about passive tags is that they can last a very long time. Active tags, on the other hand, are more advanced. They often include small batteries and can broadcast stronger signals that can cover greater distances. An active tag acts like a beacon of sorts. Tags are also referenced by the distance they can span. Vicinity RFID can be read from 20 to 30 ft. away while proximity RFID must be scanned from a close proximity. Near-field communication (NFC) is an example of the latter.

Near-field communication

NFC is a specialized subset of RFID. NFC is a short-range secured wireless connection, usually within 2 in. (4 cm) where two devices can share small amounts of data with one another. One of the devices has a readable (and sometimes writable) electronic tag that the other device can understand and communicate with. An example would be an access control card to get in to a secured area or a payment reader.

Expert Systems

An expert system is a computer-based system that simulates the judgement and decision-making ability of humans with expert knowledge or experience in a particular field. They are designed to solve complex problems by reasoning via if-then rules or similar. In essence, making a computer intelligent, or expert, is done by providing it with massive amounts of high-quality information about a specific problem area, which it can then use for informed decision-making.

Expert systems are made up of four major components. First, the knowledge base; second, production rules such as "if/then" statements; third, an inference engine that controls how the rules are applied toward the facts; and fourth, a user interface. For computer systems, knowledge is navigated in the form of rules. The most common is an "if/then" sequence. In this situation, the "if" part gives a set of logical conditions. If the conditions are met, the "then" part can be concluded. The art of designing and developing expert systems is known as knowledge engineering. The system's capabilities can be enhanced by adding to the knowledge base or to the set of rules it uses.

Edward Feigenbaum is known as the father of expert systems. As the Association for Computing Machinery notes, "Feigenbaum was interested in the problem of 'induction', and in particular how to get computers to create theories from data—theories that not only explained the particular data on which the theory was based, but could also make predictions about new data" (Nilsson, 2012). Due to this interest, and at the suggestion of friend Joshua Lederberg, Feigenbaum et al., beginning in 1965, developed a computer program that could guess the geometrical structure of complex chemical compounds given their chemical formulae and their mass spectrogram data which was called Hueristic Dendral. Hueristic processes are those that enable a person to discover or learn something for themselves. Dendritics are items that have a branched form resembling a tree.

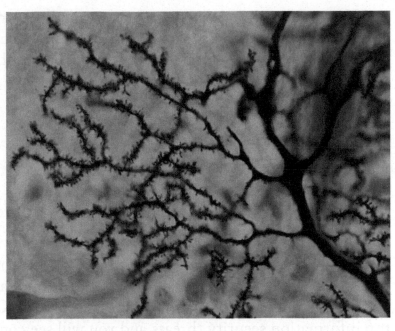

© Jose Luis Calvo/Shutterstock.com

Due to the success of Hueristic DENRAL, Feigenbaum became even more convinced about the importance of endowing computer programs with knowledge, as compared to reasoning, in the form of rules and procedures to guide the problem-solving process. Feigenbaum is also credited with the phrases "knowledge is power" and "expertise" as well (Nilsson, 2012).

Information Technology/Systems Security

IT security has always been an issue, but with the increasing diversity of use, amount of information, complexity of systems, and sheer number of users, it continues to rise in relevance and importance. There are actually a number of areas that fall under IT security such as:

- Secure central servers and end to end security for services
- Strong authentication on networks (including wireless)
- Storage, backup, and information retrieval
- Physical security
- Identity management
- And security of applications, data, and databases

In a nutshell though, it is controlling access to information or systems; so only authorized users can access it. Of key importance is the integrity of the system as a whole.

We could add to the above education and training, communication, and prescriptive guidance since humans using our systems are actually one of our biggest security risks. Consider this list from *CIO* of the six biggest business security risks:

1. Disgruntled employees
2. Careless or uninformed employees
3. Mobile devices (bring your own device)
4. Cloud applications
5. Unpatched or unpatchable devices
6. Third-party service providers (Lonoff Schiff, 2015).

As you can see from this list, there is much to be said about both our users and systems we do not directly control. Almost every item on the list is either based on users (or past users) or devices the IT staff do not control or administer.

Compare the earlier list of business security to Continuity Central's list of 2016's top information security threats and you will see common threads:

❖ Breach tsunami (bevy of breaches)
❖ Cyber warfare
❖ Internet of things
❖ Bring your own device
❖ TOR: the deep/dark web

Note: TOR is a hidden service protocol used by anonymity networks which was originally an acronym for The Onion Router.

❖ Unexposed vulnerabilities
❖ Mobile payment systems
❖ Individual cloud storage (Gill, 2016).

Again there is little that IT staff can directly control or mitigate via systems in their control. Much of it has to do with our increasing dependence on cloud-based technology and other internet-based systems. Based on these considerations, a key resource for assisting improved security is to have an IT security policy that is communicated, understood, and followed.

What are some key components to IT security in this day and age that doesn't involve humans and their tendencies or systems outside our control? Perhaps most important is the understanding of encryption—or the encoding of data. This way, even if a person gets access to the data somehow they cannot convert it to useful information.

Encryption

Encryption is a way to enhance the security of a message or file by scrambling the contents so that it can be read only by someone who has the right encryption key to unscramble it. There are two prime categories of encryption for computer systems. One is symmetric key encryption in which the two communicating computers must be known beforehand and then must each be given a matching (symmetric) key that they can use to decode the messages.

The other is public key encryption, also known as asymmetric key encryption. This method uses two different keys—a private key and a public key. The private key is known only the receiving computer, while the public key is given by that computer to any computer that wants to communicate securely with it. To decode an encrypted message, the computer must use the public key, provided by the originating computer, and its own private key (Tyson, n.d.)

Digital Certificate Management

A digital certificate authenticates the web credentials of the sender and lets the recipient of an encrypted message know that the data is from a trusted source (or a sender who claims to be one). A digital certificate is issued by a certification authority (CA) so a person who is sending an encrypted message obtains a key from them. The CA then issues the digital certificate with the applicant's public key along with other pertinent information. Digital certificates are also known as public key certificates or identity certificates (Technopedia, n.d.)

Strong Encryption

Strong encryption, also known as strong cryptography, or cryptographically strong, is cryptographic systems that are highly resistant to cryptanalysis. Cryptanalysis is the study of analyzing hidden aspects of information systems to breach security systems and gain access to the contents of encrypted messages, even without a cryptographic key.

While strong encryption is desirable if you are the possessor of information you wish to remain hidden in all circumstances, it is also effecting law enforcement and intelligence communities. There has been strong debate, for example, about a February 2016 court order that sought to compel Apple, Inc. (Apple) to assist the Federal Bureau of Investigations (FBI) in unlocking an iPhone used by one of the San Bernardino attackers. Apple refused. After months of work, the FBI was able to find a method for breaking into the phone without Apple's assistance (Brandom, 2016).

© Twin Design/Shutterstock.com

Social Media

Social media is any computer-mediated tools that allow individuals to create, share, or exchange information, interests, ideas, or media with others in virtual communities or networks. Within these forms of communication, social networking occurs. Social networking is the creation and maintenance of personal and/or business relationships online. In some cases, the items shared are totally public, and in other cases they are relatively private. One might consider it to have levels of private, semiprivate, semipublic, and public. E-mail would be an example of a private social media service, while YouTube videos, generally at least, would be considered public.

According to eBizMBA, following are the most common social networking sites for June 2016 in the United States: Facebook (1,100,000,000 unique monthly visitors), YouTube (1,000,000,000 unique monthly visitors), Twitter (310,000,000 unique monthly visitors), LinkedIn (255,000,000 unique monthly visitors), Pinterest (250,000,000 unique monthly visitors), Google Plus + (120,000,000 unique monthly visitors), Tumblr (110,000,000 unique monthly visitors), Instagram (100,000,000 unique monthly visitors), and Reddit (85,000,000 unique monthly visitors) (2016).

While there are a plethora of social networking sites available, the above lead the pack, at least for mid-2016. Facebook, YouTube, and Twitter, actually, have been in the top spots for years.

Let's take a moment to summarize these top 10.

Facebook: Facebook, located at https://www.facebook.com/, is a free-to-users (although for profit via advertising and the like) social networking site that allows users to create profiles, upload photos or videos, send or post messages, and network with others. It offers a variety of privacy options for its members and also offers a variety of ways to share information or resources including groups, events, pages, marketplace, and more. Members need to approve social media connections. It was launched in 2004 and is headquartered in Menlo Park, California.

YouTube: YouTube, located at https://www.youtube.com/, is a free video-hosting site that allows users to view, store, share, rate, and comment on video content. Almost all content has been uploaded by individual users instead of media corporations. It was launched in 2005 and is headquartered in San Bruno, California.

Twitter: Twitter, located at https://twitter.com/, is a free social networking microblogging service that permits registered users to broadcast short posts called tweets. Members can follow other members and can reply to tweets. Users can also add hashtags, a meta tag of sorts that helps create a connection between posts, in the form of #keyword. Anyone can follow anyone on public twitter. It was launched in 2006 and is headquartered in San Francisco, California.

LinkedIn: LinkedIn, located at https://www.linkedin.com/, is a social networking site intended specifically for the business community. The site allows registered users to create networks of people, called connections, they know or work with. It also permits users to create a detailed profile page that they can opt to share publically at varying levels. While the basic membership is free, LinkedIn charges for advanced access and resources. It was launched in 2003 and is headquartered in Mountain View, California.

Pinterest: Pinterest, located at https://www.pinterest.com/, is a social curation (collaborative sharing) website for sharing and categorizing media (usually photos) found online. It is essentially a visual bookmarking site. Users "pin" images, video, and other objects on their pinboard. Registration is necessary for use. It was launched in 2010 and is headquartered in San Francisco, California.

Google Plus (+): Google Plus, located at https://plus.google.com/, is a social networking site developed by Google. Its goal is to offer an enhanced social media experience which is closer to real life. Some features include grouping relationships into circles, multi-person instant messaging, text and video chat via hangouts, location tagging, and the ability to upload photos that are only visible to specific individuals. It was launched in 2011 (replacing Google Buzz) and is headquartered in Menlo Park, California.

Tumblr: Tumblr, located at https://www.tumblr.com/, is a free social networking site that permits registered users to create their own customizable blogs, known as tumblelogs, which they can post multimedia content (text, images, videos, audio, links) to. Some consider it a microblogging platform due to its simplicity. It was launched in 2007 and is headquartered in New York City, New York.

Instagram: Instagram, located at https://www.instagram.com/, is a free photo sharing and social network platform that is owned by Facebook (as of 2012). It allows members to upload, edit, and share photos or videos with other members. It was launched in 2010 and was acquired by Facebook in 2012. It is headquartered in Menlo Park, California.

Reddit: Reddit, located at https://www.reddit.com/, is an entertainment and social news website where information is socially curated (collaborative sharing) and promoted by registered members, known as redditors. The site consists of subcommunities, known as subreddits, which have specific topics. It is essentially an online bulletin board system where submissions are voted up or down in an effort to organize posts. It was launched in 2005 and is headquartered in San Francisco, California.

One cannot deny the role of social media in our everyday lives. Consider these statistics from Pew Research Center: Nearly two-thirds of American adults (65%) use social networking sites, up from 7% when Pew Research Center began systematically tracking social media usage in 2005. Young adults (ages 18–29) are the most likely to use social media—fully 90% do. Still, usage among those 65 and older has more than tripled since 2010 when 11% used social media. Today, 35% of all those 65 and older report using social media, compared with just 2% in 2005 (Perrin, 2015).

Social Media and the Workplace

What we may not realize is the role of social media in our work lives. When it comes to hiring, about 45% of employers look to social media like Facebook or Twitter to screen job candidates. Added to that, almost half of these have found content that caused them not to hire the candidate. The top reasons to pass on a candidate included candidate posted provocative or inappropriate photographs or information; candidate posted information about them drinking or using drugs; candidate bad-mouthed their previous company or fellow employee; candidate demonstrated poor communication skills; candidate posted discriminatory comments related to race, gender, religion and so on; and candidate lied about qualifications (Grasz, 2014).

Another factor to consider is if your employer can reprimand or even fire you based in things you do or say on social media. There are hundreds, if not thousands, of cases where employees have been fired, rightly or wrongly, based on actions they took in social media. In June of 2016, for example, a Bank of America employee was fired after a racially offensive Facebook post went viral (Hoff & King, 2016). In the same month, Google fired an employee for criticizing the company in a private Facebook group (Leswing, 2016) although that employee immediately took Google to court (Bergen, 2016) and an Australian airport fired an employee for Facebook posts that supported the Islamic State of Iraq and Syria (ISIS) although that employee as well has chosen to contest the termination (William, 2016). Whether you can be fired or not is based largely on the policies in place and contracts signed upon acceptance of the job. There have been cases, when employers have been legally required to rehire workers who vented in social media (Hill, 2011) but this is not commonly the case. The tendency is for employers not to tolerate workers who vent via social media and company policies are now being written to address this fact (Murphy, 2016; Hamer, Michaelson, & Robertson, 2013).

Open Access

Open access has been a cornerstone of Internet-based technology. As things stand now, the Federal Communications Commission (FCC) requires open Internet, commonly known as net neutrality. As they state:

An Open Internet means consumers can go where they want, when they want. This principle is often referred to as Net Neutrality. It means innovators can develop products and services without asking for permission. It means consumers will demand more and better broadband as they enjoy new lawful Internet services, applications and content, and broadband providers cannot block, throttle, or create special "fast lanes" for that content. The FCC's Open Internet rules protect and maintain open, uninhibited access to legal online content without broadband Internet access providers being allowed to block, impair, or establish fast/slow lanes to lawful content.

The Rules

Adopted on February 26, 2015, the FCC's Open Internet rules are designed to protect free expression and innovation on the Internet and promote investment in the nation's broadband networks. The Open Internet rules are grounded in the strongest possible legal foundation by relying on multiple sources of authority, including: Title II of the Communications Act and Section 706 of the Telecommunications Act of 1996. As part of this decision, the Commission also refrains (or "forbears") from enforcing provisions of Title II that are not relevant to modern broadband service. Together Title II and Section 706 support clear rules of the road, providing the certainty needed for innovators and investors, and the competitive choices and freedom demanded by consumers.

The Open Internet rules went into effect on June 12, 2015. They are ensuring consumers and businesses have access to a fast, fair, and open Internet.

The new rules apply to both fixed and mobile broadband service. This approach recognizes advances in technology and the growing significance of mobile broadband Internet access in recent years. These rules will protect consumers no matter how they access the Internet, whether on a desktop computer or a mobile device.

Bright Line Rules:

* No Blocking: broadband providers may not block access to legal content, applications, services, or non-harmful devices.
* No Throttling: broadband providers may not impair or degrade lawful Internet traffic on the basis of content, applications, services, or non-harmful devices.
* No Paid Prioritization: broadband providers may not favor some lawful Internet traffic over other lawful traffic in exchange for consideration of any kind—in other words, no "fast lanes." This rule also bans ISPs from prioritizing content and services of their affiliates.

To ensure an open Internet now and in the future, the Open Internet rules also establish a legal standard for other broadband provider practices to ensure that they do not unreasonably interfere with or disadvantage consumers' access to the Internet. The rules build upon existing, strong transparency requirements. They ensure that broadband providers maintain the ability to manage the technical and engineering aspects of their networks. The legal framework used to support these rules also positions the Commission for the first time to be able to address issues that may arise in the exchange of traffic between mass-market broadband providers and other networks and services. (Please note: this summary provides only a high-level overview of some key aspects of the Open Internet Order. More thorough analysis is available in the Fact Sheet and in the Order itself.)

This means that Internet service providers (including governments) cannot discriminate based on attributes of the user, the content, the site, the platform or similar. It both enables and protects free speech. It does not guarantee that the information we post or find is necessarily accurate or timely, but it does guarantee our right to communicate freely online.

Open Educational Resources

The term open access, or at least the first word, open, is now also used in many other contexts to identify resources that can be modified or repurposed by the general public. One example can be found with Open Educational Resources (OER), which is freely accessible, openly licensed resources (documents, media, etc.) that are useful for teaching, learning, or research. Perhaps, the best place to find OER materials is via the OER Commons, located at https://www.oercommons.org/.

Open Source

The term open source is used to denote software or similar for which the original source code is made freely available for redistribution and modification by any user. Generally, for something to qualify as open source in the programming world it should meet the following criteria: (1) free access, use, modification, and redistribution, (2) include source code level access, (3) permit derivations (allow for modifications and derived works), (4) offer equal opportunity for use and access, and (5) must, to the degree possible at least, be technology neutral. The Open Source Initiative includes additional criteria and expands on some of the aforementioned criteria at https://opensource.org/osd-annotated.

The idea of open access to materials, particularly in the arena of open source, as revolutionized IT programming as well as the Internet and tools therein as we know them. Through open source, individuals work able to build on and expand the work of others. They were able to push technologies and code in new directions. They were able to both make mistakes and find mistakes, causing programming as a whole to bump along in a positive direction for the better of all. Some of the many examples of open source software include: Apache HTTP server, MySQL, Python, PHP, and Moodle to name a few.

Copyright and Intellectual Property

Standard copyright rules exist on the Internet just as they do in any other medium. Generally speaking, we have well-established rules and norms for individual, joint, institutional, and vendor intellectual property ownership. We also have reasonably well-established academic norms for attribution. However, the devil is in the details (Cate, 2009). We can now distribute rights almost any way we want to distribute them. Copyright and intellectual property policy and enforcement includes a number of areas such as patents, trademarks, trade secrets, right of publicity, and copyrights.

As the United States Copyright Office (2012) states:

Copyright is a form of protection provided by the laws of the United States (title 17, U. S. Code) to the authors of "original works of authorship," including literary, dramatic, musical, artistic, and certain other intellectual works. This protection is available to both published and unpublished works. Section 106 of the 1976 Copyright Act generally gives the owner of copyright the exclusive right to do and to authorize others to do the following:

- ❖ To reproduce the work in copies or phonorecords;
- ❖ To prepare derivative works based upon the work;
- ❖ To distribute copies or phonorecords of the work to the public by sale or other transfer of ownership, or by rental, lease, or lending;
- ❖ To perform the work publicly, in the case of literary, musical, dramatic, and choreographic works, pantomimes, and motion pictures and other audiovisual works;
- ❖ To display the work publicly, in the case of literary, musical, dramatic, and choreographic works, pantomimes, and pictorial, graphic, or sculptural works, including the individual images of a motion picture or other audiovisual work; and
- ❖ In the case of sound recordings,
- ❖ To perform the work publicly by means of a digital audio transmission.

In addition, certain authors of works of visual art have the rights of attribution and integrity as described in section 106A of the 1976 Copyright Act. For further information, see Circular 40, Copyright Registration for Works of the Visual Arts.

It is illegal for anyone to violate any of the rights provided by the copyright law to the owner of copyright. These rights, however, are not unlimited in scope. Sections

107 through 121 of the 1976 Copyright Act establish limitations on these rights. In some cases, these limitations are specified exemptions from copyright liability. One major limitation is the doctrine of "fair use," which is given a statutory basis in section 107 of the 1976 Copyright Act. In other instances, the limitation takes the form of a "compulsory license" under which certain limited uses of copyrighted works are permitted upon payment of specified royalties and compliance with statutory conditions."

While there are regulations such as fair use, which allows for limited use of copyrighted materials without permissions from rights holders, particularly in the realm of education, in most cases copyright rules must be adhered to in the United States.

Creative Commons

What if you want to use music or images or whatnot from someone else in your own work but don't want to have to contact them or get release forms? In order to not plagiarize materials, we need to ensure adequate copyright release and attribution for resources we use. This is where creative commons (CC) and the public domain come in—these items already have some level of copyright release in place. Instead of focusing on copyright issues and limitations, this presentation will focus on items placed in whole or in part into the public domain.

Ownership and rights issues relating to resources available on the Internet can be a source of angst, confusion, and legal battles. In part, this is due to the automatic copyright many individuals receive, including the United States residents, upon creation of an original work. Now, however, there are options available for relaxing these rights. The service primarily used for these purposes both nationally and globally is CC (http://creativecommons.org/).

CC (http://creativecommons.org/) is a nonprofit organization that has established alternative copyright licenses for the public, which are known as CC licenses. These licenses vary based on certain features: attribution requirements, share alike stipulations, noncommercial or commercial permissions, and whether the works can be altered.

The CC license options:

❖ Attribution: This license lets others distribute, remix, tweak, and build upon your work, even commercially, as long as they credit you for the original creation. This is the most accommodating of licenses offered, in terms of what others can do with your works licensed under Attribution.

❖ Attribution Share Alike: This license lets others remix, tweak, and build upon your work even for commercial reasons, as long as they credit you and license their new creations under the identical terms. This license is often compared to open source software licenses. All new works based on yours will carry the same license, so any derivatives will also allow commercial use.

❖ Attribution No Derivatives: This license allows for redistribution, commercial and noncommercial, as long as it is passed along unchanged and in whole, with credit to you.

❖ Attribution Non-Commercial: This license lets others remix, tweak, and build upon your work noncommercially, and although their new works must also acknowledge you and be noncommercial, they don't have to license their derivative works on the same terms.

❖ Attribution Non-Commercial Share Alike: This license lets others remix, tweak, and build upon your work noncommercially, as long as they credit you and license their new creations under the identical terms. Others can download and redistribute your work just like the by-nc-nd license, but they can also translate, make remixes, and produce new stories based on your work. All new work based on yours will carry the same license, so any derivatives will also be noncommercial in nature.

❖ Attribution Non-Commercial No Derivatives: This license is the most restrictive of our six main licenses, allowing redistribution. This license is often called the "free advertising" license because it allows others to download your works and share them with others as long as they mention you and link back to you, but they can't change them in any way or use them commercially.

To learn about licensing your own work, visit http://creativecommons.org/choose/.

CC frees materials from automatically applied copyright restrictions by providing free, easy-to-use, flexible licenses for creators to place on their digital materials that permit the originator to grant rights as they see fit (Fitzergerald, 2007; Smith & Casserly, 2006). As the CC website notes, "Creative Commons provides free tools that let authors, scientists, artists, and educators easily mark their creative work with the freedoms they want it to carry. You can use CC to change your copyright terms from 'All Rights Reserved' to 'Some Rights Reserved'" (Creative Commons, 2012). This holds promise for OER movements because it helps control the costs and legal issues revolving around offering materials freely online (Caswell et al., 2008). Currently, over 30 nations now have CC licenses although it has only been in place for four years (Smith & Casserly, 2006).

Creative Commons Zero

Creative Commons Zero (CC0) essentially permits originators of materials of varying sorts to opt to put those materials into the public domain—waiving all copyright and intellectual rights. Most of CC licenses (discussed earlier) do not release materials into the public domain (Creative Commons, 2012). The public domain is "the realm of material—ideas, images, sounds, discoveries, facts, texts—that is unprotected by intellectual property rights and free for all to use or build upon" (Center for the Study of the Public Domain, n.d.). Items go into the public domain when rights have expired, been forfeited, or are not applicable for some reason.

In the case of CC0, the author chooses to put the materials in the public domain. As Creative Commons (2012) states, "Copyright and other laws throughout the world automatically extend copyright protection to works of authorship and databases, whether the author or creator wants those rights or not. CC0 gives those who want to give up those rights a way to do so, to the fullest extent allowed by law. Once the creator or a subsequent owner of a work applies CC0 to a work, the work is no longer his or hers in any meaningful sense under copyright law. Anyone can then use the work in any way and for any purpose, including commercial purposes, subject to other laws and the rights others may have in the work or how the work is used. Think of CC0 as the 'no rights reserved' option."

The ability for originators of works and materials to place these items into the public domain affects not just that individual but also all others who might make use of the resources or be affected by others who make use of the resources. In academics, this could have an impact on virtually all disciplines and areas of emphasis. It also has direct ties to consideration of what constitutes appropriate technology. In technology management, a technology is appropriate when "its intended positive consequences outweigh its unintended negative consequences" (Markert & Backer, 2010). The appropriateness of the technology is evaluated on technical, cultural and economic factors (Markert & Backer, 2010).

Right to Privacy

Privacy is something we all seem to want in some cases and all seem to be willing to give up in others. "Privacy is an important, but illusive concept in law. The right to privacy is acknowledged in several broad-based international agreements. Article 12 of the Universal Declaration of Human Rights and Article 17 of the United Nations International Covenant on Civil and Political Rights both state that, 'No one shall be subjected to arbitrary interference with his privacy, family, home or correspondence, nor to attacks upon his honour and reputation. Everyone has the right to the protection of the law against such interference or attacks'" (Stratford & Stratford, 1998). Stratford and Stratford (1998) note, "The term "privacy" does not appear in the U.S. Constitution or the Bill of Rights. However, the U.S. Supreme Court has ruled in favor of various privacy interests-deriving the right to privacy from the First, Third, Fourth, Fifth, Ninth, and Fourteenth Amendments to the Constitution."

Respect for privacy has played an important role in technology since its inception. This includes the right to privacy of users to the right to privacy of those in our larger communities. Now, however, IT evolution is making not only our ability to maintain respect for privacy harder, but seems to be altering our perception of the relevance and meaning of privacy itself. Gone are the days when we keep matters of the home in the home. Gone are the days when our children's trials and tribulations were discussed primarily at the family dinner table. Equally, gone, it seems, are the days when we as a society find issues like these problematic. While we still have power struggles between issues of freedom and control, the shape and context of these issues are changing. Now, we as a society debate about issues such as whether police should be allowed to add tracking devices to automotives belonging to criminals without their consent. We determine this should require a warrant. We debate about the implications of acquiring drones for personal video and recording use and debate about if Google Glass is going to make recording and posting of all our day to day events as common as the events themselves. With these issues we are less sure. Generally, we have preferred to have a say (if not perform the action ourselves) when information is posted that is identifiable as our own. Generally, though, we also seem to think it acceptable to post to identifiable information about others without asking them if we find it interesting, funny, or otherwise appealing to others in our circles of online friends and acquaintances. We seem to have the restrictive rules we want others to follow and then less restrictive rules we want to follow ourselves. We are, perhaps, victims of what is known as othering—where we classify others as "not one of us" and hold them to different standards or expectations.

Freedom of Speech/Expression

Freedom of speech (expression) in the United States is protected by the First Amendment to the US Constitution and by many state constitutions as well as and state and federal laws.

First Amendment to the US Constitution (Religion and Expression)

The First Amendment to the US Constitution is part of the Bill of Rights and was adopted in 1791. It provides that: Congress shall make no law respecting an establishment of religion, or prohibiting the free exercise thereof; or abridging the freedom of speech,

© MyImages-Micha/Shutterstock.com

or of the press; or the right of the people peaceably to assemble, and to petition the Government for a redress of grievances.

It should be noted that court decisions have expanded this concept to include not just verbal communication but also nonverbal expressions such as wearing a symbol, dance movements, or silent vigils. We understand that individuals and communities both large and small have a right to communicate. This is one area where issues of privacy versus freedom may become hazy if not downright contradicting. There may be a tense balancing of security (such as national security) and personal freedoms, for example. One simply has to look at the Uniting and Strengthening America by Providing Appropriate Tools Required to Intercept and Obstruct Terrorism Act (USA Patriot Act) of 2001 to find a situation such as this.

USA Patriot Act

In 2001, President George W. Bush signed in to law the Uniting and Strengthening America by Providing Appropriate Tools Required to Intercept and Obstruct Terrorism Act of 2001, also known as the USA PATRIOT act. As the United States Department of Justice (2016) summarizes:

1. The Patriot Act allows investigators to use the tools that were already available to investigate organized crime and drug trafficking.
    a. Allows law enforcement to use surveillance against more crimes of terror.
    b. Allows federal agents to follow sophisticated terrorists trained to evade detection.
    c. Allows law enforcement to conduct investigations without tipping off terrorists.
    d. Allows federal agents to ask a court for an order to obtain business records in national security terrorism cases.

2. The Patriot Act facilitated information sharing and cooperation among government agencies so that they can better "connect the dots."
3. The Patriot Act updated the law to reflect new technologies and new threats.
    a. Allows law enforcement officials to obtain a search warrant anywhere a terrorist-related activity occurred.
    b. Allows victims of computer hacking to request law enforcement assistance in monitoring the "trespassers" on their computers.

4. The Patriot Act increased the penalties for those who commit terrorist crimes.
    a. Prohibits the harboring of terrorists.
    b. Enhanced the inadequate maximum penalties for various crimes likely to be committed by terrorists.
    c. Enhanced a number of conspiracy penalties

    d. Punishes terrorist attacks on mass transit systems.

    e. Punishes bioterrorists.

    f. Eliminates the statutes of limitations for certain terrorism crimes and lengthens them for other terrorist crimes.

While parts of the Patriot Act are permanent, some parts of the Patriot Act were later extended in 2011 by President Barak Obama for four years—roving wiretaps, searches of business records, and conducting surveillance of individuals suspected of terrorist-related activities. These expired in 2015 but were generally replaced by the USA Freedom Act in 2015. That said, however, some see components of these acts as an abuse of privacy rights (Abrams, 2011).

Destruction of Information

Sometimes, the information we get on the Internet is just plain wrong. At times, this may be due to unintentional error. At other times, however, it is quite intentional. In mid-2015, Kentucky Fried Chicken, a worldwide fast food restaurant chain headquartered in Louisville, Kentucky, ended up suing several Chinese companies for formulating and spreading social media rumors that their chickens had eight legs and that their food was maggot-infested, patently false claims (Bever, 2015). This poses the question, when do you have the right to have information taken off the Internet? Added, to that, how do you do it? The online privacy blog from Abine has a good list of seven rules to get you started:

1. Walk before you run. Deletion must be done from the original source before Google will notice.
2. You'll hit some red lights. In most cases, websites don't have any duty to remove anything.
3. Keep it up. Persistence pays, so if you hit a wall, go around it.
4. Delete things from Google Search with Google's URL Removal Tool
5. If you can't delete something bad, bury it with something good
6. Report legal violations to the search engines
7. If something is truly defamatory, get a lawyer (Downey, 2012).

In some cases, a person or entity might want to be forgotten altogether or at least entirely forgotten as it relates to a certain incident. Due to this, the concept now known as the right to be forgotten came in to play around 2006.

Right to be Forgotten

As data are increasingly stored, shared, and queried, society needs to assess its viewpoints on what can or should be stored, where, and by whom. It also needs to consider who has access to this information and in what circumstances. As a part of that, is consideration on when a person or entity has a right to be forgotten. For example, if faulty information is made public, does the person or persons affected have the right to demand its' removal? If so, in what circumstances and who is responsible for the removal? This might sound easier than it is.

The European Union is moving to make it clearer. In April of 2016, the European Commission approved the General Data Protection Regulation (GDPR) which will come in to force in 2018. As Carol Magill (2016) of *The Irish News* notes, "It's a major shake-up: more than 200 pages of major reforms will introduce concepts such as the consumer's 'right to be forgotten'; raise levels of verification for opt-in consent; demand that companies store consent permissions; and make unapproved data unusable. Companies that don't comply could be fined up to 4 per cent of their global turnover, or €20 m."

The role of media in our private lives can take varying forms. As Alessandro Mantelero notes in Computer Law and Security Review, "the media affect private life in two different ways: by revealing events or information that should remain private, and thus violating the individual right to privacy, or by publicizing events whose social or political relevance prevails over their private nature. In the second case, the limitation to the right to privacy is conditioned by time: i.e., once the period of time in which interest in a specific private even is justified by its impact on the community has elapsed, the individual has the right to regain his anonymous life and privacy" (2013, p. 2) He continues, "this conception of the right to be forgotten is based on the fundamental need of the individual to determine the development of his life in an autonomous way, without being perpetually or periodically stigmatized as a consequence of a specific action performed in the past, especially when those events occurred many years ago and do not have any relationship with the contemporary context" (2013, p. 2)

With the above in mind, let's consider a case. In 2014, a top European court told Google it must delete "inadequate, irrelevant, or no longer relevant" data from its results when a member of the public requests it (Travis & Arthur, 2014). This is based on a case ruling against Google Spain by the European Union's court of justice, which was brought by a Spanish man, Mario Costeja González, "after he failed to secure the deletion of an auction notice of his repossessed home dating from 1998 on the website of a mass circulation newspaper in Catalonia" (Travis & Arthur, 2014, p. 1). The court judges ruled that "Google has to erase links to two pages on La Vanguardia's website from the results that are produced when Costeja González's name is put into the search engine" (Travis & Arthur, 2014, p. 1). What this ruling does is make "clear that a search engine such as Google has to take responsibility as a 'data controller' for the content that it links to and may be required to purge its results even if the material was previously published legally." (Travis & Arthur, 2014, p. 1).

At the point of this writing, Google is appealing to France's highest court over this legal ruling that could for it to censor its search results not just in the European Union, but worldwide. As The Guardian describes, "The search firm has filed an appeal with the Conseil d'État, the French court with the final say over matters of administrative law, in an attempt to overturn a ruling from the country's data protection authority (CNIL), which would greatly extend the remit of the so-called 'right to be forgotten'" (Hern, 2016).

Part of the issue is that not only do search terms need to be removed in the country or region where the complaint originated, but instead it must be removed from all searches on all domains.

Cases such as this pose many questions as to where responsibility lies, how long information is relevant, what is considered relevant, and who gets to determine outcomes.

There are other cases where truth is a matter of opinion. Then, the question becomes one of whose truth prevails if neither can be proven or are subjective in nature. Consider multiple examples where a company has attempted to sue someone for a bad review they placed on line. Some examples of this include:

* Etica Pizzeria who threatened to sue Julian Tully for a negative review he put on Facebook and TripAdvisor when he wrote the following: "Stay away from this place. I went there with some friends tonight who are all seasoned foodies and were treated in a fashion that I didn't think was possible. For 7 people we got a tiny amount of food (waiting more than 50 minutes in between the portions) and when we tried to complain in reasonable way we literally got told 'we have had our fill' and 'we shouldn't go out for dinner if we can't afford it.' Then they called the cops on us because we walked out. Avoid like the plague! (Unless you liked to be judged by a bunch of people who can't run a business in an 'ethical' way)." (Quinn, 2015).

* A Staten Island woman is ordered to pay $1000 due to online reviews she posted about the company Mr. Sandless, which she had hired to refinish the floors in her living room in 2015 for $695 dollars. Here, as an example, is what she wrote on Yelp: "this guy mat the owner is a scam do not use him you will regret doing business with this company I'm going to court he is a scam customers please beware he will destroy your floors he is nothing by a liar he robs customers, and promises you everything if you want s—then go with him if you like nice work find another he is A SCAM LIAR BULL—ER" (Mai & Gregorian, 2015). As the judge saw it, the terms scam, con artist and robs implied criminal wrongdoing rather than someone who failed to meet the terms of the contract (Mai & Gregorian, 2015). Note: The company originally sued for $750,000 (Regan, 2014).

Does it make matters different if a customer is paid to write a positive review, whether accurate or not? This is something Amazon.com has been fighting for years. At the time of this writing, Amazon has begun suing sellers for buying fake reviews as an effort to combat fake reviews on its platform (Conger, 2016). Previously it pursued those who posted the reviews, "Amazon says that, since early 2015, it has sued over 1,000 people who posted fake reviews for cash. Now, the company is going after the retailers themselves. Amazon said that it intends to eliminate incentives for sellers to buy fake reviews for their products." (Conger, 2016). Now, they want to eliminate the incentives for sellers to participate in the practice whether by suspending or shutting down their accounts to initiating lawsuits.

Self-Destruction

The earlier has been focusing on having external systems or individuals remove information. Let's consider if either users or systems have the capacity to permanently destroy information themselves. If one considers the top 10 most popular Defense Advanced Research Projects Agency (DARPA) stories of 2013, one will see that item eight on the list is "Vanishing Programmable Resources (VAPR) program with the aim of revolutionizing the state of the art in transient electronics or electronics capable of dissolving into the environment around them. Transient electronics developed under VAPR should maintain the current functionality and ruggedness of conventional electronics, but, when triggered, be able to degrade partially or completely into their surroundings. Once triggered to dissolve, these electronics would be useless to any enemy who might come across them. ..." Actually, one does not have to go as far as DARPA to obtain examples of individuals being able to remotely lock or disable a technology or communication tool. Simply look at the current state of smart phones. A person can currently disable a lost or stolen phone with ease. They can erase all the data on the phone and render that data unrestorable.

At the same time, we are creating technology with the capability to self-destruct, we are also creating unprecedented surveillance technology with which to attain and store data for unspecified periods of time. As Leckie and Bushman (2009) note, "the traditional forms of surveillance contrast in important ways with what can be called the new surveillance, a form that became increasingly prominent toward the end of the twentieth century. The new social surveillance can be defined as, scrutiny through the use of technical means to extract or create personal or group data, whether from individuals or contexts. Examples include the following: video cameras; computer matching, profiling, and data mining; work, computer, and electronic location monitoring; DNA analysis; drug tests; brain scans for lie detection; various self-administered tests; and thermal and other forms of imaging to reveal what is behind walls and enclosures. The use of technical means to extract and create the information implies the ability to go beyond what is offered to the unaided senses or what is voluntarily reported. Much new surveillance involves an automated process and extends the senses and cognitive abilities through the use of material artifacts or software."

Role of Informatics

When you consider the role of the informal Internet and the communication channels therein, a great number of individuals worldwide now have a voice as well as a channel for its use. However, it is not always without personal cost. In some cases in the global sphere, individuals put themselves at personal risk in order to share information, resources, or thoughts. Still, though, at least they now have a means to do so when historically they had none. As the United Nations Educational, Scientific, and Cultural Organization (UNESCO) notes in its background document distributed for a session on Ethics of the Information Society,

> Access to information and the capacity to be able to enjoy the "right to communication" are essential to the achievement of greater equity in a global society. Information and communication are both "resources" whose ethical usage and distribution can create the conditions for democracy and greater well-being. Following from this, information and communication are certainly "natural" (i.e. what we do), but neither should be considered neutral by virtue of their respective status as facts. Communication and the free circulation of information are essential to the realization of "democratic selfhood" and the formation of the global public sphere. However, when refracted through the information society, such a public sphere appears radically reconfigured. While such a sphere is certainly defined by communicative action, it is not necessarily founded upon any consensual framework of liberal reason. Indeed, it is precisely where those who are shut out of this sphere find their own voices and channels of communication.

When you consider the role of significantly more data acquisition, mining, distribution, assessment, and use, it is clear to see that informatics is changing our world. It is changing how companies do business, how trend analysis is accomplished, how educational attainment is accomplished and measured, how we perceive the world around us, and how we perceive information itself. As UNISCO continues, "A term that began readily circulating in the mid-1970s, by 'infosphere' and information society we understand a world where the boundaries between the human, technology, and the media are dissolved, a world where 'information' functions as a type of substance that animates, orders, and delimits human activity as both a field of possibility and constraint."

The Science behind the Technology

While encryption and cryptographic keys were discussed earlier, another component to computer-based systems communications is the standards and protocols used. Let's briefly consider network protocol standards.

Network Protocol Standards

When it comes to overall models for communication and their protocols, two types come in to the discussion. First, is the Transmission Control Protocol/Internet Protocol known as TCP/IP. It is the basic communication language or protocol of the Internet and was created by the Department of Defense. Next, is the Open Systems Interconnection Basic Reference (OSI) model. The latter is an abstract description for network protocol design, developed as an effort to standardize networking.

As the OSI model is broader and includes the concepts of the TCP/IP model, this is the one we will discuss now.

Open Systems Interconnection basic reference model

The OSI model is a layered, abstract description for communications and computer network protocol design. It is a set of standard specifications that allows various computer platforms to communicate openly with one another and is primarily concerned with the interconnection between systems and the manner in which these systems exchange information.

As can be seen in above diagram, there are seven layers in the OSI model. These are described by Webopedia below:

1. Physical (Layer 1):
   OSI Model, Layer 1 conveys the bit stream—electrical impulse, light or radio signal—through the network at the electrical and mechanical level. It provides the hardware means of sending and receiving data on a carrier, including defining cables, cards, and physical aspects. Fast Ethernet, RS232, and ATM are protocols with physical layer components.
   Physical examples include Ethernet, FDDI, B8ZS, V.35, V.24, RJ45.
2. Data Link (Layer 2):
   At OSI Model, Layer 2, data packets are encoded and decoded into bits. It furnishes transmission protocol knowledge and management and handles errors in the physical layer, flow control, and frame synchronization. The data link layer is divided into two sub layers: The Media Access Control (MAC) layer and the Logical Link Control (LLC) layer. The MAC sub layer controls how a

computer on the network gains access to the data and permission to transmit it. The LLC layer controls frame synchronization, flow control, and error checking. Data Link examples include PPP, FDDI, ATM, IEEE 802.5/802.2, IEEE 802.3/802.2, HDLC, Frame Relay.

3. Network (Layer 3):
    Layer 3 provides switching and routing technologies, creating logical paths, known as virtual circuits, for transmitting data from node to node. Routing and forwarding are functions of this layer, as well as addressing, internetworking, error handling, congestion control, and packet sequencing.
    Network examples include AppleTalk DDP, IP, IPX.

4. Transport (Layer 4):
    OSI Model, Layer 4, provides transparent transfer of data between end systems, or hosts, and is responsible for end-to-end error recovery and flow control. It ensures complete data transfer.
    Transport examples include SPX, TCP, UDP.

5. Session (Layer 5):
    This layer establishes, manages, and terminates connections between applications. The session layer sets up, coordinates, and terminates conversations, exchanges, and dialogues between the applications at each end. It deals with session and connection coordination.
    Session examples include NFS, NetBios names, RPC, SQL.

6. Presentation (Layer 6):
    This layer provides independence from differences in data representation (e.g., encryption) by translating from application to network format, and vice versa. The presentation layer works to transform data into the form that the application layer can accept. This layer formats and encrypts data to be sent across a network, providing freedom from compatibility problems. It is sometimes called the syntax layer.
    Presentation examples include encryption, ASCII, EBCDIC, TIFF, GIF, PICT, JPEG, MPEG, MIDI.

7. Application (Layer 7):
    OSI Model, Layer 7, supports application and end-user processes. Communication partners are identified, quality of service is identified, user authentication and privacy are considered, and any constraints on data syntax are identified. Everything at this layer is application-specific. This layer provides application services for file transfers, e-mail, and other network software services. Telnet and FTP are applications that exist entirely in the application level. Tiered application architectures are part of this layer.
    Application examples include WWW browsers, NFS, SNMP, Telnet, HTTP, FTP (Beal, 2015)

This model is used as a conceptual framework. It does not actually perform any functions in the network process nor is it tangible. Instead, it helps us better understand the complex interactions that occur.

Radio Waves and Signals

Radio waves and signals also play a key role in IT systems as was noted in the communication technology segment above. Let's take a simple example and consider RFID tags we discussed earlier in this chapter. These tags use a transceiver that generates a week signal that has a range of only a few feet to a few yards. This signal is necessary to wake or activate the tag. The signal itself is a form of energy that can be used to power the tag. Next, the transponder, which is a part of the RFID tag, converts that RF into usable power and sends and receives messages. When the transponder is hit by the radio waves, the waves go up and down the length of the transceiver, oscillating. The transponder, now waken up, sends all the information it has stored on it. As Guy McDowell (2009) explains in lay terms, "Imagine having friend who was hypnotized to wake up upon hearing a code word, tell you his name and phone number, and then fall immediately back to sleep. Yet he wouldn't do this no matter what else you said. Funny, yes it is. That's pretty much what RFID technology does."

Career Connections

The US Bureau of Labor Statistics has written a special report on careers in the growing field of technology services. The following is a summary of some of their findings:

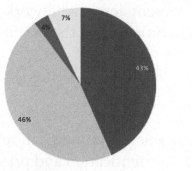

Source: U.S. Bureau of Labor Statistics, Current Employment Statistics.

- Custom computer programming services
- Computer systems design services
- Computer facilities management services
- Other computer related services

Employment distribution of computer systems design and related services, 2011

Computer systems design and related services can be broken down into four subindustries: custom computer programming services, computer systems design services, computer facilities management services, and other computer related services (see chart 1). The first two are the largest, and account for almost 90% of all IT services employment. Custom computer programming services includes establishments that write, test, and modify software for a particular client. This software includes computer programs, webpage design, and database design. Computer programming services also provide support to clients after the newly designed software is implemented.

Employment and output in computer systems design and related services are projected to grow rapidly over the next decade, outpacing similar professional, scientific, and technical industries and the economy as a whole. Between 2010 and 2020, output in computer systems design and related services is expected to grow at an average annual rate

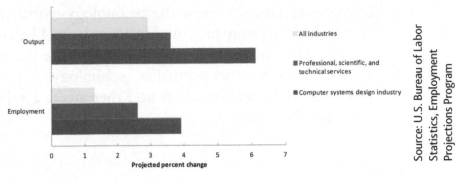

Source: U.S. Bureau of Labor Statistics, Employment Projections Program

Projected average annual percent change in output and employment in selected industries, 2010–2020

of 6.1%, compared with 3.6% for the broad industry category—professional, scientific, and technical services—and 2.9% for all industries

Employment and output in the computer systems design and related services industry is expected to grow rapidly as firms and individual consumers continue to increase their use of IT services. Cloud computing is one area that is expected to contribute to growth in this industry (Csorny, 2013).

Modular Activities

Discussions

* Post an original discussion in the online discussion board on the following topic: When it comes to domain names (website URLs), companies can now purchase a generic top-level domain (gTLD) so instead of having to use something like .com or .net you could purchase something like .car or .nonprofit or whatnot. That said, the starting price is $185,000. What do you think might be some benefits or drawbacks to having gTLDs? (200 words minimum). Next, comment on the post(s) of a minimum of one other student in a thoughtful and academic way that enhances the conversation. See rubric for grading and assessment measures. You can learn more on the web at place such as: http://newgtlds.icann.org/en/ or http://en.wikipedia.org/wiki/Generic_top-level_domain or https://www.netnames.com/gtld/

* Technology Fair: Find a current or upcoming technology (in the current news) that is of interest to you and research it. After researching a minimum of three sources (each less than a year old) pertaining to the specific technology of your choice (without duplicating other student topic choices) inspired by current events and/or the preceding lessons, share with the class the following: (a) Name and detailed overview of the technology, (b) diagram/images or other visuals of the technology,

(c) demonstration of how the technology works, and (d) affect or potential affect this technology may have on society at large. Use your creativity to share this information with the class. Treat it as if you were the teacher as you are teaching your classmates about this particular technology. After you have completed your own post, read the posts of others and then make a substantial/meaningful response to your favorite three.

Tests

❖ Online graded quiz on overall chapter content, written in a multiple-choice format. When submitted, we recommend giving the correct answer along with the page number in which it is found for questions students did not answer correctly.

❖ Complete one of the network exams found at http://www.examcompass.com/comptia/network-plus-certification/free-network-plus-practice-tests. When complete, print your results to a PDF file and submit. You will not be graded on how well you did on the practice test, but instead on your attempt to take it in order to get a better feel for things networks administrators might need to know.

❖ Use the website Socrative found at http://www.socrative.com/ (free of charge) to set up a live interactive quiz where students can instantly see the overall results for the class. Ask the following questions:

  ➢ I believe increasing use of the Internet has had a _____ influence on education (multiple choice: good influence, bad influence, no influence)

  ➢ I believe increasing use of the Internet has had a _____ influence on personal relationships (multiple choice: good influence, bad influence, no influence)

  ➢ I believe increasing use of the Internet has had a _____ influence on economy (multiple choice: good influence, bad influence, no influence)

  ➢ I believe increasing use of the Internet has had a _____ influence on politics (multiple choice: good influence, bad influence, no influence)

  ➢ I believe increasing use of the Internet has had a _____ influence morality (multiple choice: good influence, bad influence, no influence)

Follow this with a class discussion about the results.

Research

❖ Research report on computer/networking technology: Choose a computer/networking technology interest to you and write a three-page informative report about it.

❖ Imagine you just started a new a five-person (full time) shop that locates and maps property lines and landmarkers. You have zero IT currently for your company. Research the IT-related systems and items you would acquire with a rationale as to why. Also give a rough budget estimate. You must reference your research sources. You will be graded on the thoughtfulness, validity, relevance, and clarity of your answers along with being assessed on your level of research.

❖ Network Protocols: Create a detailed diagram/poster that explains network protocols (or a specific type of network protocol if you choose). Web searching will prove helpful for this assignment although what you turn in must be original work (no plagiarism)

❖ Interview a person who works in an IT-related field. Ask the following questions at a minimum and then come up with four (4) questions of your own:
  ➢ What are the import important technologies used in your job or field?
  ➢ What are challenging aspects of your job?
  ➢ What are your favorite parts of your job?
  ➢ What would you recommend to someone going in to this field?

❖ Submit both the questions and the answers.

❖ In this chapter, we discussed Dr. Edward Feigenbaum, known as the father of expert systems. He was one of a number of scientists and mathematicians to win what is known as the A.M. Turing award. Go to the website http://amturing. acm.org/ to see a list of others who have received this award. Choose one from the alphabetical listing and write a 200-word summary of their role in IT and communications.

❖ Research one of the following higher education academic areas:
  ➢ Information systems
  ➢ Information technology
  ➢ Network administration
  ➢ Computer science
  ➢ Computer engineering
  ➢ Software engineering

This might include review of offerings or degrees at your institution or at another location, contacting a professional/faculty in the field, assessing higher education level academic events or educational resources, and so on. Post a 200-word minimum summary of your findings. Make sure to include your source(s).

Design and/or Build Projects

❖ RFID technical report: You work for a large company that has decided it wants to put RFID tags on all items that cost over $100 both for merchandise going out but ideally also supplies coming in. They need out to write a technical report that explains how it works and why it is important to the other employees. Your boss recommended you look at Walmart's RFID system as an example. Your report is to be at least two written pages. As many people are visually oriented, diagrams or images are recommended as well (diagrams are in addition to the two-page written report).

❖ Using a website such as http://www.cyberpowerpc.com/, http://pcpartpicker.com/, or http://magicmicro.com/ creates a customized personal computer that you feel would best fit your current needs as a student. Submit the specifications and cost and then give a short summary of why you elected to choose some of the options you did.

Assessment Tasks

❖ Wikipedia page: Wikipedia is a free encyclopedia, written collaboratively by the people who use it. Your task for this assignment is to make a substantive change to a Wikipedia page on any topic of your choice and post what you changed here. Substantive changes include things such as: adding to the depth and detail of the content in a meaningful way, correcting errors with citations to your sources, or improving the timeliness and accuracy of information. To learn how to edit a Wikipedia page, visit https://en.wikipedia.org/wiki/Help:Editing and/or https://en.wikipedia.org/wiki/Wikipedia:Tutorial/Editing

❖ Cloud-based applications and storage: You are the IT manager at a small company (10–20 employees total; total sales under $500,000 a year) that sells yarn dog toys. Your supervisor has requested that you do an assessment (benefits and drawbacks) of moving the company from your local network and desktop-based software to a cloud-based system and offer a recommendation with rationale. You are to write up a minimum two-page report in any format you desire to get your assessment and recommendations across to upper management. You will be assessed on the accuracy, quality, professionalism, and depth of your report.

Terms

Creative Commons (CC)—Public copyright license that enables the free distribution of an otherwise copyrighted work. A CC license is used when an author wants to give people the right to share, use, and build upon a work that they have created.

CC0—CC0 is a single purpose tool, designed to take on the dedication function of the former, deprecated Public Domain Dedication and Certification

Cryptanalysis—The study of analyzing hidden aspects of information systems to breach security systems and gain access to the contents of encrypted messages, even without a cryptographic key.

Dendritics—Items that have a branched form resembling a tree.

Digital certificate management—The process of managing digital security certificates.

Digital certificate—An electronic document used to prove ownership of a public key, which permits users to share information securely over the Internet.

Encryption—Also known as cipher text, this is a process of encoding a message; so it can be read only by the sender and intended recipient. Encryption systems often use two keys, a public key, available to anyone, and a private key that allows only the recipient to decode the message.

Expert system—A computer system that emulates the decision-making ability of a human expert.

Generic top-level domain (GtLD)—A generic top-level domain (gTLD) is an internet domain name extension with three or more characters. It is one of the categories of the top-level domain (TLD) in the Domain Name System (DNS) maintained by the Internet Assigned Numbers Authority. Examples of standard top-level domains are .com, .net, and .edu. In the image below it is where you see the letters TLD.

Hueristic processes—Those that enable a person to discover or learn something for themselves.

Informatics—The science of processing data for storage, retrieval, and dissemination; information science.

Open Systems Interconnection Basic Reference (OSI) model—The OSI model is a layered, abstract description for communications and computer network protocol design. It is a set of standard specifications that allows various computer platforms to communicate openly with one another and is primarily concerned with the interconnection between systems and the manner in which these systems exchange information

Network protocol—Formal standards and policies comprised of rules, procedures, and formats that define communication between two or more devices over a network. Network protocols govern the end-to-end processes of timely, secure, and managed data or network communication.

Radio frequency identification (RFID)—A technology that incorporates the use of electromagnetic or electrostatic coupling in the radio frequency (RF) portion of the electromagnetic spectrum to uniquely identify an object, animal, or person.

Transmission Control Protocol/Internet Protocol (TCP/IP)—The language a computer uses to access the Internet. It consists of a suite of protocols designed to establish a network of networks to provide a host with access to the Internet.

References

Abrams, J. (2011). Patriot act extension signed by Obama. *Huffpost Politics*. Retrieved from http://www.huffingtonpost.com/2011/05/27/patriot-act-extension-signed-obama-autopen_n_867851.html

Anderson, M. (2015). Technology Device Ownership: 2015. Pew Research Center. Retrieved from http://www.pewinternet.org/2015/10/29/technology-device-ownership-2015/

Beal, V. (2015). The 7 Layers of the OSI Model. *Webopedia*. Retrieved from http://www.webopedia.com/quick_ref/OSI_Layers.asp

Bergen, M. (2016). A Nest employee was fired after sharing internal content on Facebook, now he's taking Google to court. *CNBC Technology*. Retrieved from http://www.cnbc.com/2016/06/03/a-nest-employee-was-fired-after-sharing-internal-content-on-facebook-now-hes-taking-google-to-court.html

Bever, L. (2015). KFC in China has not produced an eight-legged chicken. It's going to court to prove it. *The Washington Post*. Retrieved from https://www.washingtonpost.com/news/morning-mix/wp/2015/06/02/kfc-has-not-produced-an-eight-legged-chicken-its-going-to-court-to-prove-it/

Brandom, R. (2016). Apple's San Bernardino fight is officially over as government confirms working attack. *The Verge*. Retrieved from http://www.theverge.com/2016/3/28/11317396/apple-fbi-encryption-vacate-iphone-order-san-bernardino

Caswell, T., Henson, S., Jensen, M., & Wiley, D. (2008). Open educational resources: Enabling universal education. *International Review of Research in Open and Distributed Learning, 9* (1). Retrieved from http://www.irrodl.org/index.php/irrodl/article/view/469/1001

Center for the Study of the Public Domain. (n.d.). About us. Retrieved from http://www.law.duke.edu/cspd/about.html

Cisco. (2016). Cisco visual networking index: Global mobile data traffic forecast update, 2015–2020 white paper. *Cisco*. Retrieved from http://www.cisco.com/c/en/us/solutions/collateral/service-provider/visual-networking-index-vni/mobile-white-paper-c11-520862.html

Columbus, L. (2015). Roundup of small & medium business cloud computing forecasts and market estimates, 2015. *Forbes*. Retrieved from http://www.forbes.com/sites/louiscolumbus/2015/05/04/roundup-of-small-medium-business-cloud-computing-forecasts-and-market-estimates-2015/#3d3eafc81646

Conger, K. (2016). Amazon sues sellers for buying fake reviews. *Tech Crunch*. Retrieved from http://techcrunch.com/2016/06/01/amazon-sues-sellers-for-buying-fake-reviews/

Creative Commons. (2012). CC0. Retrieved from http://creativecommons.org/choose/zero/

Csorny, L. (2013). Careers in the growing field of information technology services. *Beyond the Numbers. U.S. Bureau of Labor Statistics*. Retrieved from http://www.bls.gov/opub/btn/volume-2/pdf/careers-in-growing-field-of-information-technology-services.pdf

Downey, S. (2012). How to delete things from the internet: A guide to doing the impossible. *Abine. The Online Privacy Blog*. Retrieved from https://www.abine.com/blog/2012/how-to-delete-things-from-the-internet/

eBizMBA. (2016). Top 15 most popular social networking sites June 2016. *eBizMBA*. Retrieved from http://www.ebizmba.com/articles/social-networking-websites

File, T., & Ryan, C. (2014). Computer and internet use in the United States: 2013. *U.S. Census Bureau*. Retrieved from http://www.census.gov/content/dam/Census/library/publications/2014/acs/acs-28.pdf

Fitzergerald, B. (2007). *Open content licensing (OCL) for open educational resources*. Organisation for Economic Co-operation and Development, Brisbane, Australia.

Gill, T. (2016). 2016's top information security threats. *Continuity Central*. Retrieved from http://www.continuitycentral.com/index.php/news/technology/729-2016-s-top-information-security-threats

Grasz, J. (2014). Number of employers passing on applicants due to social media posts continues to rise, according to New CareerBuilder Survey. *CareerBuilder*. Retrieved from http://www.careerbuilder.com/share/aboutus/pressreleasesdetail.aspx?sd=6%2F26%2F2014&id=pr829&ed=12%2F31%2F2014

Hamer, S., Michaelson & Robinson. (2013, July 29). Creating an effective workplace social media policy. *Bloomberg BNA*. Retrieved from http://www.bna.com/creating-an-effective-workplace-social-media-policy/

Hern. A. (2016). Google takes right to be forgotten battle to France's highest court. *The Guardian*. May, 19. Retrieved from https://www.theguardian.com/technology/2016/may/19/google-right-to-be-forgotten-fight-france-highest-court

Hill, K. (2011). Judge forces employer to rehire workers who vented on Facebook. *Forbes Tech*. Retrieved from http://www.forbes.com/sites/kashmirhill/2011/09/07/judge-forces-employer-to-rehire-workers-who-vented-on-facebook/

Hoff, V., & King, M. (2016). Bank of America employee fired after racially offensive Facebook post goes viral. *NBC 11 Alive*. Retrieved from http://www.11alive.com/news/local/bank-of-america-employee-fired-after-racially-offensive-facebook-post-goes-viral/228520320

International Telecommunication Union. (2015). *ICT facts and figures. The world in 2015*. Retrieved from https://www.itu.int/en/ITU-D/Statistics/Documents/facts/ICTFactsFigures2015.pdf

Leckie, G., & Buschman, J. (2009). *Information technology in librarianship*. Westport, CT: Libraries Unlimited.

Leswing, K. (2016). Google fired an employee for criticizing the company in a private Facebook group, report says. *Business Insider*. Retrieved from http://www.businessinsider.com/former-nest-employee-files-nlrb-complaint-against-google-2016-6

Lonoff Schiff, J. (2015). Six biggest business security risks and how you can fight back. *CIO. How-to*. Retrieved from http://www.cio.com/article/2872517/data-breach/6-biggest-business-security-risks-and-how-you-can-fight-back.html?page=2

Magill, C. (2016). Act now on general data protection regulation. *The Irish News*. Retrieved from http://www.irishnews.com/business/2016/06/07/news/act-now-on-general-data-protection-regulation-546566/

Mai, A., & Gregorian, D. (2015). Staten Island woman ordered to pay $1,000 fine for bashing floor refinishing business on Yelp. *NY Daily News*. Retrieved from http://www.nydailynews.com/new-york/s-woman-pay-1g-fine-bashing-business-yelp-article-1.2370681

Mantelero, A. (2013).The EU Proposal for a general data protection regulationand the roots of the 'right to be forgotten'. *Computer Law & Security Review, 29*(3), 229–235. Retrieved from http://www.academia.edu/3635569/The_EU_Proposal_for_a_General_Data_Protection_Regulation_and_the_roots_of_the_right_to_be_forgotten_

Markert, L. R., & Backer, P. R. (2010). *Contemporary technology*. Tinley Park, IL: Goodheart-Willcox Company.

McDowell,G.(2009).How does RFID technology work? *[Technology Explained]. Make Use of.*Retrieved from http://www.makeuseof.com/tag/technology-explained-how-do-rfid-tags-work/

Mell, P., & Grance, T. (2011). *The NIST definition of cloud computing.* United States Department of Commerce National Institute of Standards and Technology. Retrieved from http://nvlpubs. nist.gov/nistpubs/Legacy/SP/nistspecialpublication800-145.pdf

Mohamed, A. (2009). A history of cloud computing. *Computer Weekly.* Retrieved from http://www. computerweekly.com/feature/A-history-of-cloud-computing

Murphy, H. (2016). Employers need not tolerate workers screaming on the electronic street corner. *The National Law Review.* Retrieved from http://www.natlawreview.com/article/ employers-need-not-tolerate-workers-screaming-electronic-street-corner

Nilsson, N. (2012). Edward A ("Ed") Feigenbaum. *A.M. Turing Award.* Retrieved from http:// amturing.acm.org/award_winners/feigenbaum_4167235.cfm

Pardo, J., Flavin, A., & Rose, M. (2016). *2016 top markets report cloud computing. A market assessment tool for U.S. exporters.* International Trade Commission, U.S Department of Commerce Economics and Statistics Administration. Retrieved from http://trade.gov/topmarkets/pdf/ Cloud_Computing_Top_Markets_Report.pdf

Perrin, A. (2015). Social media usage: 2005-2015. *Pew Research Center Internet, Science & Tech.* Retrieved from http://www.pewinternet.org/2015/10/08/social-networking-usage-2005-2015/

Pew Global. (2015). Global computer ownership. *Pew Research Center Global Attitudes & Trends.* Retrieved from http://www.pewglobal.org/2015/03/19/internet-seen-as-positive-influence-on-education-but-negative-influence-on-morality-in-emerging-and-developing-nations/ technology-report-15/

Pew Research Center. (2014). Cell phone and smartphone ownership demographics. *Pew Research Center Internet, Science & Tech.* Retrieved from http://www.pewinternet.org/data-trend/mobile/ cell-phone-and-smartphone-ownership-demographics/

Pew Research Center. (2015). Internet seen as positive influence on education but negative on morality in emerging and developing nations. *Pew Research Center Global Attitudes & Trends.* Retrieved from http://www.pewglobal.org/2015/03/19/internet-seen-as-positive-influence-on-education-but-negative-influence-on-morality-in-emerging-and-developing-nations/

Quinn, L. (2015). Pizza shop threatens to SUE customers who left negative review that claimed he waited almost an hour for his food and warned people to 'stay away from this place'. *Daily Mail News.* Retrieved from http://www.dailymail.co.uk/news/article-3277626/Pizza-shop-threatens-SUE-customers-left-negative-review-claimed-waited-hour-food-warned-people-stay-away-place.html#ixzz4B1vBnc00

Rainie, L. (2013). Cell phone ownership hits 91% of adults. *Pew Research Center.* Retrieved from http://www.pewresearch.org/fact-tank/2013/06/06/cell-phone-ownership-hits-91-of-adults/

Regan, T. (2014). Negative online review lawsuit under way. *Remodeling.* Retrieved from http:// www.remodeling.hw.net/products/technology/outcome-uncertain-for-the-negative-review-lawsuit?utm_source=newsletter&utm_content=jump&utm_medium=email&utm_campaign= RDU_012714&day=2014-01-27

Smith, M. S., & Casserly, C. M. (2006). The promise of open educational resources. *Change, 38* (5), 8–18.

Stratford, J. S., & Stratford, J. (1998). Data protection and privacy in the United States and Europe. *IASSIST Quarterly, 22*(3), 17–20. Retrieved from http://iassistdata.org/downloads/ iqvol223stratford.pdf

Travis, A., & Arther, C. (2014). EU court backs 'right to be forgotten': Google must amend results on request. *The Guardian.* May 13. Retrieved from https://www.theguardian.com/technology/2014/ may/13/right-to-be-forgotten-eu-court-google-search-results

Tyson, J. (n.d.). How encryption works. *How Stuff Works Tech*. Retrieved from http://computer. howstuffworks.com/encryption.htm

U.S. Food and Drug Administration. (2014). *Radio Frequency Identification (RFID). Radiation-emitting products*. Retrieved from http://www.fda.gov/Radiation-EmittingProducts/ RadiationSafety/ElectromagneticCompatibilityEMC/ucm116647.htm

United States Census Bureau. (2013). *Computer and internet use in the United States: 2013*. U.S Department of Commerce Economics and Statistics Administration. Retrieved from http:// www.census.gov/content/dam/Census/library/publications/2014/acs/acs-28.pdf

United States Copyright Office. (2012). *Copyright registration for works of the visual arts*. Retrieved from http://www.copyright.gov/circs/circ40.pdf

United States Department of Justice. (2016). *The USA PATRIOT Act: Preserving life and liberty*. Retrieved from https://www.justice.gov/archive/ll/highlights.htm

William, T. (2016). Australian airport employee fired for Facebook posts supporting ISIS. *Business 2 Community*. Retrieved from http://www.business2community.com/social-buzz/australian-airport-employee-fired-facebook-posts-supporting-isis-01562855#QxlWFlHML0rIhrC2.99

World Bank. (2105). *Mobile cellular subscriptions (per 100 people)*. Retrieved from http://data.worldbank. org/indicator/IT.CEL.SETS.P2?order=wbapi_data_value_2014+wbapi_data_value+ wbapi_data_value-last&sort=asc

Chapter 6

Aerospace and Transport

(Continued)

(*Continued*)

- ❖ The science behind the technology
- ❖ Career connections
- ❖ Modular activities
  - ➤ Discussions
  - ➤ Tests
  - ➤ Research
  - ➤ Design and/or Build projects
  - ➤ Assessment tasks
- ❖ Terms
- ❖ Further Reading

Chapter Objectives

- ❖ Student will gain a rudimentary knowledge of the timeline and events important to the history of human flight.
- ❖ Students will be able to identify the different categories of aircraft and basic uses of each.
- ❖ Students will be able to demonstrate knowledge of the four forces of flight.
- ❖ Students will gain knowledge of how different aircraft are controlled about the vertical, lateral, and longitudinal axis.

Overview

Transportation is incredibly important in our modern world. The ability to move people from place to place for business and leisure, and to carry cargo and goods from the manufacturer to the markets of the world, is what makes the global economy function. Transportation is an enormous industry and covers a broad spectrum from water and land to air and space. This chapter dives deep into aerospace technologies, realizing that other modes are essential and important as well.

As long as people have walked the earth they have been fascinated with flight. They watched birds and insects, bats and seeds, and wondered what it would be like to fly. There were many reasons people wanted to take to the air. Some imagined the freedom it might bring to be among the clouds. Others just wanted a quick way to get from place to place. While still others considered the view of the ground they would have or the advantages they could gain militarily. All of these reasons appear in some fashion in the legends and mythologies of the day.

Over 5000 years ago, ancient Egyptians crafted images of the god Isis with folded wings. Around 2200 BC, the Chinese Emperor Shun was said to have built a bird

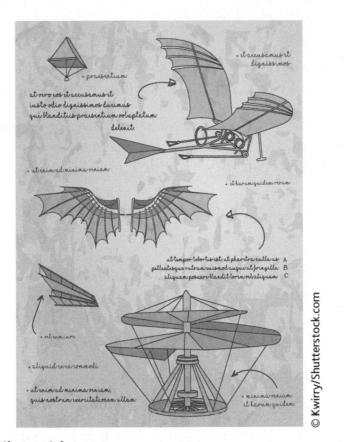

Figure 1.   Da Vinci's ornithopter.
Leonardo Da Vinci's design of a human powered flying machine

suit that included a parachute. The Greeks and Romans had mythological flyers such as Hermes or Mercury who flew with winged sandals and the Chinese had Ki Kung Shi (1700 BC) who was said to have built an invisible flying chariot. Ki, they said, was forced to destroy his craft because it was thought to be too powerful for any one man to possess. The story of the world's first astronaut, albeit a fictional one, was that of the Chinese adventurer Wan Hu. Legend has it that he strapped 47 rockets to a chair, lit the fuse and was never seen again. The popular TV show Mythbusters tried to duplicate Wan Hu's flight and it did not turn out well for their test dummy Buster.

One of the first on record to take flight seriously from a scientific standpoint was Leonardo Da Vinci. Although it took almost 300 years for his information to be published, it showed he was thinking of many of the problems faced by early aviators. His "Ornithopter" had flapping wings that were powered by the pilots pumping legs. There is no way that it could have flown, but it shows that Da Vinci was approaching the problem like a scientist.

Da Vinci had been dead for over 260 years when two French paper makers thrilled the world with the very first person carrying flight. In the next section, we talk about what led up to their success, various aspects of aerospace and aerospace technology, and the science behind it.

Introduction

An aircraft is "any device used or intended to be used for flight in the air." Here is a brief overview of how the Federal Aviation Administration (FAA) classifies aircraft:

❖ Airplane: The airplane category is divided into single-engine land, multiengine land, single-engine sea, and multiengine sea classes

❖ Rotorcraft: The rotorcraft category is divided into helicopter and gyroplane classes

❖ Glider

❖ Lighter than air: The lighter-than-air category is divided into airship and balloon classes

❖ Powered lift

❖ Powered parachute: The powered parachute category is divided into powered parachute land and powered parachute sea

❖ Weight-shift-control: The weight-shift-control category is divided into weight-shift-control land and weight-shift-control sea

We now discuss several of these classifications.

Lighter than Air (*Balloons, Dirigibles and Blimps*)

In the seventeenth century, scientists were studying the atmosphere, learning its make-up and characteristics. Italian physicist Evangelista Torricelli (1608–47) built the first mercury-filled barometer and discovered that the atmospheric pressure changes from day to day. A German scientist Otto Von Guericke (1602–86) experimented with vacuums and also built a small barometer that he used to predict the weather. Blaise Pascal (1623–62) was another mathematician and physicist from France who, building on Torricelli's work, discovered that as one goes up in altitude, the density of the air goes down. This all becomes important as we discuss our next subject, Balloons.

Balloons

For a balloon to take flight, it must trap something less dense by volume than the atmosphere around it in the envelope of the balloon. So, just like in a bottle of salad dressing where the less dense oil wants to float above the vinegar, the less dense air or other element in the balloon, wants to rise above the atmosphere near the ground. An Italian Jesuit priest named Francesco Lana De Terzi (1631–87) was one of the first to consider this concept and designed a flying machine lifted by large copper balloons in which all the air had been removed. A little misguided, he didn't consider that copper balloons thin enough to float in the surrounding atmosphere would simply be crushed by pressure if you removed the air inside. He was on the right track; however, and it wasn't too long until another Jesuit named Bartolomeu Lourenco de Gusmao (1685–1724) discovered that hot air rises. He is thought of as the father of the hot air balloon. Although he designed an airship, he never built one. Things started to come together in the late eighteenth century that would make lighter than air flight possible. British chemist and Scientific Philosopher Henry Cavendish discovered what he called "inflammable air" and that it was lighter than the air we breathe. We now know it as hydrogen.

With all of this science waiting to be put together, it was by pure experimentation that people finally flew. Joseph-Michael (1740–1810) and Jacques-Etienne (1745–1799) Montgolfier were French paper makers. One day in 1782, Joseph noticed that his wife's blouse would billow and fill with pockets of air as it dried by the fire. Not having read any of Gusmao's work, he assumed that there was something in the smoke that caused it to rise. The brothers immediately started calling it "Montgolfier gas"

Figure 2.   Montgolfier balloon.

and went about testing small cloth balloons. In June of 1783, the brothers publically demonstrated the flight of a 38-foot paper-lined balloon. It flew over 3 km (1.2 mi.) and stayed aloft for more than 10 minutes. King Louis the 16th and the Academy of Sciences in Paris wanted a demonstration; so the pair went about building a 37,000 cubic foot balloon out of taffeta and paper. They demonstrated it on September 19, 1783 by carrying a cage with a sheep, a rooster, and a duck aboard. Next, they built a 60,000 sq. ft. balloon about 70 ft. high and 50 ft. in diameter. In this balloon, Pilatre de Rozier and the Marquis d'Arlandes armed with wet sponges attached to long sticks to put out burning spots on the balloon, were the first to fly free of a tether in November 1783. They were the toast of Paris and started a flying revolution. In the same year, scientist Jacques Charles launched more balloons with the help of brothers Anne Jean and Nicholas Louis Robert, these were the first to fly using hydrogen as the lifting agent. The balloons were made using a rubbery paint over a cloth envelope. It was just a month after the Montgolfier's flew that Jac Charles and Nicolas Robert flew their 380 m³ balloon over 25 mi. and were the first to observe a sunset twice by varying the altitude of the balloon.

In September 1783, the Montgolfier brothers demonstrated the flight of their 60,000 sq. ft. balloon with Pilatre de Rozier and the Marquis d'Arlandes aboard.

Structure

Aside from the material, the balloon envelope is made of and the method of heating the air, hot air balloons have changed little since the days of the Montgolfier's. The structure of a hot air balloon consists of a basket or gondola that carries the

Figure 3.   The three main parts of a modern balloon. The "envelope" to enclose the hot air, the "burner" and tanks to make the air hot, and the "basket" to carry the people.

payload / pilot and passengers, the burner, which heats the air in the balloon and the balloon itself, which is called the envelope. Modern balloon envelopes are made of either rip stop nylon or Dacron. Nylon is lighter and stronger than Dacron but Dacron withstands higher temperatures and is less affected by moisture. Dacron is a brand name for a synthetic fiber made by DuPont chemical. Its chemical name is polyethylene terephthalate and is widely used for sails, balloon envelopes, and camping gear because it is strong, light weight, abrasion resistant, and doesn't absorb water. The burner of a balloon is used to heat the air in the envelope so that the balloon rises. It is operated by hand and burns propane stored in tanks a little larger than the tank on your patio grill.

Gas balloons use Jacques Charles idea of enclosing a gas of lighter weight by volume in the envelope, instead of heating ambient air to make it less dense. Early gas balloons used hydrogen because it had been discovered and was easy to produce. It is a very dangerous element to use because of its flammability, so modern gas balloons almost always use helium. Gas balloons are used mainly for scientific and exploration purposes because they can fly higher and stay aloft longer. They are much more expensive than your average sport balloon.

The structure of a gas balloon is similar to that of a hot air balloon. They have an envelope and gondola, but instead of a burner they control altitude through the use of something that adds weight and stability such as water, lead weights, or sand known as "ballast," and a gas valve to release the gas to maintain or vary the altitude of the balloon. The envelope itself is made not of Dacron but of a heavier nonporous material that will hold the gas for days or weeks. Gas balloons can either be zero pressure where the gas in the balloon expands as it rises or Super pressure where the envelope does not expand any more after it is filled. The zero pressure balloons are more traditional, while the super pressure balloons work better at night because the pressure maintains the size of the balloon instead of shrinking like a zero pressure balloon does because of the lack of sun radiation.

Propulsion and control

The big issue with balloons for transportation is that they are at the whim of the air mass in which they fly. If the air mass is flowing east, you are going to fly east. A balloonist can control his/her altitude and that is about it. A good balloonist can get a little directional control if there are winds blowing in different directions at different altitudes. The pilot can let the balloon flow in one direction for a while and then change his/her altitude to be moved by the air going in another direction. In most aircraft, fuel is used to propel the aircraft forward. In a balloon, the fuel is used to increase altitude. In doing so, if you are lucky you get into a wind that takes you where you want to go. If the air isn't going where you want it, you're not going to get there.

Balloons today and in the future

Today, hot air ballooning is a sport. Every year you can see hundreds of balloons in festivals from Albuquerque to Telluride, Cambridge to Las Vegas. Enthusiasts come together to show off their balloons and their skills, but no one uses their balloon to get there!

On the other hand, gas balloons are used extensively to collect weather data and to skirt the edge of space. Explorers have set around the world records not for speed but for a balloon. One thing a gas balloon can do well is go up and many believe that in the future, scientists may use balloons as a cheap way to lift objects into space. In the mid-1990s the National Aeronautics and Space Administration (NASA) experimented with what they called "ultra-long –duration –balloons" (ULDBs). The purpose was to develop a balloon that could carry very heavy loads at a constant altitude above 99% of the atmosphere for 100 days or more. In the past, small super pressure balloons—balloons that don't shrink and lose lift as they cool—were able to fly around the world several times. Larger heavy load carrying balloons had also been developed to carry weather and surface observation gear, but these could only stay aloft for a couple of days unless launched in the Antarctic night. With newer high strength materials NASA hopes to build a ULDB that could monitor weather or earth's surface for many days. They have been experimenting with pumpkin-shaped balloons because a high weight-carrying balloon needs to have a thicker and exponentially larger envelope. Thickness adds weight and surface area adds weight exponentially. The pumpkin shape holds promise but has instability issues that need to be worked out. NASA is also considering using the idea of the ULDB as a way to explore Mars. The balloon would inflate as it entered Mar's atmosphere and after jettisoning its inflation tanks, could maintain an almost constant altitude over Mars both day and night.

Balloons are also used for commercial and entrepreneurial enterprise. In 2012, 8 million people watched on YouTube as skydiver Felix Baumgartner freefell from 127,852 ft. above the earth. He exceeded the speed of sound and broke many long-held records. He accomplished this by jumping from the specialized gondola of a gas balloon.

Dirigibles and Blimps

An Airship or Dirigible is any lighter than air vehicle that can be steered under power. Some assume the word dirigible comes from the word rigid, such as rigid airship, but it actually comes from the word "diriger," a French verb meaning, "to steer." *Blimps, rigid,* and *semi-rigid* designs are all dirigibles, and all dirigibles are airships. The difference between a blimp and a rigid airship is that a rigid airship has a frame that holds the shape of the ship with gasbags inside the frame. A blimp gets its shape from the gas pressure itself and will lose its shape when deflated.

From the moment the Montgolfier's launched their first balloon, enthusiasts realized that balloons had control limitations. It is great to fly, but how do we go where we want to go? Enter John Baptist Musnier (1754–93), a French mathematician and revolutionary. Musnier, after seeing the flights of the Montegolfiers and Charles, proposed a cigar-shaped craft 84-m long driven by three propellers that were powered by 80 men. His invention, now known as a dirigible, was never completed, but other inventors such as Charles and the Robert brothers used his ideas to continue improving the designs. In 1850, Frenchman Pierre Jullian flew a model of an airship driven by a clockwork motor at the hippodrome in Paris. Henri Gifford, another French inventor and engineer, took note of the design and in September of 1852, flew the first steam powered airship that was controllable. His cigar-shaped craft had a three horsepower engine that weighed upwards of 350 pounds including the boiler. Its small engine could not overcome the wind when he demonstrated it in 1853, but in calm winds, he did much better. It wasn't until the 1880's when two gentlemen from the French Corps of Engineers, Charles Renard and Arthur Krebs, built a 50-m long ship powered by electric motors, that airships became truly controllable. The airship came into its own between 1890 and 1903 when Brazilian-born aviator Alberto Santos Dumont (1873–1932) flew his many airship designs publically in Paris France. In 1901, he flew his sixth design around the Eiffel tower three times and then moved on to a field where he made a flawless landing. During this same time period, designer Ferdinand Von Zeppelin started building the first of his giant, rigid airships in what is now Germany. His airships had rigid aluminum frames covered in doped (plasticized lacquer applied) fabric. Inside the huge envelope were several enormous bladders filled with hydrogen gas. Control of the vehicle was very much the same as modern day blimps with externally mounted engine-driven propellers and control surfaces mounted at the rear providing pitch and yaw control.

Airships had a brief popularity as a military vehicle in WWI. Germany used them as a means to deliver bombs over Great Britain until the Allies learned that all they needed to do to destroy a "Zeppelin" was to fire several incendiary or tracer bullets into the envelope. This didn't dissuade passenger service after the war however. Passengers flew all around the world with great views and staterooms in the sky. It all came to an end with the fiery crash of the LZ-129 Hindenburg over Lakehurst, New York, in 1937. Many thought it was due to sabotage but conventional thought is that a static charge lit the very flammable covering, and then the flames moved to the huge bags of hydrogen. Interestingly, the spires on buildings in New York like the Empire State building and Chrysler building were originally meant to be airship-mooring masts.

Structure

Airships are lighter than air craft with externally mounted engines that drive propellers, control surfaces (elevators and rudders) at the tail, and gondola attached to the underside of the envelope.

Propulsion

Propulsion for airships is derived from multiple power plants, typically internal combustion engines, mounted at gondola level that can be independently adjusted to propel the craft in the direction desired. Many of the newer yet smaller blimps in service today use directional ducted fans that can help increase the maneuverability of the typical blimp by rotating and sending the thrust in different directions. A directional-ducted fan is simply a propeller or fan inside an enclosed housing or nacelle. The fan is rotated at a high speed, usually by an electric motor so quite a bit of thrust is produced. The entire nacelle can be angled to point the thrust in the desired direction. Figure 5 shows the ducted fans on the "Met Life" blimp.

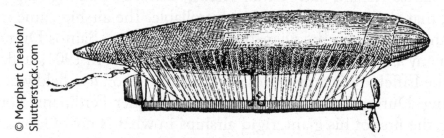

© Morphart Creation/ Shutterstock.com

Figure 4.   A vintage diagram of a great airship of the 1920's Notice the internal structure and how the gasbags don't form the shape but are inside the frame.

© Bloomberg/Contributor/Getty

Figure 5.   A Propulsor duct or fan on the Airlander 10 Hybrid Airship. These ducted fans can help steer the blimp in the desired direction.

Figure 6. The Goodyear blimp is an Icon. If you look closely, you can see the forward and aft ballonets.

Unlike balloons that can only control buoyancy, an airship can also turn left or right, climb and descend, and propel itself against prevailing winds. Buoyancy control is maintained by adjusting ballast, usually water, and by adjusting the amount of lifting gas, usually helium, in the envelope or gas bladders. This is accomplished by the use of gas valves that let the helium out. Venting helium is not the best option so designers also include airbags called ballonets. A ballonet is a bag or bladder in the interior of an airship that can be inflated and deflated to change size of the helium-filled gasbags next to them. Doing so changes the overall density of the airship compared to the outside air. Pressure can be adjusted in the ballonets to maintain the envelope shape and to adjust the relative density of the blimp. As the sun heats the blimp and the blimp starts to rise, air can be added to the ballonet to increase the blimps overall density. The opposite can be accomplished when the gas in the airship cools.

The conventional controls are similar to heavier-than-air aircraft's rudder and elevator (flight control surfaces), but there are no ailerons (hinged surface on the trailing edge of an airplane wing). Pitching up can help drive the airship up but can have the opposite effect if too much elevator is applied.

Airships today and in the future

Presently, most airships visible to the public are those used for advertising. Everyone is familiar with the Goodyear blimp with its giant electric billboard; there is also the Met Life blimp and many smaller blimps decorated with the names of companies such as "Atomic," and "Direct TV." What aren't quite as visible are the ideas that scientists, private enterprise, and the military are currently researching. Some of these include airship hybrids for heavy lifting. These hybrids would be a combination of helicopter and blimp capable of carrying aloft large loads previously impossible. Rocket scientists are considering using airships to fly in the upper atmosphere of Venus. Some entrepreneurs are considering new methods for transporting passengers. In 2015, Lockheed Martin unveiled their LMH-1 heavy lifting hybrid at the Paris Airshow. LMH-1 is heavier than air vehicle, so it can rest on the ground instead of using a mooring mast, but 80% of the lift in flight comes from the helium enclosed in the envelope. It is designed to carry a load of 20 tons. The British aerospace company Hybrid Air Vehicles touts, "We produce less noise, less pollution, have a lower carbon footprint, longer endurance and better cargo-carrying capacity than virtually any other flying vehicle." Perhaps someday one of these airship companies will produce a passenger-carrying hybrid and start an "air cruise" service. Perhaps the day of the airship is still to come. The LMH-1 is heavier than air. It will rest on the ground, but in flight, 80% of the load is carried by helium.

Airplanes

Few things, except maybe the light bulb or home computer, have altered our world quite as much as the invention of the airplane. The airplane has shrunk our world by a factor of ten and changed the way we move people and goods around the world. It has influenced agriculture, and even medicine, and has changed the very way we fight wars. Many things had to come together to make the airplane possible, and that is what we talk about in this next section.

An airplane is "an engine driven fixed wing aircraft that is heavier than air that is supported in flight by the dynamic reaction of air against its wings" (FAA).

Leonardo Da Vinci designed pedal-powered flying machines or "Ornithopters." He had plans for an early idea for a parachute and a vehicle somewhat similar to a helicopter. He also wrote about things such as center of gravity, streamlining, and center of pressure. His focus on flapping wings never achieved flight, but he was ahead of his time. Sad for the world, Leonardo's plans weren't widely distributed for over 300 years; by then, others were already surpassing his designs.

One of these pioneers was an Englishman named George Cayley (1773–1854). His work culminated a year before his death when his coachman reluctantly became the first person to ever take flight in a person-carrying glider. Cayley also invented many of the aeronautical terms used today to identify the forces of flight; lift, drag, thrust, and weight which is discussed in the science behind the technology section. His names for power plant and propeller still persist. Cayley also coined the terms *biwing* and *triwing*, and invented a curved or cambered upper wing surface. Probably, most importantly, Cayley established the idea of a "fixed wing" as a means to acquire lift.

The last part of the nineteenth century had many inventors working on flight but none were able to sustain flight. By the last decade, several were getting close to understanding how to build a controllable glider. John Montgomery (1858–1911) built several gliders in the eighteen eighties and early nineties. He would fly them in the dark of night because, at the time, anyone seriously trying to build a flying machine was thought of as a crackpot. Due to personal circumstances, he had to wait several years to demonstrate his 1893 glider. By the time he successfully demonstrated his glider he had been left behind, but his observations about air circulations and vortices proved to be important to the science of flight. German engineer Otto Lilienthal (1848–1896) is often called the father of modern aviation. Between the years of 1890 and 1896, Lilienthal built some 18 gliders and made over 2500 flights reaching distances of over 1100 ft. or 350 m. He also documented his designs, recorded the flights and wrote articles. Later he gave several lectures about flight. Lilienthal designed a powered machine, but was killed practicing in a glider with the same control system. It is suggested by some that it was this void created by Lilienthal's death that inspired the Wright brothers to get involved with flight.

In 1899, after the death of Otto Lilienthal, two brothers and bicycle makers Wilbur (1867–1912) and Orville Wright (1871–1948) decided that building an airplane was something they could accomplish. The commonly told story is that Wilbur was talking to a customer one day and was playing with an empty inner tube box. He suddenly realized that by pinching opposite corners of the box, he could warp the upper and lower surface of the box. This is how the Wrights came up with wing warping. After contacting experts like Chanute, they eventually were able to build a glider, which incorporated all of the knowledge obtained by 19th century inventors. This 1901 glider using all of the Lilienthal data acquired from Chanute didn't fly as well as their 1900 glider. They realized that even though Lilienthal kept good records his data were incorrect. At this point, they built a wind tunnel and started testing different wing shapes. They discovered that a long wing with a narrow chord was much more efficient than a short fat wing. They also discovered that none of the engine manufacturers of the day could produce an engine light enough and powerful enough for their needs. They built their own engine from aluminum and drove their handmade propellers with a chain-drive system right from their bicycle shop.

During this same time period, one of the United States most prominent scientists Samuel Pierpont Langley (1834–1906) risked his reputation by trying to build a person-carrying airplane based on successful steam and rubber band powered models he had built in years previous. Using 50,000 dollars granted by congress, Langley built the Aerodrome A and attempted to fly it off a barge in the Potomac River. His last attempt was a failure on December 8, 1903 when his assistant Charles Manly (1876–1927) nosed up and then crashed in the ice-filled Potomac.

Just nine days later, the Wrights having repaired some troubling prop shafts completed the world's first controlled, sustained, and powered flight. The problem, that came later, was that no one believed them. They did all of their work in utter obscurity while in Europe others such as Brazilian-born Alberto Santos Dumont already

Source: NPS

Figure 7. Wright first flight (https://www.nps.gov/wrbr/learn/historyculture/thefirstflight.htm) December 17, 1903, Orville Flies while Wilbur runs.

famous for his dirigible flights made flights in his 14-BIS aircraft in front of thousands of Parisians, although two years later. It didn't take long for the Wrights to gain respect as they showed the truly advanced nature of their 1905 flyer. It wasn't too long before others such as former motorcycle racer and lightweight engine builder Glenn Curtis (1878–1930) and the man you know as the inventor of the telephone Alexander Graham Bell (1847–1922) joined forces and made improvements to the Wrights' wing warping method of control by adding small opposite moving controls called ailerons.

With any war, or threat thereof, large advancements are made in technology, especially if the technology can be used in the war effort. By the time WWI was over, the airplane had reached a form familiar to us today, less a wing or two. In the four years of WWI, aircraft went from a top speed of 70–80 mph to a top speed of 180 mph, and from a maximum altitude of 10,000 ft. to a maximum of 24,000 ft.

After WWI, aviation almost died out in the United States. Had it not been for visionaries such as Billy Mitchell (1879–1936) who thought of things such as new war tactics, paratroopers, aerial refueling and so on, and who promoted aviation with record setting transcontinental and around the world flights, it might have been completely eliminated. Barnstormers and the advent of Airmail also helped establish aviation as something that needed to stay. Innovations also were fostered through events such as the Bendix Air Race and the National Air Races. When WWII started in 1938, the Monoplane had replaced the biplanes and triplanes of WWI; speeds and altitudes continued to increase. During WWII, engineers in both Britain and Germany developed jet engines independently of each other. British RAF officer and engineer Frank Whittle (1907–1996) is credited with being the first to invent a turbojet engine, but German physicist Hans von Ohain (1911–1998) built a turbojet independently at nearly the same time. We discuss the operation of a jet in the next section.

By the end of WWII, everything was in place for the modern aircraft we are all familiar with today. Even the problems that plagued so many working on vertical flight were worked out during the war by Russian Born American engineer Igor Sikorsky (1889–1972). By the end of WWII, the first practical helicopter went into production. What did all of these inventors and engineers figure out? That is what we discuss next.

Structure

An airplanes structure hasn't changed much since the first monoplanes were introduced after WWI. The main structure is the fuselage (see Figure 8) and contains the cockpit up front and the space for passengers or cargo behind. The wings hold the fuel in integral tanks and also have the ailerons flaps and spoilers. We talk about these in the next section. The tail section is called the empennage and consists of the vertical stabilizer and rudder and the horizontal stabilizer and elevator. The landing gear consists of the wheels, shock struts, and brakes. The nose wheel is steerable to control aircraft movement on the ground. Most transport category aircraft have retractable landing gear to reduce in-flight drag. Early aircraft structures were made of an internal frame covered in cloth, and then painted with dope, which makes the fabric shrink and become less porous. Until recently, modern aircraft were made from aluminum. Some are monocoque, and some are semimonocoque. The prestressed skin carries loads on a monocoque structure. Semimonocoque structures share the load between an internal frame and the prestressed skin. Most recent designs incorporate the use of composites. They are strong, light, and don't corrode; however, structural damage is not as readily evident as it is in metal structures, so they require different inspection techniques. Composites have increased airliner efficiency in the most recent models of Boeing and Airbus aircraft, not to mention designs for smaller general aviation aircraft.

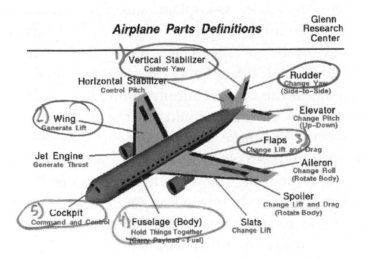

Figure 8.   Airplane diagram. The diagram depicts the components of a modern day airliner.

Propulsion

Aircraft today are either powered by reciprocating engines, like those in most cars, or turbine engines, which include turboprop, turboshaft, turbojet, and turbofan variations. Turbine engines are highly expensive but quite reliable. Piston or reciprocating engines are usually less expensive and well suited to smaller applications. For aircraft, almost all piston engines are four stroke, meaning they go through four separate cycles as they operate. The four cycles include: intake, compression, power, and exhaust. Turbine engines have all four operations going on, and even the compression ratio is close to that of piston engines. The difference is that turbine engines have them all going on at the same time. The difference in turbine engine variations is based on the application in which they are being used. Turboshaft engines are used in helicopter designs. Turboprop engines are just a jet engine geared to a propeller. This kind of application gives turbine reliability to aircraft that can operate off of runways too short for the typical jet. Turbojet engines are older designs in which all the thrust comes from the engine exhaust. Turbojets are much louder and less efficient than the newer turbofan engines. Turbofan engines get much of their thrust from the big oversized fan at the front of the engine. Part of the air passing through the fan is bypassed around the combustion section. This has the added effect of buffering the hot combustion air from the colder ambient air making the engine much quieter than the typical turbojet.

Guidance

Pilots use different types of navigation to figure out how to get from place to place. Early pilots just looked out the window and flew by reference to landmarks. This form of navigation is known as "Pilotage" and is still used today by sport pilots. In the early years of open cockpit aircraft, airmail delivery pilots used bonfires and eventually light beacons spaced every 10 mi. along the route. As airplanes got bigger and instruments came into use, various forms of ground-based navigation were used. Early ground-based AN ranging produced a solid tone on the cockpit receiver if the pilot was on course and either an A or an N in Morse code if not. Later, ground-based navigation included nondirectional beacons, where the needle in the cockpit simply pointed to the station and Very High Frequency Omni Ranging (VOR) that could tell the pilot exactly what the aircraft's bearing was to the station. Until more modern navigation systems were invented such as Inertial Navigation System (INS) and Global Positioning System (GPS), long-range navigation was still being done using star and sun sighting just like Columbus.

Modern day navigation is slowly moving to be all sky based. Airliners use accelerometers and laser-ring gyros to continuously calculate their present position. This is typically known as Inertial Navigation. This information is also continually updated with GPS and ground-based Distance Measuring Equipment (DME/DME) (See Figure 9). Many times there are two or three separate systems, so the pilots always know exactly where they are. Another huge change in aircraft control is the development of NextGen.

NextGen uses Automatic Dependent Surveillance-Broadcast (ADS-B). ADS-B is a precision satellite-based system. Unlike radar that only updates every few seconds, ADS-B updates many times a second. Radar has a lag because it takes time for the antenna to rotate and the system to process the signal return. ADS-B uses ground-based

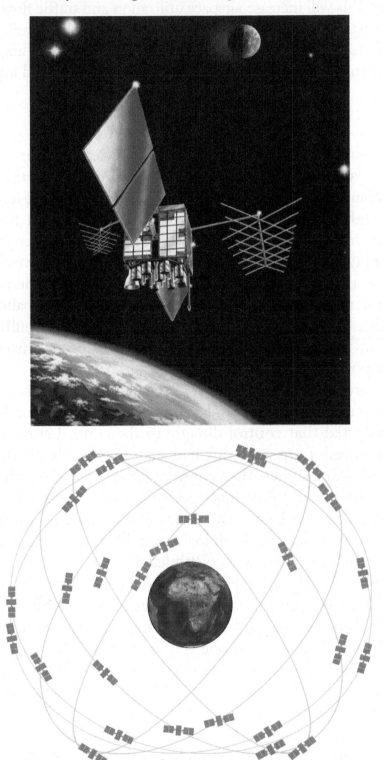

Figure 9. The Global Positioning System.

and aircraft-based receivers. Every aircraft transmits its location determined by GPS positioning to the ground and to the other airplanes in the area. Air Traffic Control (ATC) gets real-time position updates of all aircraft in the system. This makes it possible to reduce the required separation between aircraft. By 2020, all aircraft will be equipped with ADS-B in and out. This will increase airspace utilization and traffic flow in the future.

The GPS consists of 24 satellites spaced 60° apart in four planes, four satellites per plane angled at 55° to the equator in a semi-synchronous orbit. If an airplane's receiver can receive signals from four satellites, it can find its position and altitude.

Control

So how do you fly an airplane or helicopter? All heavier than air aircraft have a way to control the three axes: pitch, roll, and yaw. A pilot rolls about the longitudinal axis using ailerons. When a pilot turns the yolk or pushes the stick left, the left aileron goes up and the right aileron goes down creating more lift on the right wing and reducing lift on the left, so the airplanes rolls to the left. Airplanes yaw about the vertical axis. If a pilot pushes on the left rudder pedal, the rudder moves left, causing more camber or curve on the right side of the vertical stabilizer and the tail of the airplane moves right and the nose moves left. Finally, the airplane pitches about its lateral axis. If a pilot pulls back on the yolk or stick, the elevator goes up pulling the tail down causing the nose of the aircraft to go up. All rotations occur through the center of gravity, which is a point on the aircraft that it would balance.

Traditionally, aircraft are controlled from the cockpit directly by the pilot through a system of cables, pulleys, pushrods, and so on. When the pilot made a control movement, a cable connected that control directly to the control surface and the aircraft rolled, pitched, or yawed. Larger aircraft had such huge air loads that direct connection became unfeasible. Designers then connected the cables from the cockpit controls to hydraulic valves that would port high-pressure fluid to hydraulic actuators similar in theory to the actuators on heavy equipment like you see at construction sites. The hydraulics can easily overcome the high loads and move the rudder, ailerons, or elevators making larger and larger airplanes possible.

With the advent of computer technology, aircraft designers found that they could save many hundreds of pounds of weight by replacing the heavy control yoke and the cables running between the hydraulic actuator valves and the cockpit controls with an electronic control stick similar to the one you use to play Microsoft Flight Simulator and running a set of small wires to computers at each control surface to control the hydraulic actuator valves. This system is called "fly by wire" and has the advantage of being much lighter. With fly by wire controls, system parameters or "laws" can be set in the computer software to prevent pilots from stalling or exceeding a preset angle of bank. Two of the first aircraft to be designed with fly by wire were the General Dynamics F-16 and the Airbus A320. By replacing the control yoke alone in the

A320 saved hundreds of pounds. The next obvious evolution in control systems was to replace the wires between the control stick and the control surface computers with a fiber-optic cable. Still in testing by military designers "fly by light" includes a little more weight saving because signals can be multiplexed, meaning several different signals can be carried by a single fiber optic cable. There may also be a maintenance saving as well.

As mentioned earlier, a helicopter rotor is just a huge spinning wing, but it is not as simple as it may seem. Rules about angular momentum apply meaning that engineers had to figure out where a force needed to be applied to get the helicopter to react as expected. There are three sets of controls in a helicopter if you don't include power. The collective, on the side of the pilot seat, the cyclic, which looks a great deal like the stick in an airplane, and finally the antitorque pedals that look like the rudder pedals in an airplane. In a helicopter, all of the axes, less yaw are controlled by manipulating the pitch of rotor blades at different places on their rotation. To pitch forward, the cyclic is pushed forward, to roll right the cyclic is pushed right. The collective is used to pitch all of the rotor blades at the same time, which will increase the helicopter's altitude as long as power is increased accordingly. The antitorque pedals are used to counteract the natural tendency of the helicopter to rotate in the opposite direction of the rotor. As power is increased, the antitorque pedal is applied to prevent rotation.

Aircraft Today and in the Future

Today, aircraft are used everywhere and in many different fields. Some of the applications include: airline transportation, corporate and private transportation, aeromedical transport, domestic and international freight, fire spotting and fire suppression, traffic watch, navigation checks, weather modification, aerial photography, aerial advertising, seismic testing, backcountry skiing, scenic tours, fish spotting, logging, aerobatic competitions and demonstrations, air races, powerline and pipeline patrol, and film production.

One area of great use of aerospace is in the military. WWI General, Billy Mitchell is quoted as saying "If you control the air you cannot be beaten; if you lose the air, you cannot win." Militaries of the world have continued to innovate and build the cutting edge technologies since the first Great War. During WWII, superchargers and monoplane designs pushed the speed envelope. After WWII, jet aircraft exceeded the speed of sound. In the 1950s, jets became bigger and faster; the F86 went head to head with the Russian made Mig 15.

In the 1950s, Americans in F86 aircraft fought Chinese pilots in Russian built Mig15s. The Mig 15 was a superior fighter but the Americans were better trained.

In the 1970s, more powerful jet engines made the F15 able to accelerate straight up, and variable geometry wings made it possible for aircraft to sweep their wings for supersonic flight and then bring them forward for slower flight and landing. Two examples of this were the F14 Tomcat and the B1B Lancer.

Figure 10.   F86 Sabre Jet.

Figure 11.   F14 Tomcat and B-1B.

In the 1980s stealth technology was developed. Special coatings, angle surfaces that deflected radar away from enemy radar and weapons pods that retracted into the aircraft so no sharp edges were present, made the F117 Nighthawk and B2 Bomber almost invisible to radar.

Stealth Aircraft are shaped with round corners or angles and special coatings to deflect or absorb radar. This gives them a very small radar return.

Recent designs have included the F22 Raptor, cancelled by the federal government; it could accelerate through the speed of sound without the use of afterburner. The workhorse single seat fighter the F16 has been upgraded with newer engines that can now give it almost vertical climb capability, but it may soon be replaced by the F35 Joint Strike Fighter (Lightning). It can take off and land vertically or on a short runway depending on the configuration.

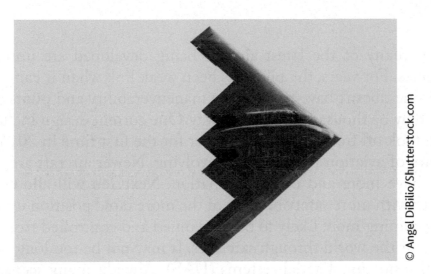

Figure 12.   The B2 Spirit.

Figure 13.   The F35 Lightning has several versions for different branches of the military.

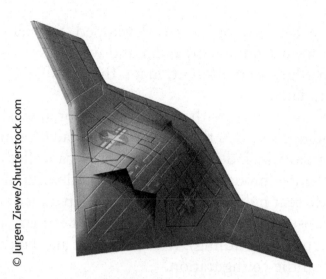

© Jurgen Ziewe/Shutterstock.com

Figure 14. The X47B is an unmanned aerial system being developed by the US Navy.

Currently, many of the latest aircraft being developed are unmanned and are known as drones. For years, the pilot has been weak link when it came to g-loads. An unmanned drone doesn't have to limit the maneuverability and pilots aren't in danger because they may be thousands of miles away. One current design includes Northrop's X47B, which took off from an aircraft carrier for the first time in 2013.

The nature of aviation is continually evolving. Newer aircraft are more and more efficient and have more and more automation. NextGen will allow airplanes to fly closer together with more safety because of the more rapid position updating. Military aircraft are becoming more likely to be unmanned and controlled from the ground on the other side of the world through satellites. It may not be too long before deliveries are made with unmanned aerial systems (UAS). Already, many scenes in movies are filmed with UAS carrying sophisticated cameras. Some companies are still working on transonic suborbital air transports that will cut travel time to a fraction of what it is today. It is hard to figure out what will and will not come to fruition but it will be fun to be involved.

Spacecraft

The Chinese invented rockets. They used them for fireworks and were the first in recorded history to use them as a weapon. They called the rockets fire sticks and used them in the thirteenth century against the Mongols. These early rockets sounded a lot like fireworks but were not all that effective as a weapon. Early in the nineteenth century, British inventor William Congreve (1772–1828) made improvements to military rockets. He standardized the formula used for the gunpowder and the length of the guide sticks so that the rockets were more accurate and by then could fly a distance of 9000 ft. Another Brit named William Hale (1797–1870), further improved Congreve's designs. By removing the guide sticks and angling the exhaust ports, Hale was able to make the rocket shell spin adding stability and accuracy. It was the red glare produced by these early rockets that inspired the words "rocket's red glare" in the US national anthem. Hale's rockets were used in the war of 1812 and were observed by composer Francis Scott Key. An English inventor Edward Boxer (1822–98) designed the first multistage rockets in 1865. Using the idea from other rocket inventors, he built a small two-stage rocket that could carry a small line to ships in peril. This made it possible for rescuers to drag larger lines from the distressed ship to get people off. Over 5000 sailors have been saved using the "Boxer" rocket as it is known.

The father of modern rocketry is American physicist Robert Goddard (1882–1945). After many tests and experiments, he built the first liquid-fueled rocket in 1926. His work was widely read by German rocket scientists Warnher von Braun (1912–77) and Hermann Oberth (1894–1989). During WWII, building on Goddard's research, they helped design the V1 and V2 rockets that the Nazi's used to terrify Londoner's from the sky. These same rocket scientists—hundreds in number—from the secret Peenemunde rocket base, immigrated to both the Union of Soviet Socialist Republics (USSR) and United states after the war and became the bases of both countries intercontinental ballistic missiles and space programs. The space programs of the U.S. and USSR are two of the most exciting and incredible endeavors ever taken on by human kind. If you are interested, the internet is filled with video and websites dedicated to nothing but space travel. We will simply talk about the basics.

Figure 15.   Robert Goddard (http://www.nasa.gov/centers/goddard/about/history/dr_goddard.html) Robert H. Goddard was the father of modern rocketry.

Rockets and Satellites

Rockets are used for many purposes today. The tasks are usually divided into military and nonmilitary. A rocket that delivers a warhead is called a missile and has a closed ballistic trajectory. A trajectory is simply the path a projectile takes to get somewhere. A missile is launched from one spot on earth and lands on another spot and follows the laws of gravity so we use the term "ballistic" to describe the trajectory. A rocket used for peaceful purposes such as moving astronauts to the International Space Station (ISS) or delivering a satellite into orbit is called a launch vehicle. Launch vehicles are used many times a year to resupply ISS to put commercial communications satellites into orbit and to launch probes on missions to investigate the solar system. Each rocket is configured to achieve a particular velocity, which will put the final stages containing the payload at the required altitude.

To clarify, a satellite is not part of a rocket; they are just one type of payload. The term satellite is simply a name for an object of smaller mass that orbits an object of larger mass. Normally we think of satellites as artificial or manmade; however, in broader terms you will hear astronomers speak of the "satellites of Mars" when they are talking about its moons. In order for an orbit to be achieved, a satellite must reach a velocity that prevents it from reentering the atmosphere. In other words, a satellite must move fast enough that the curve of the earth arcs away as it continues to fall in earth's gravity field. In essence, a satellite simply falls around the earth. In order to place a satellite in orbit, rocket scientists must calculate the velocity required for the orbital altitude desired. To give you an idea, for a satellite to reach a low earth orbit around 100 nm, it must attain a velocity of 17,500 mph. That is approximately 22 times the speed of sound. Rockets at present are the only way to accomplish this.

Structure

Most modern rockets big and small share the same common components and structures of other aircraft.

The airframe is the streamlined structure that contains all of the other systems. It needs to be light enough to take flight, but strong enough to support all of the other components on the launch pad and during massive acceleration.

Guidance

The guidance system is the brain of the rocket and may be something as simple as a gyroscope package that keeps the rocket stable. Gyroscopes are discs or wheels spun at a high velocity. Once they are spun up, they exhibit rigidity in space. In simple terms, it is easy to rotate them within the plane they are spinning but they resist being turned outside of that plane. If a guidance system has three gyros, one for each axis the resistance can be used to tell when a rocket has moved off course. Modern gyros use light instead of spinning mass. A laser ring gyro actually measures the phase of two laser beams as they

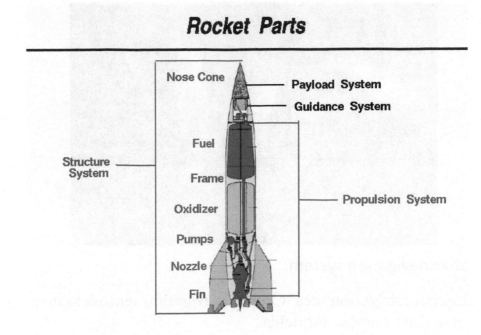

Figure 16. Rocket parts.
Rockets Systems include: Airframe (or Structure) Propulsion, Guidance and Control.

are reflected in opposite directions around a "laser ring" usually a triangular device with mirrors at the corners (see Figure 17). The counter rotating beams are directed to a photo sensor. If the beams are in phase, it means there is no rotation; if they are out of phase, it shows up precisely by how out of phase the beams are. Advanced inertial navigation systems use laser-ring gyros with accelerometers that sense acceleration in any direction to constantly update their location. Some will back up position with global positioning. This works great near earth, but for deep space, spacecraft need a whole new way of navigating.

There is a reason people always talk about rocket science in the way they do. It is very technical. For space travel, the first thing rocket scientists need to determine is where they want to go. This may not be as easy as you think because not only is the earth rotating and orbiting the sun, so are most of the destinations in the solar system. Once the future location of the destination is determined a planned trajectory can be calculated. During flight, the problem is determining if the actual trajectory differs from the plan. Once a rocket leaves the pad, it cannot be easily observed so an orbit determination team is constantly using various means to determine where its trajectory was, where it is now, and where it will be. The techniques include use of synthetic aperture radar and reconstruction of "propulsive maneuvers" to identify actual trajectory. After the position of the craft is determined, the team in charge of "orbit control" can calculate the required burns and maneuvers to meet the planned trajectory. Precise changes in velocity are made by burning the rocket engine for the calculated time to correct or alter the rockets path. If this doesn't sound complicated enough, add the fact that over time, sunlight itself can cause a spacecraft to move off course.

Figure 17.   Relative navigation system.

Modern spacecraft navigation uses "Optical" navigation sensors to map the spacecraft's destination against known star fields.

Control

The control system of most rockets includes the hydraulic lines, pumps, fluid and linkage, as well as the control surfaces to actually alter a rockets attitude and trajectory. Some rockets use small fins that can be moved just like the control surfaces of an airplane and operate under the same aerodynamic laws. Most large rockets, however, will "gimbal" or move the engine nozzle to change a rocket's attitude (the orientation of an aircraft with respect to the horizon). Once in space much of the control is based on how long the rocket engine is burned and in what direction.

A rocket has the same three axes as any other flying machine, vertical, longitudinal, and lateral; it is just hard to see a rocket roll about its longitudinal axis because it is a big cylinder. The space shuttle made seeing the roll axis easier on launch because the orbiter was hanging on the side of the huge external propellant tank.

Propulsion

A rocket engine is really nothing more than a reaction engine. Newton's third law states that "for every action there is an equal and opposite reaction," and that is exactly how a rocket engine works. The action is to eject fiery exhaust out of the rocket nozzle and the reaction is the rocket lifting from the launch pad. Newton's second law says that Force is equal to Mass multiplied by Acceleration ($F = ma$). Rocket scientists can figure out how much force a rocket engine produces by calculating the mass and velocity of the gasses ejected by a rocket engine.

The most common types of rocket engines in wide use today are either liquid fueled or solid fueled. Both types require propellants. All propellants have a reducer (fuel)

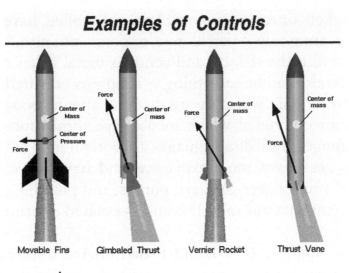

Figure 18.　Rocket control.
Rockets control direction in several ways as shown above.

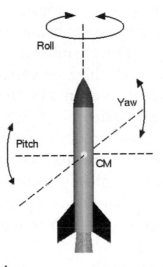

Figure 19.　Three axes on a rocket.
Rocket has the same three axes as an aircraft Roll, Pitch and Yaw.

and an oxidizer, which can be either Oxygen in liquid form or something that releases oxygen in the chemical reaction of combustion. Oxidation is the chemical reaction of oxygen molecules binding to the molecules of another element. It can be slow like that of rust or extremely rapid like a massive explosion. Rocket engines produce oxidation in the extremely rapid category.

Propellants that have the reducer and oxidizer stored separately are called bipropellants. Those that have both the reducer and oxidizer stored together are called monopropellants. Some consider solid fuels a monopropellant while some simply classify them as solid. Some examples of common liquid reducers are kerosene, gasoline, and liquid hydrogen. The most common oxidizer for liquid-fueled rockets is liquid oxygen

(LOX). Solid fuel rockets or rocket motors as they are called, have solid oxidizers such as ammonium nitrate, ammonium perchlorate, or potassium nitrate suspended in a polymer binder, which is also the reducer and contains metal flakes or other modifiers to increase thrust. The binder can be something as simple as polyurethane. There are other propellants that are hypergolic, which means that they spontaneously combust when the reducer and oxidizer are combined, which are used specifically for smaller thrust uses.

There are advantages and disadvantages to both liquid and solid fuel rockets. Liquid fuel engines are heavy, more expensive, and have more moving parts. They require tanks to hold the reducer, oxidizer, pumps, and plumbing to carry the liquids to the combustion chamber. Control valves and associated electronics are also required to adjust the power.

The combustion chamber of a liquid-fueled engine is where the propellant is burned and turned from chemical energy into kinetic energy. The pressure inside the combustion chamber is extreme and translates into very high velocity exhaust gas at the bottom of the chamber. The advantage of liquid-fuel rockets is that they produce a lot of specific thrust, and the thrust can be adjusted or shutoff.

Solid-fuel rockets have the advantage of being much simpler than liquid fuel, but one disadvantage is that the combustion chamber is the airframe itself, so the structure must be strong enough to withstand the high pressures of combustion. Another disadvantage is that solid-fuel rocket motors cannot be shut down after they are lit. Thrust adjustments must be made through grain design.

Thrust is all dependent on the surface area of the solid fuel burning. Solid motor grain design can either be *regressive*—meaning to start with a lot of thrust then slowly diminish, or *progressive*—meaning to slowly increase thrust, or it can be *neutral*—meaning to remain constant. This is all done by placing the rubberlike propellant in the airframe in a shape that has more, less, or the same amount of surface area burning.

Spacecraft today and in the future

Currently, rockets are used to boost payloads for space exploration, to place government and commercial satellites into space and to deliver warheads. Rockets have launched probes to Mercury, Venus, Mars, the Moon, Saturn, and Jupiter. The rocket *Voyager* is currently outside the orbit of Neptune.

ISS is resupplied several times a year using rockets. The Hubble space telescope launched in 1990 has been updated twice and its replacement is planned for 2018. The information gained from just that one mission has changed the way we see the universe and helped determined its very age. Don't forget rocketry is also a hobby and rocket associations everywhere send their designs skyward every month.

In the future, hopefully there will be fewer weapons delivered by rockets and more space exploration. Currently, commercial operators such as Space-X are booking flights to space for a small 6-figure fee. You can also book a flight to Mars. Hopefully, we will once again return to the moon and maybe stay a while. The Phoenix probe found frozen water on Mars, which helps with one of the biggest stumbling blocks for a manned mission, water.

A manned mission would take at minimum three years due to the orbits of Earth and Mars. If water can be resupplied onsite, the mission is much more likely to occur. It is said that for every dollar spent on space exploration, five come back in spin-off technology. Can you imagine the innovation that will happen if we spend billions going to Mars?

The Science behind the Technology

It takes an incredible amount of science for transportation. Whether you are in a boat, on a car or in the air, several science principles always apply. One major piece of science is force. In fact, the four forces of flight are Lift, Weight, Thrust, and Drag (see Figure 20).

Lift counteracts the weight of gravity and load factors exerted by accelerations. Thrust is produced by the power plant or engine of the aircraft by turning a propeller or through the exhaust gasses if it is a jet. This force is directly opposite of drag. These four forces equalize and counteract each other. In level flight, at a steady airspeed, lift is equal to weight and thrust is equal to drag. So how is lift created?

Principals discovered by Bernoulli, Newton, and Coanda all help explain how a wing lifts.

As you can see in Figure 21, as air moves over the upper surface of a wing, it moves much faster than the air on the lower surface because it must follow the curve of the wing. Bernoulli's principal says that as a fluid's velocity increases, its internal pressure decreases. This relative velocity difference forms a low pressure on the upper surface of a wing compared to the pressure on the lower surface of the wing. As the angle of attack is increased, the pressure on the lower surface increases even more and some is deflected bringing Newton's third law

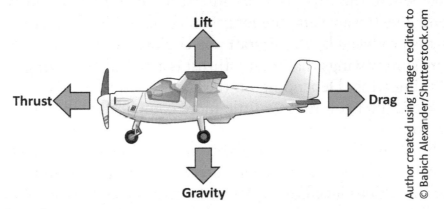

Author created using image credited to © Babich Alexander/Shutterstock.com

Figure 20.   The four forces.

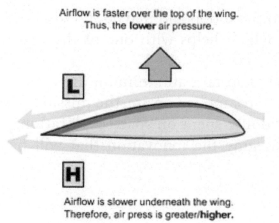

Airflow is faster over the top of the wing.
Thus, the **lower** air pressure.

L

H

Airflow is slower underneath the wing.
Therefore, air press is greater/**higher**.

Figure 21. Wing with upper camber.

into play. The Coanda effect—a fluids' tendency to attach to a smooth surface—stays in play until the wing reaches a critical angle with relation to the on rushing wind (relative wind) when this angle is exceeded this boundary layer as it is called detaches and the wing stops flying. This is what is called a stall. Not to be confused with an engine stall.

Drag is directly related to lift and airspeed, and is counteracted by thrust. There are two types of drag, Induced drag—created by the production of lift, and Parasitic drag—caused by resistance of the air as an airplane passes through it. Parasitic drag is divided into three types including Form Drag caused by the shape of the surfaces directly in the airflow; Skin Friction caused by air molecules giving up energy as they flow past rivets and surface features; and Interference Drag caused by the interference of airflow along adjacent components of the aircraft, such as struts, wing, and the tail sections.

There is a relationship of drag to airspeed. As airspeed goes up, Induced drag goes down and Parasitic drag goes up exponentially. As an airplane slows down the parasitic drag goes down, but induced drag goes up. At the very bottom of the graph you can see where the two draglines intersect. The airspeed that corresponds to this attitude will give the airplane the longest range or glide distance.

Thrust is produced by the aircraft power plant. An aircraft propeller and helicopter rotor are simply wings that rotate. Thrust is adjusted by varying the power delivered to the spinning wing. Depending on the type of propeller, this is measured in torque or RPM. Helicopter rotors are a little more complicated and are explained when we talk about control.

Weight is simply the force of gravity and must be counteracted by lift. Load factor is the extra force created when an aircraft is banked and must be counteracted by increasing the lift produced by the wing. To give you an idea, an aircraft banked to 60° produces a load factor of 2 or 2g, twice the weight of gravity.

Career Connections

Everyone is familiar with airline or air carrier transportation, but as a career one of the most fun types of flying is Private jet transport also known as Air Taxi or Charter. Have you ever wondered how the rich and famous travel? As a charter pilot, you get to fly movie stars and rock bands as well as some of the most well-connected people in the world. Instead of going to the same destinations all of the time you get to go to different spots all the time and usually where something big is taking place. Famous people also tend to have excellent catering and they share with the pilots.

If you like helping people, perhaps a career in aeromedical transport is for you. There are flight nurses and flight paramedics as well as helicopter and fixed wing pilots. Aeromedical transport can range from organ recovery carried out in a small jet to neonatal intensive care with a team moving from an airplane to helicopter. Air ambulance work is usually shift work and can be stressful but in the end there aren't many people who can say that they helped save a life today. These are just a few examples of the types of flying a person can pursue as a career. Remember that for each type of flying there are more people supporting the operation than are actually filling a pilot role.

If you are interested in cutting edge equipment, a career in aerospace or aeronautical engineering may be an option. Math is a must but all of the most innovative ideas start out as an idea somewhere. As an engineer, you may be the one to figure out how to process water for a Mars habitat or some innovative way to land on an aircraft carrier. If this is of interest to you, you may want to visit the Occupational Outlook Handbook for Aerospace Engineers found at http://www.bls.gov/ooh/architecture-and-engineering/aerospace-engineers.htm

Modular Activities

Discussions

* Post an original discussion in the online discussion board on the following topic: Discuss the ethical issues surrounding the use of UAS for military applications. What are your thoughts and why? (200-word minimum). Next, comment on the post(s) of a minimum of one other student in a thoughtful and academic way that enhances the conversation. See rubric for grading and assessment measures.
* Post an original discussion in the online discussion board on the following topic: Watch the 7:22 minute video at http://www.its.dot.gov/communications/media/15cv_future.htm and discuss what it stated here. Do you agree or disagree with this potential future of transportation? (200-word minimum) Next, comment on the post(s) of a minimum of one other student in a thoughtful and academic way that enhances the conversation. See rubric for grading and assessment measures.

Tests

❖ Online graded quiz on overall chapter content, written in multiple choice format. When submitted, we recommend giving the correct answer along with the page number in which it is found for questions students did not answer correctly.

❖ Complete the online quiz found at http://www.cnbc.com/id/38675887 and post both the question and the answer for question #10 on that site, then note the level you achieved.

Research

❖ Perform a short literature review of current safety technologies in either the automotive, aerospace, or marine industry. To accomplish this task, find at least three scholarly peer-reviewed articles on a specific safety technology for a specific industry and write an integrative assessment of your findings inclusive of the following: summary of your findings, critique of your findings, potential gaps in the research. You will be assessed on the quality, depth, accuracy, and relevance of your review.

❖ Find an online news article relating to aerospace or space exploration from within the last six months. Write a short summary in your own words (200-word minimum) and give URL location (or other reference) of article.

❖ Research current safety technologies in either the automotive, aerospace, or marine industry and write a two-page double-spaced report about your findings in APA format. Use a minimum of two references.

❖ Find an online news article relating to aerospace or space exploration from within the last six months. Write a short summary in your own words (200-word minimum) and give URL location (or other reference) of article.

❖ Research one of the following higher education academic areas:
  ➤ Aerospace engineering
  ➤ Aviation sciences
  ➤ Physics
  ➤ Atmospheric science
  ➤ Astronomy

❖ This might include review of offerings or degrees at your institution or at another location, contacting a professional/faculty in the field, assessing higher education level academic events or educational resources, and so on. Submit a 200-word minimum summary of your findings. Make sure to include your source(s).

❖ Interview a person who works in an aerospace field. As the following questions at a minimum and then come up with four (4) questions of your own:
  ➤ What role does technology play in your job or field?
  ➤ What are challenging aspects of your job?
  ➤ What are your favorite parts of your job?
  ➤ What would you recommend to someone going in to this field?

Submit both the questions and the answers.

Design and/or Build Projects

❖   Create PVC boat design document: Sketch and document how to create a simple operational floating boat that could hold two full grown adults using PVC pipe as the main component. You do not need to actually create the boat itself, just design and document it. Professionalism and detail matter. Remember: you will want to include sizes, dimensions, connection methods, steps, or similar.

❖   Create a paper/foam plate plane: Using only one or two paper or foam plates, tape, and a penny (optional), build a paper airplane that flies a minimum distance of 12 ft. while remaining right side up. While you can find templates on the Internet, you are also free (and encouraged) to experiment. Think about what makes a plane fly and you can probably create your own cool invention.

Assessment Tasks

❖   Some people feel that aerospace travel is too expensive; others feel the expense is entirely justified. Make an argument for one of these two viewpoints in a one- to two-page summary. Cite your sources in APA format.

❖   Visit http://www.nasa.gov/ and choose one of the space exploration topics under "missions." Choose from any mission on this list (including from the A–Z listing of all missions). Using the NASA website and any other resources you deem appropriate, complete the following:

➢   Name of mission
➢   Summary of the mission in your own words (cutting and pasting is not acceptable)—200-word minimum
➢   Your assessment on the value and importance of this mission with rationale for your viewpoint—200-word minimum
➢   List of references used in APA or MLA format

Terms

Ailerons—Small trailing edge control surfaces on each wing that control roll.

Airship—A powered lighter than aircraft that can be controlled.

Attitude—the orientation of an aircraft with respect to the horizon.

Ballast—A heavy material such as water, sand, lead, or iron used to add weight and stability.

Biwing—An airplane with two main wings.

Ballonet—A bag or bladder in the interior of an airship which can be inflated and deflated to change the volume of the helium filled gasbags next to them.

Blimp—An airship that holds its shape from the pressure of the gas inside and will lose its shape if deflated.

Bernoulli effect—A scientific principle that states "as the velocity of a fluid increases, its internal pressure decreases."

Camber—The curved surface of a wing. Can be upper camber or lower camber.

Coanda effect—The tendency of a fluid to attach itself to a smooth surface.

Density—The relative number of molecules in a given volume.

Dirigible—From the French word "Diriger" to steer, another word for airship.

Drag—The force opposite thrust caused by air friction and the production of lift.
    Induced—Drag created through the production of lift.
    Parasitic—Drag created when air molecules give up energy due to skin friction.

Elevator—The horizontal control surface on an aircraft at the trailing edge of the horizontal stabilizer responsible for controlling pitch.

Envelope—The membrane or fabric of a balloon that encloses the hot air or gas.

Fly by Light—A control system that transmits pilot inputs and signals between the computers responsible for control surface actuation through a fiber-optic cable.

Fly by Wire—A control system that transmits pilot inputs and signals between the computers responsible for control surface actuation through wires.

Gondola—The part of a balloon or airship, which carries the passengers and payload.

Gyroscope—A spinning mass that displays rigidity in space.

Laser Ring Gyro—A device that compares the phase of laser beams reflected in opposite directions around a small, usually triangular, ring. Motion is sensed by how much the beams are out of phase.

Lift—The force opposite weight or gravity, the aerodynamic force created by the wing.

Newton's laws of motion:
1. An object in a uniform state of motion will remain in that state until acted on by an external force.
2. Force is equal to mass times acceleration, $F = ma$.
3. For every action there is an equal and opposite reaction.

Ornithopter—From the Greek "ornithso" or bird is an aircraft that flies by flapping wings.

Payload—The part of an aircraft's load, which generates revenue such as passengers or freight. In a missile it is the warhead.

Pitch—Motion produced around the lateral axis of an aircraft or rocket.

Propeller—Essentially a rotating wing, the angle or pitch varies along its length to account for the fact that the tip moves faster than the area near the hub.

Rotorcraft—An aircraft that derives lift through rotation of a rotating wing or rotor this includes helicopters and gyrocopters.

Roll—the motion around the longitudinal axis of an aircraft or rocket.

Rudder—The control surface at the trailing edge of the vertical stabilizer of an aircraft responsible for controlling yaw.

Thrust—The force opposite of drag produced by the power plant and propeller, Jet engine or rocket motor.

Triwing—An aircraft with three main wings.

Yaw—Motion produced around the vertical axis of an aircraft or rocket.

Further Reading

Aircraft. (2016). Retrieved from http://www.faa.gov/aircraft/

Ballast | Definition of Ballast by Merriam-Webster. (2016). In *Dictionary and thesaurus | Merriam-Webster*. Retrieved from http://www.merriam-webster.com/dictionary/ballast

Ballonet | Definition of Ballonet by Merriam-Webster. (2016). In *Dictionary and thesaurus | Merriam-Webster*. Retrieved from http://www.merriam-webster.com/dictionary/ballonet

Bilstein, R. E. (2016). History of flight | aviation | Britannica.com. In *Encyclopedia Britannica*. Retrieved from http://www.britannica.com/technology/history-of-flight

FAA-Aircraft-Certification. (2016, April 23). FAA Definitions. Retrieved from http://www.faa-aircraft-certification.com/faa-definitions.html

Hybrid Air Vehicles. (n.d.). Retrieved from http://www.hybridairvehicles.com/

Merriam-Webster. (2016). Gyroscope | Definition of Gyroscope by Merriam-Webster. In *Dictionary and Thesaurus | Merriam-Webster*. Retrieved from http://www.merriam-webster.com/dictionary/gyroscope

Montgomery, J. (2015). *Aerospace: The journey of flight* (3rd ed.). Maxwell Air Force Base, Ala: Civil Air Patrol National Headquarters.

Next gen ADS-B. (n.d.). Retrieved from http://www.faa.gov/nextgen/programs/adsb/

Pagitz, M. (2007, December 15). The future of scientific ballooning. Retrieved from http://rsta.royalsocietypublishing.org/content/365/1861/3003Paris Air Show 2015: Lockheed Martin unveils new heavy-lift hybrid airship | IHS Jane's 360. (2015, June 17). Retrieved from http://www.janes.com/article/52319/paris-air-show-2015-lockheed-martin-unveils-new-heavy-lift-hybrid-airship

Administration, F. A. (2009). *Pilot's handbook of aeronautical knowledge 2008: FAA-H-8083-25A*. Chicago: Aviation Supplies & Academics.

Andrews, E. (2015, April 23). *10 Fascinating facts about the hubble space telescope—History in the headlines*. Retrieved from http://www.history.com/news/10-fascinating-facts-about-the-hubble-space-telescope

NASA-JPL. (2016). Basics of space flight section II. Space flight projects. Retrieved from http://solarsystem.nasa.gov/basics/bsf13-1.php

Shiner, L. (2002, September). How things work-Ring laser gyros. Retrieved from http://www.airspacemag.com/ist/?next=/flight-today/how-things-work-ring-laser-gyros-32371541/

Chapter 7

Location and Tracking Technologies

Outline

- ❖ Chapter learning objectives
- ❖ Overview
- ❖ Satellites
- ❖ Global positioning system
  - ➤ Adaptable navigation systems
  - ➤ Radio frequency identification
- ❖ Geographic information systems
  - ➤ Raster and vector data
  - ➤ GIS and unmanned aerial vehicles
  - ➤ GIS and privacy
- ❖ The science behind the technology
  - ➤ Latitude and longitude systems
- ❖ Career connections
- ❖ Modular activities
  - ➤ Discussions
  - ➤ Tests
  - ➤ Research
  - ➤ Design and/or Build projects
  - ➤ Assessment tasks
- ❖ Terms
- ❖ Bibliography

Chapter Learning Objectives

- ❖ Explain how satellites work.
- ❖ Examine how eccentricity and inclination influence a satellite's orbit.
- ❖ Describe the three different sections of the global positioning system.

❖ Define an adaptable navigation system.
❖ Identify how a geographic information system (GIS) works.
❖ Distinguish the differences among the three map projection systems.
❖ Differentiate the difference between raster and vector data.
❖ Describe the integration of drone technology and GIS.
❖ Summarize the ethical concerns about GIS and privacy.

Overview

"Big brother is watching you." You've probably heard this slogan or used it yourself when you realized you were on camera somewhere in public. But did you really think about what it meant, or was it just a tired cliché we all say now and then. It is so common and relevant that it could have been coined recently. However, it was from the book, *1984*, that was published in 1949. It was written by George Orwell about a dystopian society that was constantly at war. To control the people, they were put under constant surveillance by the government, which was comprised of an elite group. All individualism was squashed, and citizens could be persecuted for even thinking socially unacceptable thoughts. Orwell wrote this over half a century

©ChameleonsEye/Shutterstock.com

ago when modern surveillance technologies were in their infancy or nonexistent. Of course, World War II had recently ended. The Nazi Machine of Germany had tightly controlled its population with appalling results. Technology itself is neutral. It is the intersection of technology, culture, and government that direct how a technology will be used. Think about it. What kind of book do you believe Orwell would write today?

Imagine for a minute all the technologies that can contain electronics to track you. You wake up with your radio alarm clock, check your smartphone for messages, watch your television that may be watching you, open your computer to do research for your next paper and are followed around the internet, drive your smart car to a doctor's appointment using Global Positioning System (GPS), pick up your medicine from a pharmacist who received the prescription order online, and go to work where every keystroke on your computer is tracked and stored. During all this time, you are being identified by different cell towers as you text and talk to people. All these technologies make up a vast communication network that includes devices on the ground and in the air. It circles the globe and penetrates into deep space. Individually, many of these steps are just data. But it gets very interesting when all these technologies start talking to each other. What would they say about you?

If you've downloaded an app lately, or maybe not so lately, it probably asked for your permission to track you. If you spent any time figuring out why it would want to do so, you may have found a statement about "wanting to serve you better." If it was a navigation app, then this is likely true. Anything else is about serving the company, and not you, better. Much of the information is used for marketing purposes. Some of it is used to label people by their characteristics. How we decide to use this technology is a conversation for everyone.

Satellites

What is a satellite? If you ask an astronomer, he/she is likely to tell you that satellites are any cosmic bodies that orbit the earth or other planets. The moon is a common example. If you ask a scientist, he/she is likely to tell you that they are artificial devices put into space by humans to circle the planet. They can track weather patterns, relay messages, and take detailed pictures of Earth's surface. They are the eyes in the sky that never sleep. There are several thousand satellites in space. Many are currently being used, but a number of them are inactive. In addition to satellites, there is also a lot of spacecraft junk floating in space. This includes pieces of launch vehicles, debris from space collisions, and even paints chips. There is so much debris that National Aeronautics and Space Administration (NASA) has an Orbital Debris Program Office to deal with mission safety.

Programmer James Yoder created a website that tracks orbital objects in real time. It can be found at the following website: http://stuffin.space/. You can view the orbits of satellites and large debris by moving your mouse across the screen. If you click on one, it opens an information box about the item. If you select "Groups" at the upper left of the screen, you have other options. The first one shows the pattern of (GPS), which is helpful when you get to the following section on GPS.

There are many kinds of satellites that serve various purposes. They travel in various orbits around Earth to provide different perspectives of earth and the universe. There are three types of orbits: high earth (approximately 22,000 mi. above the earth), medium earth (approximately 1,200–22,000 mi.), and low earth (approximately 110–1,200 mi.).

The height of the orbit determines how long it takes a satellite to move around our planet. The closer the satellite is to Earth's surface, the shorter the time it takes. To see a quick video from the NASA, which compares the three orbit times, visit the following website and click on View Animation in the center of the site or on the graphic directly: http://earthobservatory.nasa.gov/Features/OrbitsCatalog/

There are two other variables in addition to height, which determine a satellite's orbit: eccentricity and inclination. See Figure 1 below from NASA on eccentricity. Eccentricity is the orbit's shape, which can be near circular (low eccentricity) or elliptical (higher eccentricity). Eccentric simply means it has an unusual shape. A satellite following an eccentric orbit does not maintain constant distance from the earth. It will be closer or farther from Earth's surface depending on where it is in its orbit.

The Figure 1 shows various levels of eccentricity (e) for satellites' orbits. A perfect circle has an e value of zero. The more eccentric (and flatter) the orbit becomes, the higher its e value. If you visit the "Stuff in Space" website posted earlier and click on a satellite orbit, you can see its eccentricity.

Inclination is the orbit's angle from the Earth's equator. In Figure 2, Satellite A has zero inclination as it is travelling with the Earth's equator. Satellite B has a 90° inclination as it is perpendicular to the equator in a polar orbit. Satellites following a polar orbit pass by both the North and South Poles. Satellite C has an approximate 45° inclination as it is about half way between the equator and North Pole. If Satellite A was traveling in the opposite direction, against Earth's rotation, it would have an inclination of 180°. The combination of height, eccentricity, and inclination control a satellite's orbit and its view of Earth.

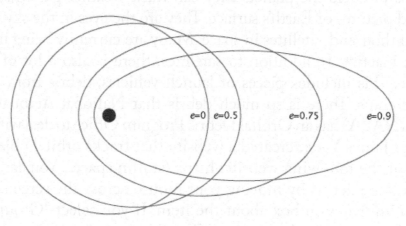

Figure 1. Various eccentricity levels for satellite orbits.

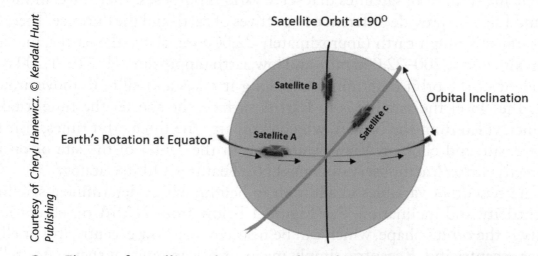

Figure 2. Figure of satellite orbits around Earth and orbital inclination.

If you see a satellite that appears to be stationary, it is in a geostationary (or geosynchronous) orbit. Satellites in this orbit always stay in the same spot over the equator. They are positioned in high-Earth orbit, about 22,000 mi. above Earth's surface, since they can follow its rotation at this height. This is why they appear to be motionless. They are traveling as fast as you. This orbit is best for communication satellites (phone, television, and radio) and those that track weather pattern, since they have an unchanging vantage point.

Satellites in medium-Earth orbit take about 12 hours to rotate around the planet. GPS satellites are positioned in this orbit. Many satellites in low-Earth orbit are used to relay weather information and scientific data. It takes them about 99 minutes to orbit the planet. If they follow the polar orbit, they are able to observe the entire Earth's surface as the planet moves below them. This is a common orbit for reconnaissance (or spy) satellites.

Sun-synchronous is a low-Earth orbit. Its pathway is inclined a few degrees beyond the poles and positioned to cross the equator at the same local time (so, no matter what time zone you are in, it will pass above at the same time). One advantage is that satellites in this orbit can take pictures of the Earth with sunlight hitting it from the same angle. This allows scientists to track geological formations (such as ice caps or rain forest coverage) over time. The Iridium satellite constellation is another low-Earth orbit system that works with many handheld devices. This network contains 66 communication satellites for voice and data and provides coverage for the entire planet. Iridium satellites fly about 485 mi. above Earth and circle it every 100 minutes.

Some satellites are made to leave the solar system altogether. Pioneer 10 and 11 were built to investigate Jupiter and were launched in the early 1970s. Voyager 1, which was launched in 1977, became the first man-made object to leave the solar system on August 25, 2012. Its twin, Voyager 2 was launched a month later. They both contained a golden record with instructions about our world, including mathematical and physical definitions as well as recordings from over 50 languages and other information. See Figure 3 for picture of record. There is also a needle and directions on how to use it in case an alien race finds it. The system is based on the old record players for those of you wondering why a needle was included. The National Oceanic and Atmospheric Administration (NOAA) even has a satellite, DSCOVR, orbiting 1 million miles from Earth that monitors deep space. It was launched on February 11, 2015, and will serve as an early warning system for solar magnetic storms. You can view daily images sent from DSCOVR at http://epic.gsfc.nasa.gov/.

Launching a satellite into space is a complex process. It is based on Newton's third law of motion, also known as "action and reaction." A massive force is created when a rocket's engine ignites and hot gases fire backward. This produces an equal force that propels the rocket skyward. Most fuel on board is used to launch the rocket fast and

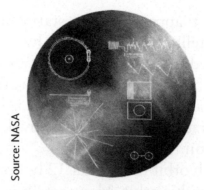

Source: NASA

Figure 3. Golden Record launched inside Pioneer 10 and 11. Figure can be found at the NASA Website located at http://voyager.jpl.nasa.gov/spacecraft/goldenrec.html

far enough to escape Earth's gravity. This is known as escape velocity, and it is achieved by traveling at least 25,000 mph.

According to the Federal Communications Commission (FCC) Satellite Learning Center, a multistage launch vehicle is needed to lift the satellite and the launch vehicles off the ground. The launcher contains rockets and fuel and weighs hundreds of tons. The amount of fuel (energy) needed depends on the desired orbit. High-Earth orbits require the most fuel. A satellite launched over the equator requires less and can also use the Earth's rotation to assist. Satellites launched at a higher incline, such as a polar orbit, also take more energy than those launched at lower inclines. Once the first-stage launch vehicle has used up its fuel lifting the payload, it is released and falls to the Earth. The second-stage launcher is then used to propel the satellite into space by igniting its own fuel supply. It is then released, but instead of returning to Earth, it is high enough to burn up in the Earth's atmosphere. The third and last, upper stage holds the satellite, which is covered by a shield called the fairing. It is the pointed top of the launcher. It protects the satellite and is a shape that offers low resistance moving through the Earth's atmosphere (ever notice that all rockets are pointy at top for this reason?). The fairing splits apart when the satellite is above the Earth's atmosphere. It also burns up instead of returning to the ground. The last set of rockets in the upper stage fire and provide the momentum to put the satellite in its place in space. It separates from the upper stage. At this point, it opens its solar panels and antennas and is ready for work.

Once a satellite is in orbit, it still needs some assistance to remain there, according to scientists at NASA's Earth Observatory. The unevenness of the Earth's terrain, along with the pull from the Sun, Moon, and even Jupiter, causes an orbit to change. High-Earth orbit satellites need to be readjusted about three or four times in their life spans. Low-Earth, Sun-synchronous orbit satellites need to be adjusted every year or two. Other low-Earth orbit satellites are pulled by drag from the planet's atmosphere, and gravity causes them to speed up. Eventually, these satellites burn up or fall to earth. Satellites can also get moved out of their orbit by colliding with space junk. NASA satellite mission controllers have had to move some satellites to avoid additional collisions.

Most satellites use radio waves to communicate. Antennas on Earth pick up the signals and process the information. To work, the satellites and antennas need to be in line of site. If a satellite does not have a clear view of a ground antenna, it will use a Tracking and Data Relay Satellite (TDRS) to pass along the information. There are nine satellites in the TDRS system. In fact, no satellite operates alone. They are all part of a larger network. One extensive NASA network is called the Near Earth Network (NEN). It has numerous tracking stations throughout the world and provides services to government agencies, international civilian space agencies, and commercial entities.

The number of satellites is immense, but listed below are some notable satellites used for different purposes:

❖ International Space Station is a joint project of space agencies in the United States, Russia, Europe, Japan, and Canada to provide long-term exploration of space. It provides the opportunity for scientists from multiple nations to research how the body, medicine, plants, technology, and so on, live, grow, and work out of Earth's atmosphere. It is located in low-Earth orbit.

❖ Geostationary Operational Environmental Satellite (GOES) program of NOAA tracks weather data, such as clouds, water vapor, wind speed, and storms and transmits them back to earth every few minutes. The program consists of both geostationary and polar-orbiting satellites to create a comprehensive global weather network.

❖ Hubble is a large space telescope. It orbits the Earth but faces outward. It has returned pictures to Earth about planets, stars, and galaxies. It is located in low-Earth orbit, so astronauts on the International Space Station can provide maintenance.

❖ COSMOS 2433 (Glonass) is the Russian counterpart to the United States' GPS navigation system and is comprised of 24 satellites deployed in three orbital planes.

❖ Space-Based Infrared System (SBIRS) is a state-of-the-art strategic satellite surveillance system designed to provide early missile warnings to the President of the United States and key defense and intelligence personnel. SBIRS is a mixture of geosynchronous satellites in highly elliptical orbits. They have the capacity to scan wide surveillance areas and a staring sensor to focus on smaller areas.

Global Positioning System

The global positioning system, or GPS, is a navigation system that is part of the global information infrastructure. GPS is a constellation of satellites located in medium-Earth orbit at approximately 11,000 mi. The satellites orbit the Earth every 12 hours and regularly transmit radio pulses.

GPS history started in the late-1950s when Soviet Union scientists launched the world's first artificial satellite, Sputnik I, on October 4, 1957. It was the size of a beach ball, weighted about 184 pounds, and took 98 minutes to orbit the Earth. It marked the start of a new technological era. It also started the U.S.-U.S.S.R. Space Race. Sputnik's launch and successful orbit took Americans off guard. They did not want their airspace vulnerable to outsiders. The United States launched its first successful satellite, Explorer I, on January 31, 1958. In July 1958, the NASA was created. Since then, its mission has been to boldly go where no man has gone before (oops, that's Star Trek).

The modern GPS system was developed in 1978 by and for the Department of Defense, but now anyone with a GPS receiver can use it. The United States is not the only country with a global satellite system. Russia has GLONASS, Europe has Galileo, and China has Beidou. GPS data are used to determine your position on earth. According to GPS.gov, there are three separate sections to this system.

1. *Space segment* (satellites): The US Airforce operates and maintains this section of the GPS navigation system. At least 24 satellites are operational around the Earth at any given time and arranged into six equally spaced orbital planes. Since a minimum of four satellites is needed to determine your geographical location on earth, this orbital satellite pattern ensures that at least four satellites are in line of site with your receiver wherever you are. Satellites that are part of the GPS system fly approximately 12,500 mi. above earth and circle it twice each day. They are powered by solar energy but contain backup batteries.

2. *Control segment* (ground stations): The US Airforce also operates and maintains this section of the GPS navigation system. It consists of a worldwide monitoring system to ensure that satellites are in their proper orbit and functioning well (see Figure 4). There is a master and alternative master control station, 11 command and control antennas, and 15 monitoring sites. The control segment monitors and analyzes satellite transmissions.

3. *User segment* (receivers): This section contains GPS receiver equipment, which can be operated by military personnel and civilians. This is where you and your GPS-enabled electronic items fit into the GPS navigation system. Most likely your phone, car, computer, and watch contain tracking technology.

To get a good overview on how GPS works, visit the following website and view the video, Space-The High Ground. You only need to view the first 4:40 minutes.

http://www.gps.gov/multimedia/videos/

To determine a location, GPS satellites transmit radio wave signals at 186,000 mi. per second to receivers on the ground. The time it takes for a signal to be received determines the distance and location. However, if only one satellite is being used, it cannot provide enough information for your exact location. For that you need at least four satellites. At first glance, it may seem that three satellites should be able to tell you your location. If you look at the left side of Figure 5, you can see that three overlapped signals do create a narrow range. Adding a fourth signal, as shown on the right side of the figure, helps pinpoint it further. Even more than four satellites can be used to create greater accuracy.

Corresponding signals from the first three satellites can generally determine your latitude, longitude, and elevation. These signals need to be simultaneous, and GPS satellites contain atomic clocks that tell time to within 40 billionths of a second. However, clocks in many GPS receivers are not that accurate. This timing error could translate into a positioning error. According to NOAA experts, a miscalculation of as

★ Master Control Station　　☆ Alternate Master Control Station
▲ Ground Antenna　　△ AFSCN Remote Tracking Station
● Air Force Monitor Station　　● NGA Monitor Station

Source: GPS

Figure 4. Image is of the Control Segment of the GPS Navigation system.

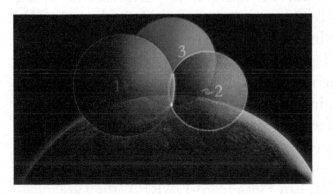

Source: NOAA

Figure 5. Using three satellites gives you two possible locations whereas adding a fourth one gives an exact location. Picture on left taken from NOAA Ocean Service Education at: http://oceanservice.noaa.gov/education/kits/geodesy/media/supp_geo09b3.html. Figure on right taken from NOAA Ocean Service Education at: http://oceanservice.noaa.gov/education/kits/geodesy/media/supp_geo09b4.html

little as one-millionth of a second between satellite and receiver clocks could cause a positioning error of up to 900 ft. The purpose of the fourth satellite is to adjust the clock error on the receiver side to correct the position calculation. The following calculation is used to calculate the distance from a satellite to receiver:

(186,000 miles/second) × (how many seconds it takes the satellite signal to reach the receiver) = how many miles the satellite is from the receiver.

The use of GPS systems has become ubiquitous in today's society. High-tech farming uses GPS and GIS to guide machinery in planting straight rows of crops, so no farmland is wasted. Herbicides and pesticides can be applied precisely using latitude and longitude data instead of keen eyesight. GPS also lends itself well to maritime use. The foundation of any shipping or fishing voyage is to know your position in a lake or ocean. This is crucial for any search and rescue mission for a vessel needing assistance. Backcountry skiers and hikers often carry GPS devices that will identify their location in the event of an avalanche or emergency. While many people want to be tracked and found, some people don't. Unfortunately, if someone knows your phone number, and your phone has a GPS receiver (and whose doesn't any more), the average citizen can never be quite sure who may be tracking them if it's turned on. Most of the time we give it away for free. When you download a new app, it will often ask your permission to track your location. Many people will simply accept this request because they want to quickly start using the new app or just find it part of the new digital normal.

Adaptable Navigation Systems

As helpful as the GPS is for the military, this capability would be lost if something happened to the satellites. Events, such as a catastrophic failure or an armed military action, could bring the system offline. After all, it is difficult to protect items in the sky, which can be vulnerable to interference by other nations—or somebody with GPS jamming technology. Plus, GPS signals are limited if users are in areas such as underground tunnels, buildings, or underwater. In 2011, the Defense Advanced Research Projects Agency (DARPA), the US Department of Defense agency responsible for developing new technology (and brains behind the development of GPS), announced a new program to address this problem. The program, titled All Source Positioning and Navigation (ASPN), is intended to be a seamless navigation tool that is not reliant on GPS. More generally, DARPA is developing adaptable navigation systems (ANS) to address three challenges. According to Mr. Lin Hass from DARPA, they are:

1. Better inertial measurement units (IMUs) that require fewer external position fixes
2. Alternate sources to GPS for those external position fixes

3. New algorithms and architectures for rapidly reconfiguring a navigation system with new and nontraditional sensors for a particular mission. (http://www.darpa.mil/program/adaptable-navigation-systems, para. 2)

It is difficult to integrate new maps and components into an existing navigation system. What they are trying to achieve is a plug-in-play approach. According to dictionary.com, plug and play is "a standard for the production of compatible computers, peripherals, and software that facilitates device installation and enables automatic configuration of the system." This is an ideal situation for a large military organization whose members use high-tech equipment in different territories and climates all over the world.

To assist in creating a robust ANS, DARPA developed a micro-technology for positioning navigation and timing (Micro-PNT). Micro-PNT are miniature sensors able to navigate precisely in the absence of GPS. They can operate in harsh environments and self-calibrate. They are also being developed to have a fully integrated miniature timing and IMUs for universal deployment. IMU can detect their own acceleration and movement, generally without the need of outside measuring devices such as GPS. Micro-PNT technology would complement existing GPS technologies that have been used in Department of Defense vehicles and weapons.

Interestingly, some solutions are not based on state-of-the-art technologies but traditional practices that have worked for millennium. The US Navy has started to train its service members to navigate using the stars. It had stopped this training over a decade ago but brought it back because of security concerns about relying exclusively on GPS.

Radio Frequency Identification

Radio frequency identification (or RFID) tags are small, flexible label that contains a microchip and transmitter. See Chapter 5 for a full discussion of the technology. An RFID receiver/scanner is used to obtain the information on the tag. Generally, they are considered short-range technology and often used for tracking inventory in-house. According to *RFID Journal*, some RFID tags have been integrated with GPS transceivers (devices that can both transmit and receive information) since 2007. One company, Zebra Technologies, has developed a real-time locating system technology for maritime shipping companies. The always-on technology integrates wireless communication with vehicle monitoring systems and RFID location devices. Managers can quickly determine the location of trucks and inventory as well as when containers were transported off ships and at what time. They can track human operators and plan their work more efficiently.

Geographic Information Systems

A geographic information system (GIS) is a dynamic, computer-based software system that displays geographically referenced data visually as a map. You may have used Google Map or similar mapping systems to find the location of a restaurant or fastest route to your favorite vacation spot. These are great tools to do simple tasks on a map and can give you a good idea of what a GIS looks like. But a true GIS can complete much more sophisticated activities and calculations. It can help doctors track a progression of a disease or even seasonal flu patterns. Firefighters can use handheld GPS units to send real-time information back to a GIS system to track forest fires as they are happening. These uses are possible because a GIS is more than just a visual map. It is an interactive visual tool that can be used to provide data to critically solve problems.

The field of geographic information science started over 50 years ago and has its roots in the physical sciences. People from various backgrounds and nationalities contributed to its overall development. Governmental agencies and large corporations were generally the first users, since the technology was expensive, and they had the resources to use it. The first GIS system was developed by the Canadian Regional Planning Information Systems Division in the 1960s. Its initial purpose was to classify and map the country's extensive natural resources. This had been a cumbersome task as Canada encompasses a large region of land, and traditional mapping techniques were inadequate. In the late 1960s, the US Census Bureau also contributed to the development of GIS while preparing for the 1970 decennial census. They wanted to create digital records of every street in the country to have the information available for easy referencing. The result was the Dual Independent Map Encoding program. Also in the late 1960s, the UK Experimental Cartography Unit published the world's first computer-generated map. However, these early systems could not edit or update the maps, which was a very time-intensive and expensive task to do by hand.

However, as computing capabilities increased at the same time that costs continued to decrease, people in other professions began using GIS technology. It still remained a highly specialized line of work with limited users. It was not until 1995 that Britain achieved the first digital map coverage in its database. Since that time, GIS software became available for desktop use and open-source applications appeared as well. The US Census Bureau, along with other government agencies, has both map shape files and demographic data available for downloading. This is an exciting time to be interested in this software as both the capabilities of the software and access of data allow even a casual user to create sophisticated maps.

In addition to government entities, other industry leaders using GIS technology include transportation authorities, utility companies, real estate businesses, retail stores, and the aviation industry, among others. GIS has become so reasonable to finance and easy to use that many small companies are using it to their advantage. It is easy to see why realtors use GIS. Their entire industry is based on location. Companies, such as Zillow, give homebuyers the ability to access data themselves, which is changing the nature of the realtor business. Utility companies, such as electric companies, can track power lines, stations, and customers. The field of transportation has been revolutionized. The convergence among wireless devices, location-tracking technologies, and spatial management tools allows businesses, such as PepsiCo, to track their fleet onscreen. Retailers can determine where their customers live and their buying behaviors. This makes a GIS a powerful marketing tool.

Politics has been one area where GIS has been used extensively. Politicians can use GIS maps to find demographic data on their constituents right down to the neighborhood level. They can also determine voter turnout at precincts and how geographic areas voted on a particular proposal. This helps candidates understand the viewpoints of citizens and where to focus their efforts on a particular issue. It has also contributed to high-tech gerrymandering. Gerrymandering is the intentional redistricting of a voting district's geographic boundaries to give a biased advantage to a person, a political party, or a particular community. This practice is not new and has been used since the early 1800s. What makes this political exercise so critical today are the abilities of powerful computers and sophisticated GISs. Political campaigners can use these technologies to make accurate predictions about voters before they even head to the polls.

Figure 6 shows a GIS map from the Centers for Disease Control (CDC) and Prevention website. It identifies heart disease death rates in the United States between 2011 and 2013 for adults aged 35 and older. The higher the rates, the darker the shade of red. Even if you are unfamiliar with heart disease, can you determine where the largest concentrations of people with heart problems live? At a glance, you can easily see they are concentrated in the South. Now, look at the Western United States. What is going on in Nevada? The map may not give you the details, but it identifies a place to do further research. If you were a cardiovascular doctor and wanted to move west, where would you locate your practice? Or supposed you were a scientist who wanted to determine what factors caused heart disease. What could be your next step after reviewing the map? You could look at other maps, such as obesity rates, meat consumption, exercise rates, or other lifestyle choices you believe may contribute to heart disease. You could also look at pollution levels or how far heart disease victims lived from a hospital. Better yet, you could layer them on top of each other and see if more than one of these factors are concentrated in a specific geographic area. The real power of a GIS system is that it has the capabilities to layer thematic maps and run spatial statistics among them to determine if an event is random or whether there is correlation.

GIS maps are not a static picture but a representation of data that are stored in a database. A database is a collection of information that is logically organized to manage, retrieve, and update information. It generally looks like a large spreadsheet but can be displayed onscreen and printed as forms, invoices, and in the case of GIS, maps. See Figure 7 for the Excel spreadsheet related to the heart disease map. The rows are the records, which are the data you are recording. The columns include the record label to explain what you are recording in your database. In this case, the file contains the field names of the state, the county, the number of heart disease death rates, and the FIPS code. The FIPS code is the designation for Federal Information Processing Standard code, and is a unique number used to identify a specific county. The first two numbers indicate the state and the last two indicate the county.

The number of records (or rows) that can be included in a database is virtually unlimited. There are 3142 records in the spreadsheet shown in Figure 7, although only seven are shown. There is one record for each county shown on the map. If you look closely at the map, you can see that each state is comprised of many counties. The shade

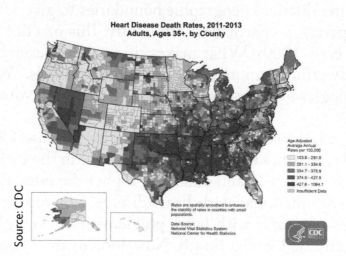

Figure 6. "Heart Disease Death Rates, 2011–2013, Adults Ages 35+, by County." CDC. June 24, 2015.

	A	B	C	D
			Heart Disease Death Rates, 2011-2013, Adults, Ages 35+, by	
1	State	County	County	FIPS Code
2	Alabama	Autauga	463.0	1001
3	Alabama	Baldwin	391.4	1003
4	Alabama	Barbour	533.1	1005
5	Alabama	Bibb	511.1	1007
6	Alabama	Blount	425.6	1009
7	Alabama	Bullock	483.2	1011

Source: CDC

Figure 7. Excel spreadsheet data that had been imported into heart disease GIS map.

of the county is dependent on the number found in Column C of the spreadsheet, which corresponds to the map's legend. If you erased all the rows in the database, the map would become white. The state and county lines would remain, since the physical map is a separate file from the data file. But it, too, has a database in the background that provides geographic coordinate system directions for how the map should be drawn. For example, data may include latitude and longitude, x and y coordinates, or other spatial indicators.

For the map file and data file to communicate with each other, they have to have a way to "talk." First, every record in a database has to have a unique identifying field. Using Figure 7 as an example again, that unique field is the FIPS code. Clearly, there are more than one row for each state, so that could not be a unique identifier. Different states have some common county names, and heart disease death rates could also be the same among counties. Therefore, you cannot be sure that either of these two fields would be unique. Only the FIPS is unique. Even if the county codes are the same (last two digits), the first two digits indicate a different state. If you looked at the database file for the map, you would also find a FIPS code identifying the county associated with its shape. By having the unique identifier in both files, the data are displayed in the correct county on the map of the United States.

You could argue that it is not even necessary to include a county name if the FIPS is all you need to plot data. This is true, but GIS maps allow you to click on a location to reveal more information about it. An information box will display the data for the fields in the underlying database. Without including the county name in the database, only the FIPS would display. Not many people would be able to identify a county only by its FIPS code. By including the name in the database, users can display it on the map.

The GIS database for this map looks like a large spreadsheet. In fact, that is often how data are added to a GIS. An existing spreadsheet can be imported into the GIS map file with information about housing, voting districts, natural resources, census information, or just about anything that is being tracked in a database. Or users can enter the information directly. Maps are imported this way as well, often through scanning and digitizing them.

Raster and Vector Data

So far you have learned about maps and the databases that provide data for them. What about the satellite images that are also overlaid onto a map to give a real picture of the Earth? In 1972, the first remote-sensing satellite for civilian use was launched. It was called Landsat 1 and is today the longest running program to record Earth's land. Of course, the satellites have gone through several versions since then and today there are Landsats 7 and 8. They operate through remote sensing, which is the science of obtaining information from a distance, often with aircraft or satellites.

Remote sensors capture data by detecting energy reflected from the Earth's surface. They can do this through active or passive means. If passive, they simply record naturally available energy reflecting off the Earth, usually from the sun. Active sensors apply direct energy, such as a laser beam, and measure the time it takes for the laser to reflect back to the sensor.

The Landsat program is now in partnership with the US Geological Survey agency. Landsat sensors can view wide geographic areas, such as highways and rivers but cannot discern images such as people or individual houses. Their resolution is coarse enough for worldwide coverage but can still distinguish changes in the environment, such as urban growth or changes in water volume. Figure 8 shows two images of Lake Meade. It is easy to identify the changes in water level between the two pictures.

Images, such as those seen in Figure 8, are not taken nor recorded like traditional pictures. They are captured as raster data. This type of data is based on the pixel format (short for picture elements) that is used to represent continuous surfaces such as Earth. If you have ever zoomed in on a picture located on a computer screen, you may have noticed how the picture becomes a series of fuzzy boxes. This is an example of raster data. According the *Landsat Science*:

Landsat sensors record reflected and emitted energy from Earth in various wavelengths of the electromagnetic spectrum. The electromagnetic spectrum includes all forms of radiated energy from tiny gamma rays and x-rays all the way to huge radio waves ...

Today, Landsats 7 and 8 "see" and record blue, green, and red light in the visible spectrum as well as near-infrared, mid-infrared, and thermal-infrared light that human eyes cannot perceive (although we can feel the thermal-infrared as heat).

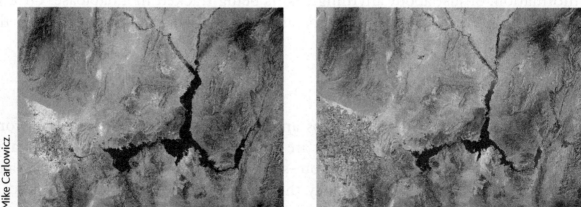

NASA Earth Observatory images by Joshua Stevens, using Landsat data from the U.S. Geological Survey. Caption by Mike Carlowicz.

Figure 8.   The pair of pictures above is of Lake Meade, which is located at the Hoover Dam. The image on the left was taken at its highest point, on May 15, 1984. The image on the right was taken at its lowest point on May 23, 2016.

Landsat records this information digitally and it is downlinked to ground stations, processed, and stored in a data archive. (NASA, Landsat Science, para. 1–2, retrieved from http://landsat.gsfc.nasa.gov/?page_id=9)

In raster data, physical space is represented by rectangular shapes, usually squares. When it is processed by GIS software, each square, or cell, is assigned geographic properties that when put together create an image.

Vector data are used to capture lines as points connected by straight lines. Areas surrounded by vector lines are called polygons. In Figure 6, map of heart disease death rates, the physical representation of counties are examples of vector data. GIS files often have layers of maps with both raster and vector data. These can be laid on top of each other so you can view an area, such as state, on a satellite map.

GIS and Unmanned Aerial Vehicles

One future direction for GIS is the use of unmanned aerial vehicle (drone) technology. They are discussed at length in the *autonomous and semi-autonomous technologies* chapter to come. That said, Drone2Map for ArcGIS is a new geographic information application released to the public in February 2016. Free trials of the software are available through Environmental Systems Research Institute's (ESRI) beta program at www.esri.com/drone2map. According to their website (http://www.esri.com/products/drone2map), users can "turn your drone into an enterprise GIS productivity tool … in minutes, not days." Drone2Map takes drone-captured pictures and allows users to create panoramic, 360° images. It is being marketed for inspecting assets, such as oil pipelines that are difficult to reach, or monitoring large landmass areas, such as where a natural disaster has taken place or environmental changes may be occurring. OpenDroneMap is an open source version for processing civilian unmanned aircraft system (UAS) imagery.

Both drone applications allow you to track areas that are difficult to see or access from the ground but can be easily seen through the eyes of a drone.

Can you think of other areas that may be difficult to see but easily accessed by a small, quite airplane that's easy to control? A place like your kitchen window. This technology is a paparazzi's dream! They don't even have to take videos. All they need to do is take a series of pictures that can be stitched together to look like an original image.

GIS and Privacy

One major concern of GISs technology is how easily it can be used to intrude on one's privacy. It is quite clear what line has been crossed if your neighbor is hovering a drone over your house. However, GIS can be used in other ways. While it is easy to find your house on a GIS map, it is the attribute data associated with your house, and your neighbors' houses that makes the technology so desirable to politicians, government officials, and businesspeople.

The availability of sociodemographic data from census bureaus, police departments, school testing companies (e.g., SAT), realtors, voting polls, and so on, can be downloaded and combined to reveal neighborhood characteristics. There is even a term for this ability: geodemographics. People who work in this field can use population data and their characteristics to accurately predict consumer behavior based on where people live. They can then divide areas into market segmentations based on common interests. Generally, this is a great way to spend marketing dollars on people who may be interested in your product. For example, you would not want to send information on tee-ball registration to a neighborhood whose median age is 56. Or market a high-end car to families who are just starting out and have a median income of $26,000. Market segmentation data are also helpful in determining where to open a new business, since it combines geographic location with sociodemographic information. As early as 2003, one computer specialist, George Beekman, found there were over 15,000 marketing databases containing personal information on about 2 billion consumers.

Today, many people disclose information about themselves all the time. They allow tracking when downloading apps and free software, complete surveys to be entered in a drawing, use loyalty cards at grocery stores, and are followed around online by cookies. Every time a credit card is used, it leaves a digital footprint. Although consumers do not give explicit consent to using this information for marketing purposes (who reads the fine print anyway?), much of this information can be bought and sold to companies who may combine it with other information. One concern is that it can all be used for social profiling. Will you be branded by where you live, where you travel, and what you buy?

The Science Behind the Technology

GIS, as discussed in length above, often work in conjunction with GPS satellite data. Many devices can send information back to a GIS system instantaneously. This real-time tracking is helpful, such as following a forest fire and determining which direction it is moving. The ability to locate and track is all about being able to measure distances. In order to take measurements accurately, it is important to understand the system of latitude and longitude and map projections.

Latitude and Longitude Systems

The latitude and longitude system is an international, numerical system. If you know the latitude and longitude of a position, you can find it no matter your language or nationality. Latitude lines are horizontal and run parallel to the equator. They indicate how far north or south you are from it. Therefore, the equator is 0° latitude. Latitudes move 90° North and 90° South.

Longitude lines are vertical and run perpendicular to the equator. These lines are referred to as meridians and indicate how far east or west you are from the prime meridian, located over the British Royal Observatory in Greenwich, England. This is the point of 0° longitude. Earth is a circle, so it is 360°. Therefore, there are 180 longitudinal lines East, of the prime meridian and 180 longitudinal lines West of the prime meridian. For more precise measurements, longitude lines are divided into 60 minutes and again into 60 seconds. According to the World Atlas website, their office in Galveston Island, Texas, is located at 29°, 16 minutes, and 22 seconds north of the equator, and 94°, 49 minutes, and 46 seconds west of the Prime Meridian. That's very specific. However, the absolute location of their office is 29°16'N, 94°49'W. Absolute locations are used in coordinate systems, which represented the geographic location of a feature. Different coordinate systems are needed because the Earth is not flat. It would be much easier to measure if it was.

Latitude and longitude lines are equidistant only at the equator. As longitude lines move toward either the North or South Pole, they start to converge. The GPS satellite system needs to know the position of the latitude, longitude, and elevation to accurately determine a receiver's location, and it does this by using four satellites as noted earlier in the chapter. A GIS does not use latitude and longitude measurements directly. It has an internal map file that is based on a coordinate system. According to ESRI, a major international supplier of GIS software, the two most common coordinate systems in a GIS are:

❖ A global or spherical coordinate system such as latitude–longitude. These are often referred to as geographic coordinate systems.
❖ A projected coordinate system such as Universal Transverse Mercator (UTM), Albers Equal Area, or Robinson, all of which (along with numerous other map projection models) provide various mechanisms to project maps of the earth's spherical surface onto a two-dimensional Cartesian coordinate plane. Projected coordinate systems are referred to as map projections. (http://resources.esri.com/help/9.3/arcgisengine/dotnet/89b720a5-7339-44b0-8b58-0f5bf2843393.htm#TypesCoordSys)

A projected coordinate system is not visually accurate since it is taking a round map and laying it flat. Have you ever tried to take a globe map and lay it flat? It is impossible to do. At some point, the paper has to be teared and adjusted to make it square. There are three types of map projections and ways to attempt to make the round Earth flat. In each instance, though, there is some distortion of the Earth's features.

1. Cylindrical: Image a round Earth in a cylinder, with all its features being projected on the cylinder and then unwrapped to a flat image. (See Figure 9.)
2. Azimuthal or Planar: Image a round Earth projected onto a flat piece of paper that it is touching. These images are of a limited area such as the poles.
3. Conic: Imagine a round Earth projected in a cone and then unwrapped.

When using a GIS, it is important to know what coordinate system you are using and to define it within the map. If two datasets for the map do not match, they will not align and cause inaccuracies within your map. You can see by looking at Figure 9, how difficult it could be to try and align maps of different projections.

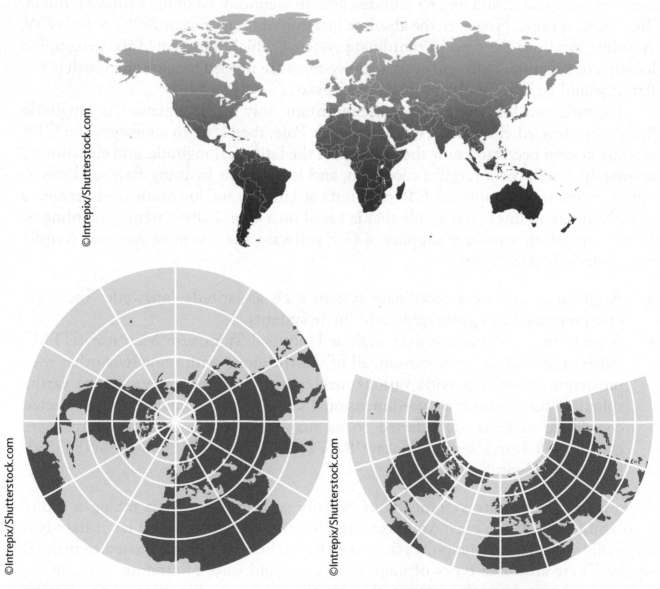

Figure 9.   A cylindrical, azimuthal, and conic map projection side by side.

Career Connections

According to Tech Republic, "programmers and developers can find high-paying projects in many industries, but one field that's drawing increasing attention is that of geologic information systems (GIS)" (Where are the jobs in the GIS field?, http://www.techrepublic.com/blog/career-management/where-are-the-jobs-in-the-gis-field/). This is because more manufacturers, marketers, and businesspeople are finding value in adding GIS to their operations. Jobs include geological surveyors, environmental scientists, and forestry technicians. The Bureau of Labor Statistics shows that geographers earn an average of $74,000 per year and the job outlook has remained somewhat steady. Cartographers and photogrammetrists, people who collect, measure, and interpret geographic information for mapping purposes, earn an average of $61,000 per year, and the job outlook is expected to grow 29%, which is much faster than normal. To get a job in this profession, you need a bachelor's degree in cartography, geography, or geomatics (a fairly new discipline increasingly found at universities, including Utah Valley University). To become a satellite engineering analyst, you need a bachelor's degree in engineering, physics, or related discipline.

Modular Activities

Discussions

❖ Post an original discussion in the online discussion board on the following topic: Some say there are situations where location tracking goes too far either in what is tracked, how long information is stored, or based on who has access to the information. What are your thoughts and why? (200 words minimum). Next, comment on the post(s) of a minimum of one other student in a thoughtful and academic way that enhances the conversation. See rubric for grading and assessment measures.

❖ Watch the TEDxUVU video, Storing and Tracking Your Life using GIS Technology, at https://www.youtube.com/watch?v=PjA8y-MioM0. Post an original discussion about your thoughts on a content topic, for example, do you think this technology improves the political process, does mapping crime data make people safer or increase social profiling, will society become more or less homogeneous using GIS capabilities? In your initial thread, do addition research on your topic to support your opinion. Next, comment on the post(s) of a minimum of one other student in a thoughtful and academic way. Use information from the book or additional research to support your comment and add to the discussion.

❖ Take a 16-minute tour of the International Space Station at http://www.n2yo.com/satellite-news/Narrated-tour-of-the-International-Space-Station/3463.

What kind of research is being done? Have students find something of interest to them in the video and provide an update. They could include the history of the topic, what countries are involved in the project, what has been found so far, and what are future implications.

❖ Listen to the broadcast of "U.S. Navy Brings Back Navigation by the Stars for Officers" on NPR at http://www.npr.org/2016/02/22/467210492/u-s-navy-brings-back-navigation-by-the-stars-for-officers. Ask students what they think about military personnel learning an ancient form of navigation after relying on state-of-the-art technology for generations. Do they think this practice will continue? Are there other traditional skills that would be useful to know if technology doesn't work—in the military, home, and society? What skills are we currently using (e.g., handwriting) that are also being phased out? Is this a good thing for our society or are there unseen consequences?

❖ Have students visit a lifestyle segmentation site, such as Claritas at https://www.claritas.com/sitereports/Default.jsp. They can review what information is available on consumers and how it is obtained. Have them first write a paragraph on their findings and what they think of this information—both the pros and cons. Then, they can make a comment on another students posting but providing additional information they have found.

❖ Have students try to determine all of the databases in which their names may belong, how this information may be used, and whether there are any privacy laws that protect their information.

Tests

❖ Online graded quiz on overall chapter content, written in multiple-choice format. When submitted, we recommend giving the correct answer along with the page number in which it is found for questions students did not answer correctly.

Research

❖ NASA for students. View this site for assignment ideas: http://www.nasa.gov/audience/forstudents/postsecondary/index.html

❖ In the section on Satellites, a website is highlighted called "Stuff in Space." This is an interesting, interactive website that is updated daily. Have your students spend time on it and identify a few items they find of interest and write a research paper on them, including their purpose, history, nationality, and contribution to science.
   ➢ http://stuffin.space/

❖ Have students write a modern version of the book, *1984*. They could include current political situations as the backdrop, with real, leading-edge technologies as the main characters. It would be fiction but based in reality. The technology uses would be plausible based on their research of them and creative ideas of how they could be used.

* Have students research and discover what satellites are now orbiting earth in one of the three orbits (high, medium, and low). They can include the history, purpose, and future implications of these satellites.

* Have students research the Golden Record aboard Voyager 1 and translate the pictures. Also have them explain how scientists attempted to include and explain items that may end up in the hands of aliens who do now know how we communicate.

* Have students locate and visit a ViewSpace location at: http://hubblesite.org/explore_astronomy/visit_viewspace/. For example, Utah has two locations (Salt Lake City and Provo) where the public can visit for free or a small fee. Have students write a two-page paper on what they saw, the science behind the technology, and how the information can be used to improve the lives of people on Earth (e.g., other planets to live on, information to help improve the environment, warnings about threats to Earth such as meteors).

* Have students do research at the CDC or similar to find GIS maps. Have them write a two-page narrative about a map they select, including source of data, what is being tracked, and what they have learned by viewing the map. One good source is the maps to track flu activity in the United States. They can view a series of the maps and determine if the flu is spreading, already widespread, regional, or local. It is found at http://www.cdc.gov/flu/weekly/usmap.htm and includes an arrow to view previous weeks. This allows students to view it over time.

* Research one of the following higher education academic areas:
  - Earth science (geology, oceanography, meteorology)
  - Geography
  - Geomatics

This might include review of offerings or degrees at your institution or at another location, contacting a professional/faculty in the field, assessing higher education level academic events or educational resources, and so on. Submit a 200 words minimum summary of your findings. Make sure to include your source(s).

Design and/or Build Projects

* Amazing Space is for grade 10–12 but can be adapted to a higher level (or left as it is as an introduction to estimating and sampling). It is an online exploration called Galaxy Hunter and uses real data from the Hubble Space Telescope: http://amazingspace.org/resource_page/235/math_estimating_sampling/topic#educator_tab

❖ Have students create their own golden record. What major scientific breakthroughs would they include, how would they record the information (e.g., a record still or some other medium), and how would they let other civilizations know how to retrieve the information from their device?

❖ ESRI, a major GIS software company, is licensed at many schools and universities. Contact your IT administrator to see if your school has a license that students can access. Or, go the ESRI student software page at http:// www.esri.com/landing-pages/education-promo for more information. You can get a free trial of Lynda.com, ArcGIS Essential Training at https://www. lynda.com/ArcGIS-tutorials/ArcGIS-Essential-Training/180108-2.html? utm_source=bing&utm_medium=cpc&utm_campaign=11-US-Search-Dev-ArcGIS&cid=11-us:en:ps:lp:prosc:s50:1963:all:bing:mbm-gis_book&utm_ content=%7Bcreative%7D&utm_term=%2BGIS%20%2BBook. Have students complete this tutorial or others you may find online.

❖ Peruse the website http://nasawavelength.org/ and find a topic of interest that relates to the use of location and tracking technology. Create a poster that explains your chosen technology (earth structure, earth and space science, engineering and technology) at a high school level. You will be assessed on the depth, accuracy, and professionalism of the information. You must cite your source(s).

❖ There is a beta site for educators to help students understand satellite data and images as well as related content, http://nasawavelength.org/

❖ The USGS archive contains archives of Landsat data located at http://landsat. gsfc.nasa.gov/?page_id=2370. Have students browse through the resources to find images in which they are interested. Have them create a PowerPoint, poster, or online media presentation of the topic, including pictures, the history, uses, and future implications.

❖ Use GIS to review the 2012 presidential election at http://www.arcgis.com/ home/item.html?id=ea94ccd1385747d4b0348769dc5e51c9. Students can click on the map to reveal a world map. The Election 2012 layer is already turned on. Students can zoom into the United States. When they click on a state it reveals population information, mediate age, percent male/female, and so on. When you hover over election 2012, the five options appear beneath: show legend, show table, change style, filter, and more options. Have students work with this map and develop a narrative about their own state or region of the country.

❖ Download free Landsat imagery at the following site: http://gisgeography.com/ usgs-earth-explorer-download-free-landsat-imagery/.

❖ Google Earth has a web page that includes beginner, advanced, and 3D buildings tutorials at http://www.google.com/earth/learn/. Have students select and complete one of the tutorials based on their interest and skill level/.

Assessment Tasks

❖ Set up a free account on Open Street Map, and open source mapping software at http://www.openstreetmap.org/#map=5/46.950/20.083. Search for locations in your area. See if you can edit one of the areas (edit is in the upper left). Write a 200-word summary of your findings and thoughts about this resource.

❖ Visit the USGS Earth Explorer found at http://earthexplorer.usgs.gov/. Type in your address and hit "show." Now click on the address/place to add coordinates to the area of interest control.

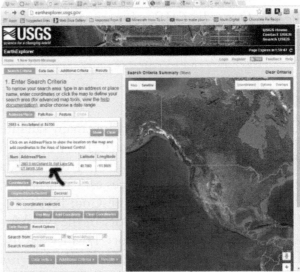

Note your location coordinates (degree/minute/second). Go to the "data sets" tab, search for "LIDAR," and click to ensure it is chosen. Lidar, an acronym for Light Detection and Ranging, is a surveying technology that measures distance by illuminating a target with a laser light. Now, go to the "results" tab. Click on the LIDAR image on the left. This will show you the topography of your area. Save a copy of the image and submit along with the details on your location coordinates.

❖ Location and tracking technology is used by law enforcement. Choose an area in which it is used by local or national law enforcement such as vehicle tracking, suspect tracking, phone tracking, or the like. Find a resource or article that addresses one of these areas and write a critical assessment of the article. Submit the following: (a) article title, location and author; (b) short summary of the article (200 words minimum); (c) an assessment of the usefulness, authority, and coverage of the article (200 words minimum).

❖ Location and tracking technology is used by the US Department of Agriculture in a variety of agencies such as forest service, agricultural research, risk management, and resource conservation. Find a resource or article that addresses one of these areas and write a critical assessment of the article. Submit the following: (a) article title, location, and author; (b) short summary of the article (200 words minimum); (c) an assessment of the usefulness, authority, and coverage of the article (200 words minimum).

Terms

ANS: A navigation system that does not rely solely on GPS.

Eccentricity: The shape of a satellite's orbit (e.g., circle or oblong).

GIS: Geographic information system is a computer-based software system used to display information visually on a map.

GPS: Global Positioning System is a constellation of satellites part of the global information infrastructure used to determine a receiver's position on earth.

Inclination: The angle of a satellite's orbit from the equator.

Landsat: Group of satellites taking pictures of the Earth using remote sensing.

Raster Data: In GIS, representation of the land in a grid of cells (also known as pixels).

Vector Data: In GIS, representation of the world using points (e.g., city location), lines (e.g. highway and roads), and polygons (e.g., counties and land parcels).

Bibliography

Cain, F. (2013, October 24). How many satellites are in space? *Universe Today*. Retrieved from http://www.universetoday.com/42198/how-many-satellites-in-space/

ESRI. Coordinate systems, map projections, and geographic (datum) transformations. Retrieved from http://resources.esri.com/help/9.3/arcgisengine/dotnet/89b720a5-7339-44b0-8b58-0f5bf2843393.htm#TypesCoordSys

Davis, G. (2011, June). History of the NOAA satellite program. *NASA*. Retrieved from http://goes.gsfc.nasa.gov/text/history/History_NOAA_Satellites.pdf

DeMers, M. N. (2003). *Fundamentals of geographic information systems* (2nd ed.). Hoboken, NJ: John Wiley & Sons.

Drummond, K. (2012, June 13). When GPS goes down, pentagon still wants a way to fight. *WIRED*. Retrieved from https://www.wired.com/2012/06/darpa-gps/

ESRI. ESRI Releases Drone2Map for ArcGIS. (2016, February 24). *ESRI* Retrieved from http://www.esri.com/esri-news/releases/16-1qtr/esri-releases-drone2map-for-arcgis

European Space Agency. (n.d.). Hubble overview. *European Space Agency*. Retrieved from http://www.esa.int/Our_Activities/Space_Science/Hubble_overview

Federal Communications Commission. (2016). How do satellites work? *FCC Satellite Learning Center*. Retrieved from http://transition.fcc.gov/cgb/kidszone/satellite/kidz/into_space.html

Franzen, C. (2015, July 8). See all the satellites and space junk circling Earth in real-time. *Popular Science*. Retrieved from http://www.popsci.com/now-you-can-see-all-space-junk-floating-around-earth-real-time

Gilman, L. (n.d.). Reconnaissance satellite. *Encyclopedia.com*. Retrieved from http://www.encyclopedia.com/topic/reconnaissance_satellite.aspx

GPS.gov. (n.d.). The global positioning system. Retrieved from GPS.gov [it is maintained by the National Coordination Office for Space-Based Positioning, Navigation, and Timing, so should that go here instead?] http://www.gps.gov/systems/gps/

Hackbarth, G., & Mennecke, B. (2005). Strategic positioning of location applications for geo-business. In J. B. Pick (Ed.), *Geographic information systems in business* (pp. 198–210). Hershey, PA: Idea Group Publishing.

Hanewicz, C. (2012). Geographic information systems and the political process. Presented at the Western Political Science Association 2012 Annual Meeting, Portland, OR. Retrieved from https://wpsa.research.pdx.edu/meet/2012/hanewicz.pdf

Harris, R., Sleight, P., & Webber, R. (2005). *Geodemographics, GIS and neighbourhood targeting*. Chichester, West Sussex, England: John Wiley & Sons, Ltd.

XSENS. IMU inertial measurement unit. Retrieved from https://www.xsens.com/tags/imu/

Landsat Science, NASA. The Landsat program. Retrieved from http://landsat.gsfc.nasa.gov/

Landsat Science. (2016, December 10) About. *NASA*. Retrieved from http://landsat.gsfc.nasa.gov/?page_id=2

Lockheed Martin. (2013, March 19). Lockheed Martin-built infrared surveillance satellite launched successfully *Lockheed Martin*. Retrieved from http://www.lockheedmartin.com/us/news/press-releases/2013/march/319-ss-sbirs.html

Longley, P. A., Goodchild, M. F., Maguire, D. J., & Rhind, D. W. (2015). *Geographic information science and systems*. Hoboken, NJ: John Wiley & Sons.

Lutwak, R. (n.d.). Micro-technology for positioning, navigation and timing (Micro-PNT). *DARPA*. Retrieved from http://www.darpa.mil/program/micro-technology-for-positioning-navigation-and-timing

Markert, L. R., & Backer, P. R. (2010). *Contemporary technology: Innovations, issues, and perspectives* (5th ed.). Tinley Park, IL: The Goodheart-Wilcox Company.

N2YO.com. (n.d.). Iridium satellites. Retrieved from http://www.n2yo.com/satellites/?c=15

N2YO.com. Most tracked satellites at N2YO.com. Retrieved from http://www.n2yo.com/satellites/?c=most-popular

NASA. (2007, October 10). Sputnik and the dawn of the space age. Retrieved from http://history.nasa.gov/sputnik/

NASA. (2016, March 4). Tracking and data relay satellites. Retrieved from https://www.nasa.gov/directorates/heo/scan/services/networks/txt_tdrs.html

NASA. (2016, March 10). History, near earth network. Retrieved from https://www.nasa.gov/directorates/heo/scan/services/networks/txt_nen.html

NASA. (n.d.). *NASA Orbital Debris Program Office*. Retrieved from http://orbitaldebris.jsc.nasa.gov/index.html

NASA. (n.d.). What is the golden record? *NASA, Jet Propulsion Laboratory*. Retrieved from http://voyager.jpl.nasa.gov/spacecraft/goldenrec.html

National Oceanic and Atmospheric Administration (NOAA). (n.d.). What is remote sensing? Retrieved from the http://oceanservice.noaa.gov/facts/remotesensing.html

NOAA Ocean Service Education. (2008, March 25). Do you know where you are? – The global positioning system. Retrieved from http://oceanservice.noaa.gov/education/kits/geodesy/geo09_gps.html

PBS. What is the purpose of the space station? Retrieved from http://www.pbs.org/spacestation/station/purpose.htm

Pick, J. B. (2008). *Geo-business GIS in the digital organization*. Hoboken, NJ: John Wiley & Sons.

Reynolds, G. (2007). *Ethics in information technology* (2nd ed.). Boston, MA: Thompson Course Technology.

Riebeek, H. (2009, September 4). Catalog of earth satellite orbits. *NASA Earth Observatory*. Retrieved from http://earthobservatory.nasa.gov/Features/OrbitsCatalog/

Roberti, M. (2013, September 17). Has RFID been integrated with GPS? *RFID Journal*. Retrieved from http://www.rfidjournal.com/blogs/experts/entry?10729

Russian Strategic Nuclear Forces. (2013, May 1). *Glonass*. Retrieved from Russian Strategic Nuclear Forces at http://russianforces.org/space/navigation/glonass.shtml

NPR. U.S. Navy brings back navigation by the stars for officers. Broadcast on National Public Radio, Morning Edition, on February 22, 2016. Retrieved from http://www.npr.org/2016/02/22/467210492/u-s-navy-brings-back-navigation-by-the-stars-for-officers.

Tech Republic. (2013, June 25). Where are the jobs in the GIS field? Retrieved from http://www.techrepublic.com/blog/career-management/where-are-the-jobs-in-the-gis-field/

Woodford, C. (2015, September 8). Space rockets. *Explain That Stuff*. Retrieved from http://www.explainthatstuff.com/spacerockets.html

World Atlas. (n.d.). Latitude and longitude facts. *World Atlas*, World Map. Retrieved from http://www.worldatlas.com/aatlas/imageg.htm

Zebra. (2012). Marine terminal automation. Retrieved from https://www.zebra.com/content/dam/zebra/product-information/en-us/brochures-datasheets/misc/marine-terminal-en.pdf

Chapter 8

Autonomous and Semiautonomous Technologies

(Continued)

(*Continued*)

- ❖ The role of autonomous and semiautonomous technology in earth science
- ❖ The science behind the technology
  - ➢ Geographic information Systems
    - ■ Global positioning systems
    - ■ Remote sensing
  - ➢ Obstacle avoiding sensors
  - ➢ Sensing and perception
  - ➢ Mobility and manipulation
  - ➢ Information exchange systems
  - ➢ Built in testing of systems
  - ➢ The physics of autonomous and semiautonomous technologies
    - ■ Mechanics
    - ■ Electricity and magnetism
    - ■ Optics
    - ■ Thermodynamics
- ❖ Career connections
- ❖ Modular activities
  - ➢ Discussions
  - ➢ Tests
  - ➢ Research
  - ➢ Design and/or Build projects
  - ➢ Assessment tasks
- ❖ Terms
- ❖ References

Chapter Learning Objectives

- ❖ Define autonomous technologies
- ❖ Describe unmanned aerial vehicles (UAV, also known as drones) along with social and legal considerations of their use
- ❖ Distinguish semiautonomous and autonomous technologies
- ❖ Summarize international defense use of semiautonomous and autonomous technologies
- ❖ Examine the role of mechatronics and robotics in semiautonomous and autonomous technologies

Overview

What if our roads were built with autonomous and robotically controlled equipment? This could include wheel loaders and trucks, excavators, bulldozers, graders, scrapers, and the like. It is something the Center for Earthworks Engineering Research (CEER) at Iowa State University (ISU) has been researching. As they see it, use of such equipment can help reduce construction costs across the country, currently upward of $180 billion a year (Hallmark, 2015).

How about an autonomous oil skimmer? Not only could it skim oil from a given area (such as where the oil spill occurred in the Gulf of Mexico in 2010, which is the largest accidental spill in world history), it could also monitor the thickness of the oil in real time and direct its attention to where the thickest amounts lay. This system could also statistically track its recovery rate and report real time performance monitoring. This is a system the Bureau of Safety and Environmental Enforcement are working on as of March 2016 (Johnson, 2016). In almost any area we can likely find uses for autonomous or semiautonomous technologies.

Unmanned Systems

There are a variety of unmanned systems (UMS) that will be discussed in this chapter that vary in their sizes, intricacy, levels of autonomy, and area of use. Generally speaking, though, UMS receive instructions of some sort from a person which it then accomplishes either with or without further human to robot interactions. They can be used in situations of varying complexity and/or environmental difficulty which will be factors for its ultimate level of autonomy.

When it comes to size, there are UMS, known as micro systems, that weigh under eight pounds yet there are other large systems that weigh over 30,000 pounds. Researchers at Korea's Seoul National University and Harvard's Wyss Institute for Biologically Inspired Engineering, for example, have emulated the biomechanics necessary for making a 0.002 oz. microbot be able to walk and jump on water without breaking the surface tension, just as the water strider insect can. They also have varying levels of cognizance (observation, perception, and knowledge) which might range from observing external data to processing of information to make actionable decisions and performing tactical behaviors. A majority of these systems, however, work in collaboration or cooperation with human operators via sharing of data remotely using technologies such as Common Relative Operational Pictures (CROP), where an identical display of relevant information is shared across systems and participants. It is through computer-based systems that operations are coordinated for mission plans, maneuvering, communication, and the like, with the human being the superior decision maker and the equipment the subordinate.

Let's consider the UMS Integrated Roadmap for fiscal year 2013–2038 written by the United States Department of Defense. Obviously their focus will be on battlefield capabilities but the general idea of varied types and styles of UMS, even in this single context, is evident. It includes not only air but also maritime (water) and ground (land). Of particular interest when it comes to the technologies used are: interoperability and modularity; communication systems, spectrum, and resilience; security (research and intelligence/technology protection (RITP)); persistent resilience; autonomy and cognitive behavior; and weaponry. The Department of Defense UMS Funding scheduled for 2014–2018 as follows (Table 1):

In the abovementioned table, RDTE stands for Research, Development, Test, and Evaluation. Proc stands for procurement, and OM stands for operations and

Table 1.   DoD Unmanned Systems Funding ($ mil/PB14)

FYDP		2014	2015	2016	2017	2018	Total
	RDTE	1189.4	1674.0	1521.4	1189.4	1087.9	6662.2
Air	Proc	1505.5	2010.2	1843.5	1870.7	2152.8	9382.7
	OM	1080.9	1135.2	1102.7	1156.9	1178.5	5654.1
Domain Total		3775.9	4819.4	4467.6	4217.0	4419.3	21699.1
FYDP		**2014**	**2015**	**2016**	**2017**	**2018**	**Total**
	RDTE	6.5	19.1	13.6	11.1	10.6	60.9
Ground	Proc	6.5	27.9	30.7	42.6	55.4	163.1
	OM	0.0	0.0	0.0	0.0	0.0	0.0
Domain Total		13.0	47.0	44.3	53.7	66.0	223.9
FYDP		**2014**	**2015**	**2016**	**2017**	**2018**	**Total**
	RDTE	62.8	54.8	66.1	81.0	87.2	351.9
Maritime	Proc	104.0	184.8	160.1	158.1	101.1	708.2
	OM	163.4	170.3	182.4	190.5	193.6	900.2
Domain Total		330.2	409.8	408.6	429.7	381.8	1960.2
FYDP		**2014**	**2015**	**2016**	**2017**	**2018**	**Total**
All	RDTE	1,258.7	1,747.9	1,601.1	1,281.5	1,185.7	7,075.0
Unmanned	Proc	1,616.0	2,222.9	2,034.3	2,071.4	2,309.3	10,253.9
Systems	OM	1,224.3	1,305.4	1,285.1	1,347.4	1,372.1	6,554.3
Domain Total		4,119.1	5,276.2	4,920.5	4,700.4	4,867.1	23,883.2

Source: DOD. http://www.defense.gov/Portals/1/Documents/pubs/DOD-USRM-2013.pdf

Note: Ground operations and maintenance (OM) is funded with overseas contingency operations funding.

maintenance. As you can see from above, while air is the leading expense at 91%, maritime is approximately 8%. While only 1% of the current budget is allocated to ground systems, that number increases steadily each year while air and maritime both decrease.

UMS used in international defense (2013) include the following:

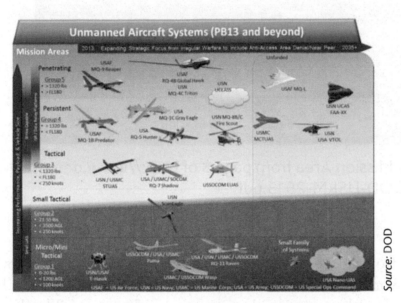

Source: DOD

Figure 1. UAS (PB13 and Beyond). (From page 6 of http://www.defense.gov/Portals/1/Documents/pubs/DOD-USRM-2013.pdf

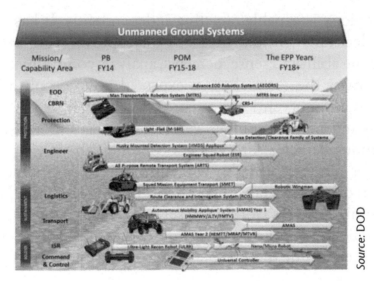

Source: DOD

Figure 2. UGS by Mission/Capability Area. From page 7 of http://www.defense.gov/Portals/1/Documents/pubs/DOD-USRM-2013.pdf

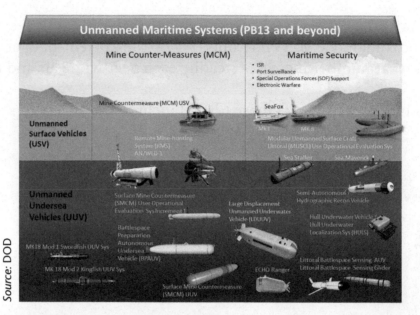

Source: DOD

Figure 3.   UMS by Mission Area from page 8 of http://www.defense.gov/Portals/1/ Documents/pubs/DOD-USRM-2013.pdf

As you can see from above, the systems are vast and varied. Each, however, serve a purpose and largely work together to create a complete overarching system. They allow for strategic planning and implementation that was previously not possible. (Figure 1, 2, & 3)

Autonomous Technologies

Autonomous technology is capable of acting independently. It has the ability to govern itself, making decisions without the need for approval, verification, or clearance. This entails having the ability to sense, perceive, analyze, communicate, plan, and make decisions. In these cases the technology is expected to accomplish its defined-scope mission without human intervention.

An example would be autonomous vehicles such as Google and Fiat Chrysler or Nissan autonomous vehicles. Proponents contend self-driving technology could drastically reduce the number of highway fatalities, since computer systems don't get distracted, text, or in other ways take their attention from the road. National Highway Traffic Safety Administration is working to set up formal guidelines for these types of vehicles to ensure reliability and consumer safety. These guidelines would incorporate issues such as whether or not these vehicles need steering wheels or foot pedals for manual override.

Vehicles are not the only machines potentially going autonomous though. The United States Marine Corps and the Department of Defense are now developing and testing autonomous robot-drone teams for potential future use on the battlefield and beyond. One aspect of these developments is known as the Unmanned Tactical Autonomous Control and Collaboration (UTACC). The purpose of UTACC is to address issues relating to massive information flow coming in to operators and users of these technologies at a level which is too cumbersome to handle. UTACC, which could involve any number of unmanned vehicles working in cooperation with each other, works to enhance mission/goal accomplishment while at the same time limiting the information load and requisite cognitive stress on the human defense teams by working collaboratively with the team in a fairly autonomous fashion. This is accomplished using distributed Real-time Autonomously Guided Operations eNgine (DRAGON) software. Consider the US Marines who are working on a ground robot with an attached air robot. Essentially, the ground vehicle launches an air vehicle to fill in gaps in its sensor imagery. This robot/drone team then sends back information to an operator who can determine how to best use the information (Gallagher, 2016).

Other areas of international defense are assessing or developing similar technologies. The US Army, for example, is looking toward fully autonomous tactical vehicles. Mark Mazzara, robotics interoperability lead for the Army's Program Executive Office—Combat Support and Combat Service Support at Detroit Arsenal, Michigan describes the incorporation of these technologies in a three phased approach: (1) driver-safety and driver-assist technologies that are upgrades to vehicles, (2) basic autonomy capabilities, and finally (3) a fully autonomous tactical vehicle (Ferdinando, 2015). Sea Hunter, a robotic warship designed, developed, and built by the Defense Advanced Research Projects Agency (DARPA) via their antisubmarine warfare continuous-trail unmanned vessel (ACTUV) program is a similar technology. Sea Hunter has revolutionized autonomous navigation and human-machine collaboration and could ultimately change the nature of US maritime operations (Pellerin, 2016).

Space exploration is yet another area being impacted by unmanned vehicle technologies. Curiosity Mars Rover which was launched by the US National Aeronautics and Space Administration (NASA) team in November of 2011 is one such example. It landed successfully on Mars in August of 2012 with the goal of determining if Mars was ever able to support microbial life. The rover is capable of travelling about 30 m (98 ft.) per hour based on external factors such as terrain and internal factors such as power levels. Its role is to study rocks, soils, and the local geologic setting in order to detect chemical building blocks of life (e.g., forms of carbon) on Mars and assess what the Martian environment was like in the past. To communicate back to earth it uses the NASA Deep Space Network (DSN) which is an international network of antennas that provide the communication links between the scientists and engineers on Earth to the missions in space and on Mars (Nasa Jet Propulsion Laboratory, 2016).

Semiautonomous Technologies

Semiautonomous technology acts independently to some degree, but requires input from another source, often in the form of human intervention. An example of this type of technology can be found in Volvo S90 cars which now come equipped with Park Assist Pilot which lets you know if there's a suitable space and can reverse into a parking bay or parallel park in a gap just 1.2 times the size of the car. The driver handles the gears, brake, and accelerator and the car's automated system handles the navigation in to the parking spot. Actually, a number of cars are now semiautonomous, including Tesla Model S, BMW 750i, Infiniti Q50S, and Mercedes-Benz S65 AMG to name a few.

Today, semiautonomous systems are used as well for dangerous underground mining work, to clear landmines, provide surveillance, convoy supplies, and acquire targets, among many other things. Byrnecut, Australia's largest underground mining contractor has used Caterpillar semiautonomous Load Haul Dump (LHD) machines for years for its efficiency, effectiveness and safety. Robotic Mine Detection Systems (RMDS) used by the military give us unprecedented capability to detect, mark, and neutralize explosive hazards at far less risk. These machines can function with standard manned operation, teleoperation with cruise control, and semiautonomous path following. There is even a semiautonomous pipe bomb end cap remover which can be used by first responders in bomb threat situations to keep humans at a safe distance while dismantling the bombs and collecting video and physical evidence.

Even medical fields are increasingly using semiautonomous technologies. One area of rapid growth is in the use of robotics in surgery. Robot surgeons, after all, more consistent in their methods and don't tremor or shake. It is unlikely; however, that we will see robot doctors replace our human surgeons any time soon. Instead, these robotic technologies can be used to assist doctors in very difficult or precise procedures.

Automation Technology

As was discussed in the advanced manufacturing and production chapter, automation and control technology play a strong role in state-of-the-art manufacturing. It is directly tied to many industrial processes. Automated systems are increasingly common because of their consistency, safety, control, and effectiveness. Automation plays equally relevant and important roles in other areas as well, including transportation, medical technology, robotics, and more for some of the same reasons.

What exactly is automation technology though? To answer that, we need to start by defining automation. Automation, according to the Merriam Webster dictionary (2016), is "(1) the technique of making an apparatus, a process, or a system operates automatically, (2) the state of being operated automatically; and (3) automatically controlled operation of an apparatus, process, or system by mechanical or electronic devices that take the place of human labor." Following that, automation technology is automated systems used to produce goods or services. Often it is classified into the areas of manufacturing and service. These types of systems are used for a number of reasons. Some more common reasons are because there is high risk of injury or fatigue to humans, a shortage of labor, a high cost of labor, a need for very high speed production, or quality control.

In automation technology there are multiple types of automation systems that can be used. They range from permanently fixed systems which are made for a very specific purpose (such as completing a fixed sequence of operations in a consistent fashion) to completely flexible or programmable automation that can be used for any multitude of tasks. Generally speaking though with greater customization capability comes lower overall production volume. The other issue is that flexible automation generally requires more manpower and human skills than a completely automated fixed system. Imagine a fixed system whose sole purpose is to create a specific widget. This is all it does and it does it very well and very quickly. While it might need occasional hardware repairs or general maintenance, it pretty much just does its job on its own without any intervention. All its parts have a very specific purpose. Now imagine you have a flexible system where you can adjust the size, shape, and even functionalities of the widget. Someone has to tell the system what it is to create and how. The system, as well, has to have enough complexity in its structure and mechanisms to handle these new requirements. In this case it is quite possible that there will be parts of the system that are not being used and other parts of the system that are being stretched to their maximum capacity. Neither system is necessarily better than the other, they just serve different situations more effectively. A part of the consideration of which system is best for a specific company or entity has, in part, to do with its intended production as single jobs, batch production, or mass production. If you are mass producing many identical items then fixed automation may be the way to go; if you are producing low volumes or even specific items (customization) then flexible automation, if any, is likely your best choice. Somewhere in the middle of these two is programmable automation where you have some flexibility but also a system intended primarily for batch production.

Unmanned Aircraft Systems

Unmanned Aircraft Systems (UAS), as the name implies, are not manned by humans. These systems are also known as Unmanned Aerial Systems. A UAS system includes not just the aircraft but also other elements such as ground control or communication systems. They are the counterpart to manned aircraft. As noted generally about UMS earlier, UAS also come in a variety of shapes and sizes to serve very diverse purposes.

The Federal Aviation Administration (FAA) defines UAS as follows:

A UAS is the unmanned aircraft (UA) and all of the associated support equipment, control station, data links, telemetry, communications and navigation equipment, etc., necessary to operate the UA.

The UA is the flying portion of the system, flown by a pilot via a ground control system, or autonomously through use of an on-board computer, communication links, and any additional equipment that is necessary for the UA to operate safely. The FAA (2016) issues an experimental airworthiness certificate for the entire system, not just the flying portion of the system.

Unmanned Aerial Vehicle

The terminology in this area is evolving, but generally speaking the term Unmanned Aerial Vehicle (UAV) or similar (unmanned aerospace vehicle, uninhabited aerial vehicle, unmanned air vehicle, unmanned airborne vehicle) has no pilot onboard. The vehicle can be a remote controlled aircraft or can fly autonomously based on either preprogrammed flight plans or dynamic automation systems. This term was more commonly used during the earlier development years of UAS systems of the late 1990s.

Drones

Drone systems, according to Merriam Webster (2016), are "an unmanned aircraft or ship guided by remote control or onboard computers." While industry may prefer the term UAV, drone is what is used by the public at large to mean the same thing. That said, there are some proponents who say drones should include not just air vehicles but also land or sea vehicles that are autonomously or remotely guided (Maartens, 2015). Consider, for example, media and marketing use of the term "submarine drone" or "water drones", neither of which, generally speaking, fly at all. In some cases it gets muddy definition-wise when we consider systems that indeed travel by air but can also travel by land or sea. One such example is the Parrot Minidrone which can do all three at the low cost of under $200 (Moynihan, 2015). There are others that think drone systems must be able to fly autonomously (Villesenor, 2012). At the time of this writing, however, the leaning is toward the definition of drone being what would equate to a UAV. For the purposes of this book, the terms will be used interchangeably.

Quadcopters

Quadcopters are simply multirotor helicopters that are propelled by four rotors. They are also sometimes called quadrotors or a quadrotor helicopters. While in almost all cases a quadcopter is an UAV it does not necessarily have to be. It is possible to have a manned quadcopter, although quite uncommon due to safety and the difficulty of scaling a quadcopter to the size necessary to carry human passengers. An added clarification, while almost all quadcopters are UAV not all UAV are quadcopters. Some UAV look more like planes, for example. (Drone & Quadcopter, 2015).

History of UAS

While pilotless aircraft has been around since World War I, the first large-scale production, purpose-built drone, known as the Radioplane, was designed by Walter Righter and then purchased and refined by Reginald Denny. At its core

© Margo Harrison/Shutterstock.com

was a radio control system made by Kenneth Case in 1937. Righter, Case, and Denny worked together to build various models which were later purchased, primarily, by the US military (Naughton, 2005).

Nowadays UAS come in many shapes and sizes. Some drones such as the militaries Predator and Reaper are as large as a full sized jet and is most often used for similar purposes, namely surveillance and combat. Other drones are small enough to fit on your desk and can be used for agricultural purposes, search and rescue, surveillance, or even delivering goods.

In the past, drone technology was only affordable by large and affluent groups, basically limited to governments and militaries. Recently however, due to technological advancements, and the miniaturization of crucial drone technology, drones in a simple yet still capable form, have become affordable and usable to the public. Their use been booming worldwide and now includes categories such as military, commercial, and recreational use. While their functionality and associated costs range greatly, their relevance is apparent in every area.

Military

In the military, UAS play a number of roles. They may be used for reconnaissance (observation, battlefield intelligence), logistics (delivering cargo), target or combat (air strikes, target acquisition). Today, drones are central to US national security strategy. They range from small hand-thrown Ravens to the US air force's RQ-4 Global Hawk, the largest in the force.

If we consider military UAS funding then we quickly can see their relevance. For the fiscal year 2017, the Department of Defense asked for an acquisition budget total of $183.9 billion. Almost one billion would go toward the Navy's MQ-4C Triton--$357 million in research, development, test, and evaluation; and another $619.7 million toward procurement (Office of the Under Secretary of Defense, 2016). For the same project, in 2016, almost the same amount was spent—$378 million for research, development, test, and evaluation; and another $548 toward procurement (Office of the Under Secretary of Defense, 2015). MQ-4C Triton is intended to provide persistent maritime intelligence, surveillance, and reconnaissance capability.

UAS (and the drones that are a part of them) have their longest history in military applications. Early drones include the Firebee drone which was used during Vietnam War to fly over North Vietnam. The aircraft was about the size of the current Predator and was originally used for day reconnaissance. Eventually, though, the Firebee was used for night photos, signal and communication intelligence, leaflet dropping, and surface-to-air missile radar detection (Garamone, 2002). The Firebees included the Ryan 147, AQM-34 and the Lightening Bug. Each was launched from the wings of a large aircraft. They flew preprogrammed routes and could also be controlled by officers onboard the originating aircraft.

As UAS technology progressed in the military it continued to be refined and strengthened. Consider the Predator drone for example. This drone is used for armed reconnaissance, airborne surveillance, and target acquisition. Its wingspan is 55 ft., length 27 ft., and height 7 ft. It can travel up to 770 mi. at an average speed of about 84 mph. The cost for four aircraft with sensors, a ground control station, and satellite link is $20 million in 2009 dollars (U.S. Air Force, 2015).

© Oleg Yarko/Shutterstock.com

Drones are perhaps most known for their use by the military over foreign lands, particularly in Afghanistan and Iraq. These drones are actually controlled by separate location such as via a satellite link from Nellis and Creech USAF base outside Las Vegas, Nevada. Drones are launched from the conflict area but control of the drones are usually handled by a remotely located controllers via video. Actually, there are multiple people involved remotely. Often one person controls (flies) the drone, while another person monitors the cameras and sensors and a third person interacts with any ground troops or commanding officers.

While drones and their parent UAS technology were initially designed by the military for international defense purposes they will undoubtedly find their place in the civilian world. Despite differences in equipment and needs, military and civilian drone-research programs have been closely linked, with advances flowing between the two sides.

Government, Non-military

The US Government also uses drones in nonmilitary settings. The Department of the Interior (DOI), which manages America's natural and cultural resources and includes the Bureau of Land Management (BLM) and National Park Service (NPS), has a drone fleet which consists of helicopters like the Pulse Aerospace Vapor 55TM or the Falcon Hover quadcopter. The Vapor carries items such as airborne lidar (surveying technology that measures distance by illuminating a target with laser light and high definition electro-optical/infrared (EO/IR) brushless gimbals. The Hover is a single person portable system which is made for easy setup and simple logistics (U.S. Department of the Interior, 2016). Some of the many functions drones serve for the United States. DOI include wildfire tracking, wildlife monitoring, hydrology, geological or geophysical surveys, and assessing volcanic activity.

Civilian/Commercial

Often the drones used in the military are of the highest complexity as well as the associated highest costs. Equally, as with many military and defense technologies, they led to growth and development in both commercial and recreational use. In commercial and civilian fields, for example, scientist Emma Marris points to drone use in areas such as wildlife research, hurricane research, climate studies, volcano monitoring, and studies of polar regions. In her eyes drones are becoming an indispensable technology. Researchers across the globe are trying to improve the autonomy, maneuverability, and endurance of UAS.

One commercial use of drones is in agriculture. One such case is AGERpoint, Inc, a Florida-based analytics company that offers remote scanning services to growers of permanent crops. Using fixed-wing UAVs, AGERpoint can scan between 900 and 3000 acres a day. They are able to offer highly accurate three-dimensional modules of the crops on the grower's property (Rusnak, 2015). We also have drones that are used for search and rescue, geological assessment, surveillance, and many other functions.

Recreational

Recreationally, drones are becoming increasingly popular. One Washington Post article notes, "The FAA conservatively estimates that, within a decade, private drones will constitute a $90 billion industry. Already, according to the Consumer Electronic Association, 2014 sales of UAVs are forecast at $84 million and 250,000 units. By 2018, CEA predicts consumer drone sales will approach $300 million and about a million units."(Downes, 2014). It should be noted that these drones are small, consumer-friendly devices that sell for less than $300. They are known as small unmanned aircraft systems (sUAS). In some cases a drone might only cost $20.

One issue, however, is that very few recreational (hobbyist) drone fliers know the basic rules for drone flight as established by congress and in some cases might not even know the rules exist at all. For example, hobbyists cannot fly higher than 400 ft., their sUAS must be in eyesight at all times, they are to stay 25 ft. away from individuals, and are not to fly within 5 mi. of an airport without prior control tower approval. The FAA has actually created an educational program called Know Before You Fly, which can be found at http://knowbeforeyoufly.org/, to try and help educate recreational users. Also, any recreational drones that weigh between .55 pounds (250 g) up to 55 pounds (25 kg) need to be registered with the FAA at https://registermyuas.faa.gov/. If the drone is over 55 pounds then the user must use traditional aircraft registration under 14 CFR Part 47 as can be found at http://www.faa.gov/licenses_certificates/aircraft_certification/aircraft_registry/UA/

Rules, Regulations, and Standards

We just discussed some rules, regulations, and standards for recreational drone users. There are a number of others to take in to consideration as they exist for our protection and the protection of others. The possibilities for UAS use in the civilian world are exciting. However, this quickly rising powerful technology has left not only the country, but the world, unsure as to where their exact place in society lies. Regulations, particularly in the United States, place strict limits on where and how researchers can use the devices. As these rules and standards evolve, drones will likely take to the skies in higher numbers.

UAS of all sizes and shapes have continually developing and refining regulations and policies coming in to play. A recommended place to start looking the current state is at the FAA UAS Regulations and Policies web site at https://www.faa.gov/uas/regulations_policies/. This site includes information for large audiences. An addition made, for example, in mid-2016 was the first ever clarification of educational use of UAS. While not an official policy, it is an interpretation of use in relation to (1) use of UA for hoppy or recreational purposes at educational institutions and community-sponsored events; and (2) student use of UA in furtherance of receiving instruction at accredited educational institutions (Govan, 2016).

Commercial small craft drone operators and others had to wait for the Department of Transportation's FAA to finalize rules for small UA well past when initial demand for such resources came in to play. The first operational rules for routine commercial use of sUAS was put in to place in late June of 2016. It was a first step to opening pathways toward fully integrating UAS in to our nation's airspace (Dorr & Duquette, 2016). These rules are known as part 107 and can be found at http://www.faa.gov/uas/media/Part_107_Summary.pdf. Until that point, launching a commercial drone operation was a challenge and had high barriers for entry. For those wanting to go into commercial drone operation before that point, there were three options: (1) operate without permission from the FAA, (2) apply for a section 333 Exemption from the FAA, or (3) wait for the FAA to finalize formal rules (Part 107) (sUAS News, 2016). With Section 333 exemption, however the volume was so high that the exemptions were taking over six months to be approved. Another major hurdle facing commercial drone operators until the release of part 107 was the legal requirement for an operator's certificate. Operators had to obtain at least a sport-pilot's license and have current flight review, along with a FAA medical certificate. Based on the passing of part 107, however, a number of these hurdles have been removed. This rule permits operators to instead pass an initial aeronautical knowledge test and recurrent test every 24 months and not need the FAA medical certificate. The aircraft must weigh under 55 pounds, be in visual line of sight only, be flown only in daylight or civil twilight (30 minutes before official sunrise and 30 minutes after official sunset), not operate over any persons not directly participating in the operation (nor under a covered structure or inside a covered stationary vehicle), have a maximum ground speed of 100 mph, fly at a maximum altitude of 400 ft. above ground level, and have minimum weather visibility of 3 mi. from the control station though.

Privacy and Surveillance Regulations

It should be noted that UAV operators of all sorts must comply with currently existing laws and regulations as well at a national and state wide level. So, if privacy is a concern, then local and national rules apply.

While there are no federal rules explicitly govern when a drone is trespassing or violating privacy, the FAA has created regulations that limit their use in some public places. For example, the aircraft must be kept in line of sight of the pilot, and cannot fly near stadiums, large crowds, or airports. For statewide regulations the situation varies rather significantly based on where are located. Let's say a recreational user of a drone is flying over a neighbor's house and is recording video of the family in their back yard. Is this permissible? Well, it depends. As Farkas (n.d.) of Nolo points out, "Different states have different laws regarding surveillance. In some states, mere visual recording is not illegal so long as the camera is on your neighbor's property. In other states, visual recording is acceptable but any audio recording is not. And in other states, all forms of recording might face criminal or civil penalties. Generally, any publically viewable areas like back yards are fair game—which is how companies like Google can record their Street View images across the United States." It is somewhat dependent on your reasonable expectation of privacy, which does not usually extend to locations in plain view but could extend to a fenced-in yard or similar.

Utah, as an example, has the following rules relating to privacy violation, in relation to audio at least, under Utah Title 76: 76-9-401 Definitions: For purposes of this part: (1) "Private place" means a place where one may reasonably expect to be safe from casual or hostile intrusion or surveillance. (2) "Eavesdrop" means to overhear, record, amplify, or transmit any part of a wire or oral communication of others without the consent of at least one party thereto by means of any electronic, mechanical, or other device. (3) "Public" includes any professional or social group of which the victim of a defamation is a member.

76-9-402 Privacy violation. (1) A person is guilty of privacy violation if, except as authorized by law, he: (a) Trespasses on property with intent to subject anyone to eavesdropping or other surveillance in a private place; or (b) Installs in any private place, without the consent of the person or persons entitled to privacy there, any device for observing, photographing, recording, amplifying, or broadcasting sounds or events in the place or uses any such unauthorized installation; or (c) Installs or uses outside of a private place any device for hearing, recording, amplifying, or broadcasting sounds originating in the place which would not ordinarily be audible or comprehensible outside, without the consent of the person or persons entitled to privacy there. (2) Privacy violation is a class B misdemeanor

76-9-403 Communication Abuse. (1) A person commits communication abuse if, except as authorized by law, he: (a) Intercepts, without the consent of the sender or receiver, a message by telephone, telegraph, letter, or other means of communicating privately; this paragraph does not extend to: (i) Overhearing of messages through a regularly installed instrument on a telephone party line or on an extension; or (ii) Interception by the telephone company or subscriber incident to enforcement of regulations limiting use of the facilities or to other normal operation and use; or (b) Divulges without consent of the sender or receiver the existence or contents of any such message if the actor knows that the message was illegally intercepted or if he learned of the message in the course of employment with an agency engaged in transmitting it.

As can be found at https://le.utah.gov/xcode/Title76/C76_1800010118000101.pdf. It should be noted though that above only applies to audio, and special attention should be paid to the sentence "wire or oral communication of others without the consent of at least one party thereto by means of any electronic, mechanical, or other device" which means no matter how many people are recorded, only one of them needs to give consent (and if the person who is recording is a part of the party they can give the consent themselves).

In Utah, among many states, acceptable drone use is currently being clarified and defined. In March, 2016, for example, a bill was proposed by Utah State Senator Wayne Harper that would "would establish criminal penalties for misusing UA and empower first responders or law enforcement to "neutralize" a drone. Examples of misusing a drone would include voyeurism, flying them within 500 ft. of a correctional facility, photography near crowds of more than 500 people and flying them within 3 mi. of a wildfire" (Risen, 2016). While it did not pass, the sentiment remains strong for some that regulations such as this should exist, just as sentiment also exists against too much regulation (Romboy, 2016). As US News and World Report states, "lawmakers seeking broader rules in 26 states including California, Florida and Arkansas have passed regulations governing the use of UA, including penalties for voyeurism or privacy violations" (Risen, 2016). Some state level examples currently in place:

* Arizona SB 1449 prohibits certain operation of UAS, including operation in violation of FAA regulations and operation that interferes with first responders. The law prohibits operating near, or using UAS to take images of, a critical facility. It also preempts any locality from regulating UAS.
* Idaho SB 1213 prohibits the use of UAS for hunting, molesting, or locating game animals, game birds, and furbearing animals.
* Indiana HB 1013 allows the use of UAS to photograph or take video of a traffic crash site. HB 1246 prohibits the use of UAS to scout game during hunting season.
* Louisiana SB 73 adds intentionally crossing a police cordon using a drone to the crime of obstructing an officer. Allows law enforcement or fire department personnel to disable the UAS if it endangers the public or an officer's safety.
* Oregon HB 4066 modifies definitions related UAS and makes it a class A misdemeanor to operate a weaponized UAS. It also creates the offense of reckless interference with an aircraft through certain uses of UAS. The law regulates the use of drones by public bodies, including requiring policies and procedures for the retention of data. It also prohibits the use of UAS near critical infrastructure, including correctional facilities. SB 5702 specifies the fees for registration of public UAS.
* Wisconsin SB 338 prohibits using a drone to interfere with hunting, fishing, or trapping. AB 670 prohibits the operation of UAS over correctional facilities (National Conference of State Legislators, 2016).

As can be seen from abovementioned examples, there remain many discrepancies in standards and regulations and there remains much to be clarified as well. Researchers, developers, and scientists have been working to improve the technology to make the devices more agile, more autonomous, and better able to work in groups. As each of these areas are refined it is likely additional rules, regulations, and standards will follow.

International Defense

Earlier in this chapter we talked about the military use of unmanned transportation systems used by the military, when an emphasis on UAV. However, there is also use of autonomous and semiautonomous technology for international defense that does not have to do with transport. The nature of war and the battlefield have undergone significant changes. Part of the change is due to the advances in technology. There have been improvements in weapons, planes, and ships, but advances in communication technology have been just as important. The "fog or war" has been partially lifted as front-line soldiers have transitioned to the new "information warriors." They have real-time data of their environment, including GPS location and continual contact with their colleagues. Another change is how war is fought. Traditionally, there were big battles with clear enemies. Today's battle is more likely to be a low-intensity conflict with an ambiguous enemy. Even if it is clear who the enemy is, war is less likely to involve face-to-face combat. Military personnel can now use computers to control weapons as well as pilotless planes.

It is possible that a number of military functions could be performed by machines acting rather autonomously, without, direct interaction from human operators, in our military future. This could range from simple logistical tasks all the way to potential application of deadly force. One area of discussion relates to the development of LAWS, an area of recent growth as well as discussion.

Lethal Autonomous Weapon Systems

According to The Conversation journal, diplomats meet in early April, 2016 for the United Nations' third Informal Expert Meeting on LAWS, also known as killer robots. Four general views were addressed:

1. Rely on existing laws—The UK's position is that existing international humanitarian law is sufficient to regulate emerging technologies in artificial intelligence
2. Ban machine learning—The French delegation said a ban would be "premature" and that they are open to accepting the legality of an "off the loop" LAWS with a "human in the wider loop." This means the machine can select targets and fire autonomously, but humans still set the rules of engagement.
3. Position 3: Ban "off the loop" with a "human in the wider loop"—The Dutch and Swiss delegations suggested "off the loop" [humans are note kept "in the loop"] systems with a "human in the wider loop" could comply with international humanitarian law, exhibit sufficiently meaningful human control, and meet the dictates of the public conscience. The UK, France, and Canada spoke against a ban on such systems.
4. Position 4: Ban 'in the loop' weapons—Pakistan and Palestine will support any measure broad enough to ban telepiloted drones. However, most nations see this as beyond the scope of the LAWS debate, as humans make the decisions to select and engage targets, even though many agree drones are a human rights disaster (Welsh, 2016).

It was also noted that, "It is generally agreed that LAWS are governed by international humanitarian law. For example, robots cannot ignore the principles of distinction between civilians and combatants, or proportionality in the scale of attack" (Welsh, 2016) and human commanders would have responsibility for their robots.

As Welch points out, there are two key areas where one might want to assert meaningful human control of autonomous weapons. First, in defining the policy rules that the autonomous weapon mechanically follows. Second, in the execution of those rules when firing (Welsh, 2016b). These areas need to be considered not just on a departmental or national level, but on a global level as well.

Fire and Forget Missiles

Fire and forget missiles are ones that do not require guidance after launch. The information about the target is programmed in to the missile before launch, inclusive of things like coordinate, radar measurements, and images. The missile then guides itself with the use of technologies such as gyroscopes, accelerometers, global positioning

systems (GPS), radar homing, and homing optics. One such missile is Britain's Royal Air Force's Brimstone missile.

Brimstone Missile

The Brimstone missile is a fire and forget missile with a high powered guidance system that allows accuracy even on moving targets. It can direct itself using either radar or can be guided by a weapons operator in the air or on the ground via lasers. Added to that, this missile is known to cause low collateral damage. The explosions can be limited to small areas and with less debris (BBC, 2015). Ben Goodlad, senior weapons analyst at IHS Jane's, said the Brimstone missile was the only one being used against Isil that could be either laser-guided onto the target by a weapons operator, or direct itself with radar guidance. He said: "Brimstone is the only weapon currently being used in air strikes that provides a fire and forget capability capable of hitting moving targets" (Farmer, 2014). This particular weapon, among others, is now being assessed for the US Navy as well.

If we look at the House Report 114–102 National Defense Authorization Act for Fiscal Year 2016 items of special interest we will see a number of autonomous or semiautonomous items on the list. These include the aforementioned Brimstone Missile, but also an Antisubmarine Warfare Continuous Trail Unmanned Vessel and unmanned carrier aviation (114th Congress, 2015). There is also development of weaponry such as multiple independently targetable reentry vehicles which are ballistic missile payloads that contain several independent warheads which maneuver using small on board rocket motors and computerized guidance systems that make use of accelerometers and gyroscopes to calculate position, orientation, and velocity of itself and its target without the need for external references or support.

Military Robots

Robots are also now playing a role in the military. For example, the army uses robotic ground systems that haul gear, navigate tunnels and rough terrain, monitor remote areas, capture and transmit images, search for roadside bombs, remove obstacles from roads, and sometimes go where no soldier can safely go. As Maj. Gen. Walter L. Davis, deputy director of the US Army Capabilities Integration Center notes, UMS are able to offer situational awareness, unmanned lethal and nonlethal fires, unattended precision target attack and acquisition, maximum standoff from threats, and can perform unmanned logistics and support services. (Pellerin, 2011).

Consider, as an example, the military's robot pack mule. Defense Advanced Research Projects Agencies (DARPA's) LS3, a semiautonomous Legged Squad Support System, can carry 400 pounds of warfighter equipment, walk 20 mi. at a time, and act as an auxiliary power source for troops to recharge batteries for radios and handheld devices while on patrol. This type of system could significantly impact the issue of loads of equipment necessary to be hauled for soldiers and marines (Cronk, 2012).

Semiautonomous gun/robot

Another area of consideration in relation to international or national defense is the role of semiautonomous weapons that are smaller in scale than missiles and instead serve purposes more similar to those of guns. Consider the Samsung SGR-1. This weapon, a robot of sorts, has heat and motion detectors to identify potential targets up to 2 mi. away. It has a 5.5 mm machine gun and a 40 mm grenade launcher although it is necessary for a human operator to give the go ahead to fire (Prigg, 2014). It is used to patrol the border between North and South Korea, installed by South Korea (Prado, 2015).

Here in the United States, if we chose to use a system such as this, would also need a human operator to give the command to fire since in 2012 the Department of Defense established a five year directive that all weapons in operation and development need to have a human in the loop. This directive have details such as "Semi-autonomous weapon systems that are onboard or integrated with unmanned platforms must be designed such that, in the event of degraded or lost communications, the system does not autonomously select and engage individual targets or specific target groups that have not been previously selected by an authorized human operator." This directive can be found at http://www.dtic.mil/whs/directives/corres/pdf/300009p.pdf

The Role of Autonomous and Semiautonomous Technology in Earth Science

Autonomous and semiautonomous technologies play a significant role in a variety of areas of science and engineering as well in international defense. Let's consider the role of some of these technologies in earth science. UAVs are playing an ever increasing role in geosciences research. They are used in areas such as study of volcanic activity, surveying of natural resources, weather study, identification and tracking of faults, assessment of earthquake damage, and so on.

One such example is the University of Costa Rica's use of an UAV, called a Dragon Eye. Dragon Eye was used to study the Turrialba volcano since 2013. The scientists, via this UAV, are able to better monitor the volcano, understand emissions effects on local ecology and agriculture, and study the dynamics of volcanic plumes in general (Oleson, 2013).

Actually, here in the United States, the US Geological Survey (USGS) even has a specific National UAS project office. This office is "leading the research of UAS technology in anticipation of transforming the Department of the Interior (DOI) approach for collecting remote sensing data. UAS technology is being investigated by monitoring environmental conditions and landscape change rates, responding to natural hazards, recognize the consequences and benefits of land and climate change, conduct wildlife inventories, and support related land management missions. The USGS is teaming with all of the DOI bureaus, academia, industry, state, and local agencies under guidance from the FAA and the DOI Office of Aviation Services (OAS) to lead the safe, efficient, cost-effective and leading-edge investigation of the potential uses for UAS technology in scientific research and operational activities of the Department." (USGS, 2016).

U.S. Geological Survey use of UAS
[image taken from https://www.flickr.com/photos/ usgeologicalsurvey/22358987095 and in the public domain]

Earth scientists are not alone, by the way. Biologists, as well, make active use of UAV systems. In particular they can be used for wildlife management, as well as in the study of agricultural growth and development. UAVs have been used to study large mammals, sea turtles, and many bird populations. According to Mike Hutt of USGS, in 2013, about half of the projects on which his office has collaborated have involved wildlife. He uses the study of cranes as an example, "The cranes spook easily when they're out feeding during the day, making conventional observations from manned aircraft difficult. But by flying a Raven at night over the roosting cranes, we could fly as low as [23 m] above them and with the thermal camera actually . . . do a very accurate count of the number of cranes as they were migrating through the wildlife refuges" (Oleson, 2013).

The Science behind the Technology

In almost all autonomous and semiautonomous systems there is an operator control interface that is used for human interaction. With these, the operator can control, communicate with, transfer data between, and plan actions with the unmanned or self-directed system. In many cases the unmanned or self-directed systems have onboard navigation using a GPS as well.

Geographic Information Systems

Geographic Information Systems (GIS) are discussed in the location and tracking technology chapter. Let's take a few minutes to review this technology again though as it certainly impacts autonomous and semiautonomous technology. GIS are designed to capture, store, manipulate, analyze, manage, and present a great variety of spatial and geographic data. Geologic data then is simply collection of information that describes objects with relation to space, usually using x and y coordinates (grid structure) and longitudes or latitudes.

Note: See the chapter on location and tracking technology to learn more about GIS, GPS, and remote sensing.

Global positioning systems

GPS are actually radio-based satellite navigation systems that supply specially coded satellite signals that can be processed in a GPS receiver, enabling it to compute its position, velocity, and time. Usually, four GPS satellite signals are used to compute three-dimensional positions (latitude, longitude, and altitude) anywhere on earth, at any time and in any weather condition—assuming there is an unobstructed line of sight to four or more GPS satellites. In space there are 24 satellites minimum orbiting the earth at an approximate height of 12,600 mi. which are used specifically for civilian GPS systems via US government (originally the Department of Defense).

Remote Sensing

Remote sensing is how the data is gathered. It is the scanning of the earth (or any object) by a high flying aircraft (including satellites) to acquire information about it. Often these objects are UAV. It is used in many fields, including, but not limited to hydrology, ecology, oceanography, glaciology, and geology.

© Andrey Armyagov/Shutterstock.com

Obstacle Avoiding Sensors

Autonomous and semiautonomous systems are also often equipped with obstacle avoiding sensors which use algorithms to avoid collisions. These are critical for autonomous vehicles such as mobile robots or intelligent vehicles. Obstacle avoidance includes not only obstacle detection but also avoidance. Generally speaking, they sense, then plan, then act. Obstacles might be detected with technology such as sonar sensors (rotating or fixed) or laser sensors. This might include not just obstacles with a visible height and mass, but also things like potholes or dead ends. It also can potentially determine things like turning radius, maximum speeds, or breaking distance. In order to accomplish this, they need to be able to obtain and interpret a variety of information from multiple sources. This might include things like sensing its current position, velocity, acceleration, and surrounding range.

Sensing and Perception

Vision and range sensors play a role in obstacle avoidance, but there are a number of other types of sensors that play critical roles in robotic perception. Some of these include acoustics (sound) and haptics (tactile). As Paul Fitzpatrick (2003) notes, "Perception is key to intelligent behavior. While the field of Artificial Intelligence has made impressive strides in replicating some aspects of cognition, such as planning and plan execution, machine perception remains distressingly brittle and task-specific."

The NASA groups sensing and perception in to the following categories:

❖ 3D Sensing: provides 3D measurements of the environment for mobility and for surface and in space manipulation.

❖ State Estimation: provides multi-sensor, vision-aided pose, and velocity estimation for mobility and for manipulation (both objects being manipulated as well as their corresponding manipulators).

❖ Onboard Mapping: provides terrain maps (topographic and trafficability) and landmark models for surface and above-surface mobility and manipulation.

❖ Object, Event, and Activity Recognition: recognizes natural and human-made objects, natural dynamic events, and human activities near robot systems

❖ Force and Tactile Sensing: senses forces, torques, and contacts of the mobility or manipulation platform with the environment or with other platforms.

❖ Onboard Science Data Analysis: Automated data analysis for decision making.

Each of the abovementioned necessitates sensors, sensing techniques, and algorithms for interpreting incoming data (NASA, 2015, p. 12). The ability to sense and perceive are essential for autonomous and robotic systems. A number of constraints in the use of robotics has to do with the limitations of current technology in these areas.

Note: See chapter on robotics and artificial intelligence to learn more.

Mobility and Manipulation

Mobility consists of the ability to move from one place to another in the environment. This includes flying, walking, climbing, rappelling, tunneling, swimming, sailing, and thrusting. For this to occur, systems must handle problems of scene modeling, classification, and recognition. The system must then be able to integrate this information to avoid collisions, navigate, and learn. NASA (2015, p. 12) groups mobility in to the following categories:

❖ Extreme-Terrain Mobility: provides mobility across terrains with challenging topographies and challenging regolith properties for bodies with substantial gravity.

❖ Below-Surface Mobility: provides access to and mobility below a solid or liquid surface.

❖ Above-Surface Mobility: provides coverage of, access to, and mobility above planetary surfaces.

❖ Small-Body and Microgravity Mobility: provides mobility across surfaces of small bodies or microgravity environments without surface contact.

❖ Surface Mobility: provides efficient mobility across nonextreme terrains or liquid surfaces.

❖ Robot Navigation: provides autonomous and supervised mobility for surface, above-surface, and extreme terrains.

When considering mobility one only has to think of NASA's Mars rovers including Spirit (2004), Opportunity (2004), and now Curiosity (2012). These rovers travel the surface of mars recording information and sending it back to earth. Curiosity, for example, seeks evidence of organics, the chemical building blocks of life.

Manipulation, on the other hand, pertains to making intentional changes to the environment or objects in that environment. This includes moving or handling objects, assembling, excavating, collecting and handling samples, and anchoring among other actions. Often manipulation technology makes use of arms or other limbs, fingers, cables, and scoops.

NASA (2015, p. 12) groups manipulation in to the following categories:

* Manipulator Components: provide key components that impact the design of manipulators to improve their performance, such as actuators, controllers, and lightweight structures.
* Dexterous Manipulation: provides a capability to grasp, change the grasp of, and smoothly articulate objects (e.g., positioning and orienting of objects), as well as manipulate interfaces on a spacecraft.
* Modeling of Contact Dynamics: Relevant to both mobility and manipulation, in particular, for limbed platforms that intentionally make and break contact.
* Mobile Manipulation: provides a capability for coordinating mobility and manipulation to expand the workspace of robotic platforms.
* Collaborative Manipulation: provides a capability to coordinate and jointly handle and manipulate objects using either multiple robots or robot-human teams.
* Sample Acquisition and Handling: provides a capability to extract and handle rock, regolith, or organic samples, at both large and small scales, for resource processing, sample analysis, or sample caching for future analysis or usage.
* Grappling: provides a capability to capture, anchor to, or interface with large structured and unstructured objects that are free-floating in space or on a planetary surface.

The ability to manipulate gives critical capabilities such as positioning of instruments in space, extracting samples, and handling objects in support of missions. Manipulation system metrics include items such as payload capacity, reach, dexterity, speed, accuracy, and repeatability (NASA, 2015, p. 35)

Computational Stereo

Computational stereo, also known as stereovision, stereoscopic vision, or computer stereo vision, is the ability to view the same physical point from at least two different viewpoints which allows depth from triangulation. Much of geometric vision is based on information from two (or more) camera locations. Computational stereo systems must be able to address issues of calibration, matching, and reconstruction.

Note: See chapter on robotics and artificial intelligence to learn more about computational stereo.

Information Exchange Systems

Information exchange is affected by factors such as proprietary interfaces, data, bandwidth, data type, frequencies and settings, and the like. However interoperability necessitates information exchange. Thus, there is desire to create information exchange requirements (IERs) and standards which are consistently used in a stable fashion. As the US Department of Defense (2013, p. 31) notes in their UMS Integrated Roadmap for fiscal years 2013–2038, "Stable IERs that address joint and Service needs, interoperability profiles (IOPs), middleware (which can translate multiple system inputs and outputs), and other areas are needed to reach the necessary level of interoperability across manned and unmanned systems." At this time, however, these requirements and standards do not solidly exist.

US Under Secretary of Defense (USD) has the following vision for interoperability standards and policy that would span all joint, interagency, intergovernmental, and multinational efforts:

> The ability of systems, units, or forces to provide services to and accept services from other systems, units, or forces and to make use of the services, units, or forces; and to use the services so exchanged to enable them to operate effectively together. An example for the use of this policy would be the condition achieved among communications-electronics systems or items of communications electronics equipment when information or services can be exchanged directly and satisfactorily between them and/or their users (U.S. Department of Defense, 2013)

Built in Testing of Systems

Often, these types of technologies have equipment and software embedded in to their systems which can perform tests to verify their mechanical and electrical functionality including areas such as sequencing, data processing, and interpretation. They contain a sort of self-health monitoring system. The practice that makes this possible is system health management. Often these systems use a monitor-and-respond type of approach based on local system thresholds (upper and lower limits). The approach is to use logic or procedures to detect, isolate, verify, and recover from faults.

The Physics of Autonomous and Semiautonomous Technologies

There are a number of areas of physics that directly relate to the development of autonomous and semiautonomous technologies. The mechanics of the systems are critical for functionality. Electricity and magnetism are needed for the functioning of the systems. Thermodynamics help address the state and stability of the system. And optics are necessary for system sensing and perception.

Mechanics

Mechanics is the study of motion and its causes. Classical mechanics is concerned with the set of physical laws describing the motion of bodies under the influence of a system of forces. Within mechanics is the study of kinematics which describes motion, and dynamics which addresses the causes of motion.

Electricity and Magnetism

Electromagnetism is the study of electromagnetic force, a type of physical interaction that occurs between electrically charged particles. Both electricity and magnetism can be static. However, when they change or move together they make electromagnetic waves. These waves are formed when an electronic field couples with a magnetic field. Waves in the electromagnetic spectrum vary in size from very long radio waves the size of buildings, to very short gamma-rays smaller than the size of the nucleus of an atom.

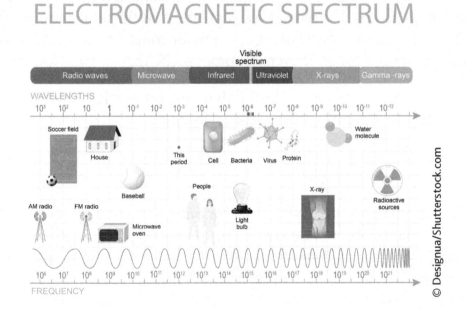

© Designua/Shutterstock.com

Optics

Optics is the branch of physics that involves the behavior and properties of light. This includes its interactions with matter and the instruments that use and detect it. Light, actually, is an electromagnetic wave (see above). Optics most commonly describes the behavior of infrared, visible, and ultraviolet light, such as how the light is dispersed.

Some models used for the study of optics include geometric optics which treat light as a collection of rays or beams, and physical optics which include wave effects.

Thermodynamics

Thermodynamics is the branch of science concerned with heat and temperature and their relation to energy and work. There are three principal laws of thermodynamics. Here is how NASA Glenn Research Center describes them:

❖ The zeroth law of thermodynamics involves some simple definitions of thermodynamic equilibrium. Thermodynamic equilibrium leads to the large scale definition of temperature, as opposed to the small scale definition related to the kinetic energy of the molecules.

❖ The first law of thermodynamics relates the various forms of kinetic and potential energy in a system to the work which a system can perform and to the transfer of heat. This law is sometimes taken as the definition of internal energy, and introduces an additional state variable, enthalpy. The first law of thermodynamics allows for many possible states of a system to exist. But experience indicates that only certain states occur.

❖ This leads to the second law of thermodynamics and the definition of another state variable called entropy. The second law stipulates that the total entropy of a system plus its environment cannot decrease; it can remain constant for a reversible process but must always increase for an irreversible process. (NASA Glenn Research Center, n.d.)

Each law leads to the definition of thermodynamic properties which help us to understand and predict the operation of a physical system.

Career Connections

Some might see autonomous and semiautonomous robotic systems as taking a way jobs, but indeed many jobs will be created in the development, refinement, and implementation of these systems. Faculty at Oxford University completed a study in 2013 which researched the future of employment. As they see it, about 47% of total US employment is at risk of being automated or replaced. Some areas they see as being at a high probably of computerization include office and administrative support, sales and sales-related jobs, service jobs, and production. Those with lower probably of computerization include education, legal, community service, arts, media; computer, engineering and science; and healthcare practitioners and technical, and management, business and financial (Frey & Osborne, 2013).

With the early mentioned, there remain a great variety of job opportunities in the areas of autonomous and semiautonomous technologies. These include physical science, engineering, and computing to name a few. Software developers and engineers, research and development, automation technicians, control specialists, process control engineers, mechanical designers, assemblers, test operators or engineers, instructional operators, site leads, project coordinators, avionics technicians, pilots, operators, and field service technicians are some of the many jobs becoming increasingly available within the area of autonomous and semiautonomous technology development. There are also jobs being created based on the accessibility of this new technology such as in aerial photography, cinematography of extreme sports, land surveying, and outdoor film.

Modular Activities

Discussions

Post an original discussion in the online discussion board on the following topic: Some say there are situations where autonomous technology goes too far based on how it is used and by whom. What are your thoughts and why? (200 word minimum). Next, comment on the post(s) of a minimum of one other student in a thoughtful and academic way that enhances the conversation. See rubric for grading and assessment measures.

* Web search the autonomous electric minibus known as Olli which is a 3D printed vehicle that was designed as a 12-seat self-driving system. What are your thoughts about Olli and why? (200 word minimum). Next, comment on the post(s) of a minimum of one other student in a thoughtful and academic way that enhances the conversation. See rubric for grading and assessment measures.
* Review the summary of small unmanned aircraft rules (part 107) at http://www.faa.gov/uas/media/Part_107_Summary.pdf. Which aspects do you see as the most important and why? Are there things you feel should be changed? (200 word minimum). Next, comment on the post(s) of a minimum of one other student in a thoughtful and academic way that enhances the conversation. See rubric for grading and assessment measures.

Tests

Online graded quiz on overall chapter content, written in multiple choice format. When submitted, we recommend giving the correct answer along with the page number in which it is found for questions students did not answer correctly.

* Use the web site Socrative found at http://www.socrative.com/ (free of charge) to set up a live interactive quiz where students can instantly see the overall results for the class. Ask the following questions:

> ➤ When I think about LAWS I feel they are [short answer]
> ➤ I believe UAS should be permitted to be used for surveillance or tracking by law enforcement [multiple choice: yes, no, only for persons with a prior criminal record, only for felony-level situations, not sure, other]
> ➤ I know someone who has a recreational drone that they use to record video outside their own property [true/false]
> ➤ I permit my cell phone to track my location [multiple choice: yes, no, sometimes, not applicable, not sure]
> ➤ I think it is acceptable for parents to use tracking on the cell phones of their children through the age of 18 [true/false]

Then, have a class discussion about the results.

Research

Research the use of autonomous or semiautonomous technology used in agriculture (robots, UAS, etc.). Write a two page paper about your findings. You need a minimum of two references and your paper must be written in APA or MLA format.

- ❖ Create an annotated bibliography of autonomous technology resources used in medicine in APA or MLA format. Include a minimum of five items on your list. You must include the following: (a) information about the article or source (author, title, publication date, etc.); a summary of the article or source written in your own words (150 word minimum); and a short assessment of the source (100 word minimum). You can learn more about annotated bibliographies at https://owl.english.purdue.edu/owl/resource/614/03/

- ❖ Develop a research question that relates directly to autonomous or semiautonomous technology. Key criteria: (1) it must be something that can actually be researched academically, (2) it must be focused so it could be covered fully in an approximately 20 page paper (you do not need to write the paper, just the research question), (3) it must be unique (something not already researched by others), and (4) it must be clear. If you are not sure where to start, do a web search for "develop a research question" to get some ideas.

- ❖ You have been asked to come up with five questions and answers for an "All about Drones" website. Write up your five questions and their accurate answers. Cite your sources for your answers.

- ❖ Interview a person who works in a field related to autonomous or semiautonomous technology. As the following questions at a minimum and then come up with four questions of your own:
  - ➤ What technologies play the greatest role your field and why?
  - ➤ What are challenging aspects of your job?

> ➤ What are your favorite parts of your job?
> ➤ What would you recommend to someone going in to this field?

Submit both the questions and the answers.

Design and/or build projects

Inside of an autonomous car: You have been asked with designing the interior of an autonomous car. Determine what you would be important and why. Create diagrams of your internal components and design with written explanations/rationale for key parts. You will be assessed on the depth, quality, professionalism, and thoughtfulness of your work.

❖ You have been asked to recommend rules and regulations for the use of semiautonomous guns in the United States for civilian and well as governmental use. What would you recommend and why? (200 word minimum)

Assessment tasks

Choose an autonomous or semiautonomous technology which is used for international defense. Write a report assessing your chosen technology. Include in this report: (a) an overview of the technology (200 word minimum), (b) relevant facts about the technology (200 word minimum), and (c) your personal assessment of the technology (200 word minimum)

❖ Review the California Department of Motor Vehicles regulations for deployment of autonomous vehicles for public operation which is found at https://www.dmv.ca.gov/portal/dmv/detail/vr/autonomous/auto. Do you think these same regulations should be approved nationwide? Why or why not? What would you change? (200 word minimum).

❖ Browse the book "College Physics" at https://openstax.org/details/college-physics (available for free download as PDF). Choose two topic areas covered in the book and address how they are relevant to the development of autonomous or semiautonomous technology. Include: (a) a short overview of the two topic areas you chose (150 word minimum) and (b) a summary of how they are relevant to the development of autonomous or semiautonomous technology (150 word minimum)

Terms

Autonomous technology—is capable of acting independently. It has the ability to govern itself, making decisions without the need for approval, verification, or clearance. This entails having the ability to sense, perceive, analyze, communicate, plan, and make decisions. In these cases the technology is expected to accomplish its defined-scope mission without human intervention.

Geographic Information Systems (GIS)—systems that are designed to capture, store, manipulate, analyze, manage, and present a great variety of spatial and geographic data.

Lethal Autonomous Weapon Systems (LAWS)—a type of military robot designed to select and attack military targets (people, installations) without intervention by a human operator. LAW are also called "lethal autonomous robots" (LAR), "robotic weapons" or "killer robots."

Quadcopter—multi-rotor helicopters that are propelled by four rotors. They are also sometimes called quadrotors or a quadrotor helicopters. While in almost all cases a quadcopter is an UAV it does not necessarily have to be.

Semiautonomous technology—acts independently to some degree, but requires input from another source, often in the form of human intervention.

Unmanned aircraft/aerial systems (UAS)—are not manned by humans. These systems are also known as UAS. A UAS system includes not just the aircraft but also other elements such as ground control or communication systems. They are the counterpart to manned aircraft.

Unmanned Aerial Vehicles (UAV)—have no pilot on board. The vehicle can be a remote controlled aircraft or can fly autonomously based on either preprogrammed flight plans or dynamic automation systems.

Unmanned systems—receive instructions of some sort from a person which it then accomplishes either with or without further human to robot interactions. They can be used in situations of varying complexity and/or environmental difficulty which will be factors for its ultimate level of autonomy.

References

114th Congress (2015). Committee Reports 114th Congress (2015–2016) House Report 114–102. Retrieved from http://thomas.loc.gov/cgi-bin/cpquery/?&sid=cp114llNjh&r_n=hr102.114& dbname=cp114&&sel=TOC_340316&

BBC. (2015). *Who, What, Why: What is the Brimstone missile?* Retrieved from http://www.bbc.com/news/magazine-34973203

Cronk, T. (2012). Robot to serve as future military's 'Pack Mule'. *DoD News. U.S. Department of Defense.* Retrieved from http://archive.defense.gov/news/newsarticle.aspx?ID=118838

Dorr, L., & Duquette, A. (2016). *DOT and FAA finalize rules for small unmanned aircraft systems.* Federal Aviation Administration. Retrieved from https://www.faa.gov/news/press_releases/news_story.cfm?newsId=20515

Downes, L. (2014). America can't lead the world in innovation if the FAA keeps dragging its feet on drone rules. *The Washington Post Online. Innovations.* Retrieved from https://www.washingtonpost.com/news/innovations/wp/2014/10/08/america-cant-lead-the-world-in-innovation-if-the-faa-keeps-dragging-its-feet-on-drone-rules/?utm_term=.1cfcda1ae474

Drone., & Quadcopter. (2015). *What is a drone? Drone vs quadcopter.* Your Guide to Hobby Drones and Quadcopters. Retrieved from http://droneandquadcopter.com/what-is-a-drone/

Farmer, B. (2014). Brimstone: British missile envied by the US for war on Isil. *The Telegraph.* Retrieved from http://www.telegraph.co.uk/news/uknews/defence/11133680/Brimstone-British-missile-envied-by-the-US-for-war-on-Isil.html

Farkas, B. (n.d.) *Can my neighbor legally point a security camera at my property?* NOLO. Retrieved from http://www.nolo.com/legal-encyclopedia/can-neighbor-legally-point-security-camera-property.html

Federal Aviation Administration (FAA). (2016). *Section 333.* Retrieved from https://www.faa.gov/uas/beyond_the_basics/section_333/

Federal Aviation Administration. (2016b). *Unmanned aircraft systems (UAS) frequently asked questions.* Retrieved from https://www.faa.gov/uas/faqs/

Ferdinando, L. (2015). *Army looks toward fully autonomous tactical vehicle.* U.S. Army. Retrieved from https://www.army.mil/article/146372

Fitzpatrick, P. (2003). *Perception and perspective in robotics. Proceedings of the 25th annual meeting of the Cognitive Science Society.* Boston.

Frey, C., & Osborne, M. (2013). *The future of employment: How susceptible are jobs to computerization?* Retrieved from http://www.oxfordmartin.ox.ac.uk/k/downloads/academi/The_Future_of_Employment.pdf

Gallagher, S. (2016). *Marines test autonomous robot-drone teams for future on battlefield.* ARS Technica Technology Lab. Retrieved from http://arstechnica.com/information-technology/2016/05/by-our-powers-combined-marines-test-teams-of-autonomous-robots-drones/

Garamone, J. (2002). From U.S. Civil War to Afghanistan: A Short History of UAVs. *U.S. Department of Defense News.* Retrieved from http://archive.defense.gov/news/newsarticle.aspx?id=44164

Govan, R. (2016). Federal Aviation Administration Memorandum. Retrieved from https://www.faa.gov/uas/resources/uas_regulations_policy/media/interpretation-educational-use-of-uas.pdf

Hallmark, S. (2015). *Studying the impacts of autonomous and robotically controlled road-building equipment.* United States Department of Transportation. Retrieved from http://www.rita.dot.gov/utc/publications/spotlight/spotlight_2015_05

Johnson, G. (2016). *Oil spill response research (OSRR) program.* Bureau of Safety and Environmental Enforcement. Retrieved from http://www.bsee.gov/Technology-and-Research/Oil-Spill-Response-Research/Projects/Project1037/

Maartens, E. (2015). *Drone vs UAV? What's the difference?* Ezvid. Retrieved from http://www.ezvid.com/drone-vs-uav-whats-the-difference

Merriam Webster (2016). *Automation.* Merriam Webster Dictonary. Retrieved from http://www.merriam-webster.com/dictionary/automation

Moynihan, T. (2015). Parrot's tiny new drones travel by land, air, and sea. *Wired Magazine.* Retrieved from http://www.wired.com/2015/06/parrots-tiny-new-drones-travel-land-air-sea/

NASA Glenn Research Center (n.d.). *What is thermodynamics?* Retrieved from https://www.grc.nasa.gov/www/k-12/airplane/thermo.html

NASA. (2015). *NASA Technology Roadmaps*. TA 4: Robotics and autonomous systems. Retrieved from http://www.nasa.gov/sites/default/files/atoms/files/2015_nasa_technology_roadmaps_ta_4_robotics_autonomous_systems.pdf

Nasa Jet Propulsion Laboratory. (2016). *Mars science laboratory curiosity rover*. Retrieved from http://mars.nasa.gov/msl/

National Conference of State Legislators. (2016). *Current unmanned aircraft state law landscape*. Retrieved from http://www.ncsl.org/research/transportation/current-unmanned-aircraft-state-law-landscape.aspx

Naughton, R. (2005). *Hargrove: The Pioneers*. Centre for Telecommunications and Information Engineering. Monash University Engineering. Retrieved from http://www.ctie.monash.edu.au/hargrave/righter3.html

Office of the Under Secretary of Defense (Comptroller)/CFO. (2015). *Program Acquisition Cost by Weapon System*. United States Department of Defense Fiscal Year 2016 Budget Request. Retrieved from http://comptroller.defense.gov/Portals/45/Documents/defbudget/fy2016/FY2016_Weapons.pdf

Office of the Under Secretary of Defense (Comptroller)/CFO. (2016). *Program Acquisition Cost by Weapon System*. United States Department of Defense Fiscal Year 2017 Budget Request. Retrieved from http://comptroller.defense.gov/Portals/45/Documents/defbudget/fy2017/FY2017_Weapons.pdf

Oleson, T. (2013). Droning on for science. *Earth: The Science Behind the Headlines*. Retrieved from http://www.earthmagazine.org/article/droning-science

Pellerin, C. (2011). *Robots could save soldiers' lives, army general says*. U.S. Department of Defense. Retrieved from http://archive.defense.gov/news/newsarticle.aspx?id=65064

Pellerin, C. (2016). *Work: Robot warship demonstrates advances in autonomy, human-machine collaboration*. U.S. Department of Defense. DoD News. Retrieved from http://www.defense.gov/News-Article-View/Article/716156/work-robot-warship-demonstrates-advances-in-autonomy-human-machine-collaboration

Prado, Guia Maarie Del. (2015). These weapons can find a target all by themselves — and researchers are terrified. *Business Insider Online. Tech Insider*. Retrieved from http://www.businessinsider.com/which-artificially-intelligent-semi-autonomous-weapons-exist-2015-7

Prigg, M. (2014). Who goes there? Samsung unveils robot sentry that can kill from two miles away. *Daily Mail*. Retrieved from http://www.dailymail.co.uk/sciencetech/article-2756847/Who-goes-Samsung-reveals-robot-sentry-set-eye-North-Korea.html#ixzz4Bmm6m600

Risen, T. (2016). *America's civil war on drones*. U.S. News and World Report. Retrieved from http://www.usnews.com/news/articles/2016-03-04/war-between-drones-privacy-escalates-in-utah

Romboy, D. (2016). Drone operators, academics urge Utah lawmakers against too much regulation. *Deseret News*. Retrieved from http://www.deseretnews.com/article/865654521/Drone-operators-academics-urge-Utah-lawmakers-against-too-much-regulation.html?pg=all

Rusnak, P. (2015). *Farm-aiding drone technology given lift with FAA approval*. Growing Produce. Retrieved from http://www.growingproduce.com/vegetables/farm-aiding-drone-technology-given-lift-with-faa-approval/

sUAS News. (2016). Commercial Drone Operations: Wait for Part 107 or Get a 333? Retrieved from http://www.suasnews.com/2016/04/wait-for-part-107-or-get-a-333/

U.S. Air Force. (2015). *MQ-1B predator*. Retrieved from http://www.af.mil/AboutUs/FactSheets/Display/tabid/224/Article/104469/mq-1b-predator.aspx

U.S. Department of Defense. (2013). *Unmanned systems integrated roadmap. FY2013-2038.* Retrieved from http://www.defense.gov/Portals/1/Documents/pubs/DOD-USRM-2013.pdf

U.S. Department of the Interior. (2016). *Office of aviation services.* DOI UAS Fleet. Retrieved from https://www.doi.gov/aviation/uas/fleet

U.S. Geographical Survey (USGS). (2016). *Unmanned aircraft systems (UAS) project office.* Retrieved from http://rmgsc.cr.usgs.gov/uas/

Villesenor, J. (2012). *What is a drone, anyway?* Scientific American Guest Blog. Retrieved from http://blogs.scientificamerican.com/guest-blog/what-is-a-drone-anyway/

Welsh, S. (2016). *World split on how to regulate 'killer robots'.* The Conversation. Retrieved from http://theconversation.com/world-split-on-how-to-regulate-killer-robots-57734?utm_source=Triggermail&utm_medium=email&utm_campaign=Post%20Blast%20%28bii-iot%29:%20Drone%20reportedly%20hits%20airplane%20in%20London%20%E2%80%94%20Laws%20for%20autonomous%20military%20robots%20%E2%80%94%20Beverly%20Hills%20wants%20to%20use%20self-driving%20cars%20for%20public%20transportation&utm_term=BII%20List%20IoT%20ALL

Welsh, S. (2016b). *We need to keep humans in the loop when robots fight wars.* The Conversation. Retrieved from https://theconversation.com/we-need-to-keep-humans-in-the-loop-when-robots-fight-wars-53641

Chapter 9

Mediated and Virtual Reality

Outline

- ❖ Chapter learning objectives
- ❖ Overview
- ❖ Smart systems and gaming consoles
  - ➢ Gaming
  - ➢ Game-based dynamics and systems
  - ➢ Stereo projection systems
  - ➢ Interaction devices
  - ➢ Collective intelligence
  - ➢ Television, movies, and video in VR
  - ➢ The issue of cost
- ❖ Immersive experience
- ❖ Augmented or mediated reality
  - ➢ Enhanced reality
- ❖ Training/simulation
  - ➢ Telepresence
  - ➢ Voice control and speech recognition
- ❖ Virtual reality
- ❖ The science behind the technology
  - ➢ Computer science
    - ■ Object-oriented programming
  - ➢ Earth science
- ❖ Career connections
- ❖ Modular activities
  - ➢ Discussions
  - ➢ Tests

(Continued)

(*Continued*)

> ➢ Research
> ➢ Design and/or Build projects
> ➢ Assessment Tasks
> ❖ Terms
> ❖ References

Chapter Learning Objectives

- ❖ Define virtual reality as compared to mediated reality
- ❖ Identify virtual reality equipment and environments
- ❖ Examine elements that constitute a virtual or mediated reality environment
- ❖ Compare different types of mediated or virtual reality type environments
- ❖ Assess various virtual or mediate reality resources and tools

Overview

Mediated and virtual reality (VR) systems have varied meanings but at their core they are computer-mediated systems that use data to manipulate or interact with the user's

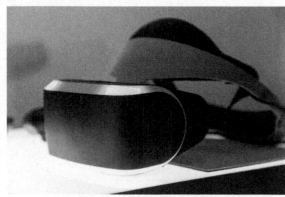

© Barone Firenze/Shutterstock.com

sensory environment, whether visual, auditory, haptic (touch), or otherwise. It could be a system as basic as a glove input device. Increasingly, mobile technologies, such as smartphones or tablet computers, play the role of offering mediated reality. This is possible because mobile technology usually incorporates not just cameras but also sensors such as accelerometers that measure acceleration, global positioning systems, and solid state compasses. Regardless of format, these personal display devices offer immersive experiences, which are increasingly common in areas such as three-dimensional (3D) gaming or training on the use of tactical equipment.

There are various technologies used for display of and interaction with information in the realm of mediated or VR. Some examples include head-mounted, smart eyeglass systems, heads up displays, or contact lenses.

- ❖ A head-mounted display (HMD) is a display device paired to a headset such as a harness or helmet that has a display optic in front of one (monocular) or both (binocular) eyes. This is the most traditional format for offering virtual or augmented reality. Samples of this type of system include Oculus Rift, HTC Vive, or Samsung Gear.

❖ Eyeglass smart systems would be something like Sony SED-E1 SmartEyeglass or Moverio BT-200 Smart Glasses, which project overlaid transparent images onto each lens, displaying various types of information in a heads-up display format. This category also includes more complex systems such as EyeTap systems, which are worn in front of the eye and are both able to act as a camera/recording system while also superimposing imagery to the natural scene available to the eye.

❖ Heads up displays project items (usually digital and transparent) at head level on to places like your windshield and include products such as the Garmin head up display or systems in 2016 Audi A7 or Mercedes S55 automotives.

❖ Contact lens systems improve sharpness of vision while enabling the wearer to view extreme detail of objects placed very near to the eye. This way the user is able to simultaneously focus on virtual content from the eyewear and on an entire spectrum of activities in the real world. An example of such a system is the Innovega iOptik contact lens.

Smart Systems and Gaming Consoles

Smart systems are able to sense the environment and put that information into action in a controlled fashion by analyzing the situation and then making decisions based on all available data. Digital devices working together as smart systems can sense, act upon, and communicate about a situation. They can recognize patterns, make predictions, and support human decision-making. In the world of mediated and VR, having systems, players, and units that can communicate with one another and share information changes the world in which the technology exists. In gaming, avatars can play together; in augmented reality that overlays information on visual elements such as street names and distance projections over the road in front of you; or in training where the decision of one participant affects all participants makes the world not only more interactive but more interdependent—as the real world is. To experience this type of smart system and immersive virtual environment, a user typically needs a high end graphic computer or system, a stereo projection system, software and interfaces, and interaction devices. Each will be discussed in more detail next.

Gaming

In the gaming world, not only is VR itself playing a role, but with the integration of 3D graphics, the experience is all the more immersive. Often times, the user has an avatar—a holographic image that portrays the person—which is used within the gaming environment. In this way, the user is immersed into a meta-universe created by the game developers.

This is important to consider because, according to the Entertainment Software Association [ESA], in 2015 there were 155 million Americans who play video games. Four out of five US households own a device used to play video games, and 51% of use households own a dedicated game console. In game playing households, there is an average of two gamers and for all Americans 42% play video games regularly (3 hours or more a week). These game players are not necessarily children. As a matter of fact, the average game player is 35 years old (2015).

If we look at frequent players, 39% play social games and 56% play with others. For most frequent gamers who play with others, they spend an average of 6.5 hours per week playing with others online and 5 hours per week playing with others in person (ESA, 2015). Interaction with others, it seems, is relevant to gaming.

When it comes to the types of games we like to play, for video games, the best-selling genres for 2014 were shooter games (21.7%), followed by sport games (13.3%). The number one selling video game overall was Call of Duty: Advanced Warfare. For computer games, the best-selling genres for the same year were strategy games (37.7%) and casual games (24.8%). The number one computer game overall was The Sims 4 (ESA, 2015).

It should be noted that gaming is not entirely recreational, as is implied in the statistics mentioned earlier. Gaming can also be used for very practical and applied purposes such as distraction, training, or rehabilitation. VR has been used to improve walking for individuals with musculoskeletal and neuromuscular conditions, in the rehabilitation of the upper extremities post-stroke, in improving balance in older adults, and in supplying controlled fitness options to patients who are critically ill.

Game-Based Dynamics and Systems

Smart gaming consoles are becoming increasingly powerful with a focus on integration with other media and increased connectivity. Due to this, the gaming experience is becoming increasingly complex and life-like. Here are a few examples of common gaming consoles along with their primary strengths:

❖ Nintendo Wii U (2013): gets gamers physically involved via motion controllers (motion, such as waving ones hand, is sensed by the controller); good for family activities; handheld controller with a 6.2-inch touchscreen that combines tablet functions with video games

❖ Sony PlayStation 4 (2013): high level of processing power, intense-action-packed game play, high-quality graphics; ability to stream video game content between devices

❖ Microsoft Xbox One (2013): high level of processing power, intense-action-packed game play, high-quality graphics

Portable gaming consoles include:

❖ Sony PlayStation Vita (2011): offers a PlayStation-like experience, impressive graphics, and ability to stream your game PS4 game via the Vita, allowing you to play your console games any time and any place with Internet

❖ Nintendo 3DS (2010): 3D graphics without the need for glasses; stereoscopic images (illusion of depth)

Note: Smart systems are discussed further in the Information Systems and Technology chapter

Stereo Projection Systems

In order to see objects in 3D, stereo projection systems, also known as computer stereo vision, stereographic systems, or stereographic projection, are often used. This involves the extraction of 3D information from two distinct digital images. By comparing information about a scene from two vantage points, 3D information can be extracted by examination of the relative positions of objects in the two panels, operating similarly to the biological process stereopsis. In this type of system, two projectors display two distinct images (from two different vantage points) shown at the average eye distance (pupillary distance) of approximately 6.2–6.5 cm. The brain then superimposes both images to a single 3D image with depth. Different filter technologies are used to separate the images for the left and right eye using either passive stereo, as is found in glasses with color or polarization filters; or active stereo, as is found in shutter classes or liquid crystal displays (LCD). Some kinds of stereo projection systems include screen projection systems (such as computer-aided virtual environments known as cave automatic virtual environment [CAVE] which will be discussed later), head-mounted displays, and desktop graphic workstations.

Interaction Devices

In most cases, virtual and mediated reality systems offer interactivity for the user. This is done with interaction devices including tracking systems that measure eye position to render images concerning point of view, voice control, force feedback, or devices such as sensing gloves. Using interaction devices, the user can have experiences such as the perception of grabbing and moving an object. This is done, in part, using proprioceptive cues. Proprioception, by the way, is the unconscious perception of movement and spatial orientation arising from stimuli within the body itself. So, for example, if you extend your finger and see a finger where you believe your finger position to be, you accept the extended finger as your own. In order to work successfully, the display must be continuously redrawn (in stereo). Since the user is constantly moving, the position must be tracked

and the objects in the environment updated. The display has to be redrawn with the new view position, new user body configuration, and new object locations as applicable. For this to occur the system must have low latency (lag time).

Collective Intelligence

Collective intelligence (CI) is intelligence that comes from shared or group participation. Via collaboration, collective effort, and competition there comes consensus or other definitive decision-making. While collective intelligence has been a mainstay of human existence generally, in this context it is specific to the online and/or gaming world. Here, groups of people and computer-based systems collectively work together via the Internet. One example most of us have likely used is Wikipedia, an open access encyclopedia that is edited by the general public. Requisite is that the groups be made of individual actors, whether they are people, computational agents, or organizations. They must demonstrate collective behavior to perceive, learn, judge, problem solve, or perform similar functions. As Thomas Malone and Michael Bernstein (n.d.) note, it is, all in all, groups of individuals acting collectively in ways that seem intelligent. Multiplayer games can function in this fashion.

Researcher Jane McGonigal studied the game "I Love Bees," which was used to create an immersive back story for Microsoft's science fiction shoot video game Halo 2. The fans collaboratively author a narrative bridge between the original Halo videogame and its sequel. She describes what she sees as core requirements for future learning systems:

> I explore the three stages of I Love Bees gameplay that ultimately produced a game-based CI. They are: 1) collective cognition, 2) cooperation, and 3) coordination. These three stages encompass, respectively, the initial formation of community, the development of distributed skill sets, and the scaffolding of group challenge that are essential elements of both massively-multiplayer game systems and the new CI knowledge networks. I also identify the three aspects of I Love Bees' game design that resulted in these distinct stages of highly collaborative gameplay: 1) massively distributed content, 2) meaningful ambiguity, and 3) real-time responsiveness. I offer these elements as a reproducible set of core design requirements that may be used to inspire future learning systems that support and ultimately bring to a satisfying conclusion a firsthand engagement with collective intelligence (2007, p. 10)

As she concludes her research she states, "As the leading edge of research, industry, politics, social innovation and cultural production increasingly seek to harness the wisdom of the crowd and the power of the collective, it is urgent that we create engaging, firsthand experiences of collective intelligence for as wide and as general a young audience as possible. Search and analysis games are poised to become our best tool for helping as many and diverse a population as possible develop an interest and gain direct experience participating in our ever-more collective network culture" (p. 38). What she noted in 2007 seems to ring very true today.

Television, Movies, and Video in VR

3D and VR home entertainment is now close to reality for many of us. As Drew Grant notes in The Observer, "many credible news outlets (CNN, The New York Times, Wired) that year heralded the era of 3D home entertainment as more than speculation, but the (very near) future" (2016). In part this prediction may be based on the fact that Toshiba, Samsung, and Sony have invested billions in creating 3D capable televisions. So now we add VR systems in to the mix. Using a headset, service providers can effectively block out all of your vision other than the screen you are to be watching. As it continues to be refined though we may see more immersive storytelling experiences that include things like motion detection (Grant, 2016). Touchstone research, in a 2015 comprehensive online survey with over 2000 participants from age 10 through 65+, found that while 60% of users were interested in gaming in VR, 66% where interested in television, movies, or video in the same (Burch, 2015).

It seems that industry is already aware of this area of potential demand. In 2015, Netflix went into a joint partnership with Oculus. This means that you can watch Netflix using VR via streaming on Oculus-made Samsung Gear VR (Matthews, 2015). As Anthony Park (2015) states in the Netflix blog, "We've been working with Oculus to develop a Netflix app for Samsung Gear VR. The app includes a Netflix Living Room, allowing members to get the Netflix experience from the comfort of a virtual couch, wherever they bring their Gear VR headset. It's available to Oculus users today. We've been working closely with John Carmack, CTO of Oculus and programmer extraordinaire, to bring our TV user interface to the Gear VR headset."

The Issue of Cost

Some see one of the greatest challenges of VR in cost. According to Touchstone research, the first wave of high end VR head mounted displays will be financially out of reach for a majority of users. Based on their results from a 2015 comprehensive online survey with over 2000 participants from age 10 through 65+, approximately 40% would buy the devices if the cost is between $400 and $599. If the cost goes up to $1000, then only 10% would make the purchase. Then again though, if younger individuals have their way, the numbers will be higher. Of the generation Z persons interviewed (born approximately 1995–2009), 70% would "definitely" or "probably" ask their parents for a VR device. (Burch, 2015)

So where are costs actually at for a typical user? Let's consider the top players for 2016 mentioned earlier—Oculus Rift, PlayStation VR, Gear VR, and HTC Vive. Daniel Howley, a reporter for Yahoo! Technology has this to say on the costs of Oculus Rift, PlayStation VR, and HTC Vive:

❖ Oculus Rift: Headset, $599; high-powered gaming PC, $950 (system with an Intel Core i5-4590 processor or better, at least 8 GB of random access memory (RAM), and—most importantly—either an Nvidia GTX 970 or an AMD R9 290 graphics card or better. In addition, you'll need a free HDMI port, three USB 3.0 slots, and 1 USB 2.0 slot)
❖ PlayStation VR: Headset, $399; PlayStation 4 console, $350
❖ HTC Vive: Headset, $799; high-powered gaming PC, $950 (similar to have for aforementioned Oculus Rift); and you need a 15-by-15-foot space (2016)

The one that Howley doesn't mention is Gear VR, Gear VR—Headset, $99; Samsung handset, $600 (compatible with Samsung Galaxy S7, S7 edge, Note5, S6 edge+, S6, S6 edge). Then again though, for the VR hardware market, the prospects are good, even if customers may feel the systems are outside of their current price point. Statista expects the VR hardware market to grow worldwide from 3.2 billion dollars in 2016 to 15.9 billion by 2020 (Statista, 2016).

Added to that, VR systems as used by commercial, industrial, or governmental entities while significantly more expensive, can also decrease the need for physical prototypes, shorten product development time, improve communication, and offer better visualization of ideas, each of which could lead to significant company savings either directly or indirectly in the long run.

Immersive Experience

Immersive experience, also known as immersive viewing or immersion, is generally seen as a system that seems to (or actually does) surround the player or audience so that they feel completely involved and encompassed by something. It is the perception of being physically present in a nonphysical world where the perception is created by surrounding the user in images, sound, or other stimuli that creates a completely engrossing environment. An extreme level would be a CAVE where projectors are directed at (or projected from) three to six walls of a room-sized cube. In this environment, the participant then wears special glasses that are synchronized with the projectors so they can actually seem to walk around an image and study it from all angles. To make this happen, sensors within the room track the viewer's position to properly align the perspective.

The name CAVE is not only an acronym, it is also meant to reference *The Allegory of the Cave* as found in Plato's *The Republic* in which the philosopher plays with ideas of perception, reality, and illusion. In this scenario, prisoners can only see one wall, illuminated from behind them by fire. All they can see are shadows, which the perceive to be real since that is their only basis for ideas of what real objects are (Grimes, 2013; Noor, 1995).

EcoMUVE, a multiuser virtual environment, is another example of an immersive experience, although not as intense. Through a primarily self-directed process of inquiry and exploration, find resources, develop hypotheses, test their interpretations, and try to resolve an ecological mystery (Walsh, 2014). As their site describes, "EcoMUVE is a curriculum that was developed at the Harvard Graduate School of Education that uses immersive virtual environments to teach middle school students about ecosystems and causal patterns. EcoMUVE was developed with funding from the Institute of Educational Science (IES), U.S. Department of Education. The goal of the EcoMUVE project is to help students develop a deeper understanding of ecosystems and causal patterns with a curriculum that uses Multi-User Virtual Environments (MUVEs). MUVEs are 3-D virtual worlds that have a look and feel similar to videogames. They are accessed via computers and, in our case, recreate authentic ecological settings within which students explore and collect information. Students work individually at their computers and collaborate in teams within the virtual world. The immersive interface allows students to learn science by exploring and solving problems in realistic environments." (ecoMUVE, n.d.).

Augmented or Mediated Reality

Augmented reality (AR) is also known as mediated or moderated reality. It is a situation in which a view of reality is modified (possibly even diminished rather than augmented) by a computer. AR is a live direct or indirect view of a physical, real-world environment whose elements are augmented (or supplemented) by computer-generated sensory input such as sound, video, graphics, or global positioning system (GPS) data.

Imagine you're a recreational stargazer using a newly acquired deep sky telescope. You look into the sky and wonder what, exactly, you are seeing. Which are planets? Are any galaxies? Where are the constellations you learned about as a kid in books? And what are Messier objects? It sure would be handy if there were something that could just tell you exactly what you were looking at. Wait! There is! As a matter of fact, there are actually a number of free mobile apps that do just this thing. Come to think of it, you don't even need a telescope if you have the app, although you might not be able to "really" see some things the app tells you are there without one. These apps use the GPS and compass in your phone or tablet or other electronic device to pinpoint your location and show you where all the surrounding celestial bodies are, even if your view is blocked. Added to that, they don't even need to have Internet access to work. You just open the app, point your smartphone at the sky, and the app will identify the planets, stars, galaxies, and objects for you (Dickerson, 2013).

Note: GPSs are discussed in more detail in the location and tracking technologies chapter.

The above is an example of augmented or mediated reality. It is simply taking reality and modifying it a bit and re-presenting it to you. In this case, it is super-imposing a computer-generated image on your view of the world, thus providing a composite view.

Enhanced Reality

Enhanced reality is a similar concept but in this case you have a live, direct view of the real world that in some way is enhanced as you view it. The view is not indirect or recreated, it is simply enhanced. Consider thermal imaging googles as an example. With these googles you are still directly seeing your own line of sight with your own eyes, it's just that the googles make it so you can now see infrared better as well and thus can identify heat given off an object (hotter objects give off more light, generally). It does this by detecting tiny differences in temperature via the infrared radiation and then creating an electronic image based on that information (Rouse, 2011).

Training/Simulation

For starters, we need to understand that simulation is a technique for practice and learning, not a technology. It replaces and/or amplifies real experiences with guided ones that are often immersive in nature (Lateef, 2010). If we turn to Merriam-Webster dictionary (n.d.), the definition of simulation is "something that is made to look, feel, or behave like something else especially so that it can be studied or used to train people." For the purposes of this chapter, we will stick to discussion of mediated and VR type systems which are used to supply simulation.

Training for real-life activities has always been an important area for virtual-reality-type systems. Some of the earliest projects to use virtual-reality-type computing for training or simulation were from the military. As of the 1950s there was a universal flight training simulation system which would be used to train bomber crews. The computer made to drive this system was known as Whirlwind I. It was initiated by the Office of Naval Research (US Navy) in 1944 and was completed in 1951. It was during this project that Jay W. Forrester, of Massachusetts Institute of Technology (MIT) invented "random-access, coincident-current magnetic storage, which became the standard memory device for digital computers, replacing electrostatic tubes. For this he was granted a patent in 1956. The change to magnetic core memory provided high levels of speed and reliability" (MIT Institute Archives & Special Collections, 2009). This patent was the first RAM that was practical, reliable, and high speed. International Business Machines (IBM) later paid MIT $13 million for the

© Halfpoint/Shutterstock.com

rights to Forrester's patent after years of litigation. The Whirlwind I project led to the development of Semi-Automated Ground Environment (SAGE) Air Defense System project in the late 1950s.

VR simulations are appealing because they provide training that is as near equal as possible without the physical risk. Practice in virtual machine (VM) systems can train individuals both at reduced cost and with more safety. When a pilot takes a misstep in a VM system, for example, there is no risk of personal injury to passengers or damage to the aircraft itself. Instead, the "game" is reset and the trainee tries again. These types of flight simulation systems are often closed mechanical ground-based systems that use visual, auditory, and motion feedback to give the sensation of flight although by the late 1970s we began to see head-mounted displays like the McDonnell Douglas Corporation's VITAL helmet (Lowood, n.d.). Flight simulation has been advancing ever since.

Not all training systems make use of real-world imagery though as flight simulators do. Fredrick Brooks used VR type systems to enhance the perception and comprehension by biochemists of complex molecules, as well systems that permit architects to walk through buildings still under design (Booch, 1999). In his eyes, the VR efforts of vehicular simulators do not represent the state of VR in general since they have very specialized properties, nor does he seen entertainment systems as representative because the VR experience itself is what is sought over the insight from the experience and users are more willing to suspend disbelief, which is not so much the case in other applications Brooks sees the increasing adoption of VR techniques and technologies as even more important than the advances in the technologies themselves (1999).

Telepresence

Telepresence involves the use of VR technology, especially for remote control of machinery or for apparent participation in distant events. It is essentially robotic remote control from potentially vast distances. The human has a sense of being at the location (virtual reality) where the remotely controlled robot (known as a telechir) resides and operates. The control and feedback are accomplished via telemetry.

Source: NASA

NASA Astronaut Tom Marshburn with R 2 Humanoid Robot Onboard the International Space Station.

Margaret Rouse (2015) describes one such system, "In a telepresence system, the telechir is often a humanoid robot, also known as an android. The control station can consist of a full body suit that the user wears. Sensors detect, and transducers reproduce, sensations of vision and sound. In some systems, tactile sensing is also possible (this is called haptics). The user wears headgear with a display and headphones that reproduce scenes and sounds as they appear at the site of the telechir. Binocular machine vision allows a sense of depth. Binaural machine hearing facilitates the perception of sounds with a sense of loudness and direction. The telechir may have one or two arms with end effectors (grippers) resembling human hands. In haptic systems, the user wears data gloves."

The telechir does not need to be a full android, however. It could be a single robotic arm or hand instead. Regardless, telepresence is useful in a variety of applications. They could be used in situations which would be high risk environmentally for humans (high heat, low oxygen, extreme cold, extreme pressure, etc.), or high risk situationally (disarming bombs, working with toxic material, working highly flammable materials, etc.), or in situations were very specialized skills are needed (such as the case with a super specialized surgeon located in another country).

Voice Control and Speech Recognition

Voice control is just what it sounds like it is. You control something with your voice. In the world of mediated or VR, this means we could like the technological system we are using to recognize speech. Speech recognition, also known as voice recognition, is the ability of a machine or program to identify words and phrases in spoken language and convert them to a machine-readable format. Not only do we want it to recognize what we said, we want it to respond to these spoken commands. This means that the machine or program needs to receive, interpret, and act upon dictation it receives. You might perhaps recall the Fijit Friends released in mid-2011 by Mattel. Fijit Friends could respond not just to being poked in certain places to be made to act, but had basic voice recognition. They could respond to key words but also interpret beats of music and start to perform dance movements accordingly. If your child asked the Fijit to "tell me a joke" it would do just that. If the child said "chat with me", to some degree the Fijit could, assuming the child used only terms listed on the instructions and/or cheat sheet. Fijit contained over 150 built-in phrases and jokes. Plus, it only cost around $30–$50 ($50 was its highest price on release but it quickly dropped to closer to $35)

Voice control has improved significantly in recent years, and it has been added to a vast array of products. I can now talk to my remote control devices, toys, appliances, car, even my home heating and air conditioning system. Plus, the prices keep dropping as the technology becomes more commonplace. What is important, though, is that the system understands me. In a number of cases, this means it needs to "learn" how I speak—my inflections and intonations, dialect, and speech patterns. This necessitates computational linguistics at increasingly complex levels. To help facilitate this, some systems (known as speaker dependent) use "training" where an individual reads specific text or isolated vocabulary into the system which it can analyze to fine-tune its recognition capabilities.

Virtual Reality

While augmented reality enhances or modifies reality in some manner, VR replaces the real world with a simulated one. VR environments are immersive and generated worlds that are so convincing that users are inclined to react the way they would in real life. In most cases, they use computer modeling and simulation to enable users to interact with an artificial 3D environment. Tools used to create the artificial environment might include items such as gloves, goggles, headsets, or bodysuits. To make this happen, however, sensory input from the outside world must be blocked or masked. Often, the senses emphasized are visual and auditory. Often, VR environments include not just immersion but also interactivity, so the participant feels they are a part of the action themselves. The synthesized world being experienced is not static, but actually responds to user inputs.

Historically, one might consider Morton Hellig's Sensorama (US patent 3.050.870), a machine invented in the last 1950s and built and patented in the early 1960s, to be an early example of VR. It offered users an immersive, multisensory (also called multimodal) experience for the user via a mechanical device that operated as a simulator, which used a 3D picture via short films as well as olfaction (smell), stereo sound, seat vibration, and vent blown air to create the illusion of being in another place.

Morton also developed an early telesphere mask that he patented in the late 1950s (US patent 2.955.156). It was a head-mounted display that provided a 3D television with wide vision and stereo sound.

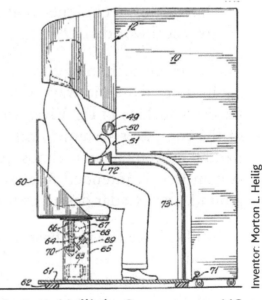

Morton Hellig's Sensorama. US Patent 3.050.870.

Inventor: Morton L. Heilig

Others may see modern computers, sophisticated graphics processors, and related technology as necessary for a true VR experience since products such as Morton's were limited to short films and the very limited experiences therein. Some view motion sensors that pick up a user's movements and adjust the screen views and similar technology in real time to be a requisite part of a true VR experience.

As Fredrick Brooks sees it, there are four technologies that are crucial for VR. From his special report *What's Real About Virtual Reality?* he writes:

Four technologies are crucial for VR:4,5

❖ the visual (and aural and haptic) displays that immerse the user in the virtual world and that block out contradictory sensory impressions from the real world;
❖ the graphics rendering system that generates, at 20 to 30 frames per second, the ever-changing images;
❖ the tracking system that continually reports the position and orientation of the user's head and limbs; and
❖ the database construction and maintenance system for building and maintaining detailed and realistic models of the virtual world.

Four auxiliary technologies are important, but not nearly so crucial:

❖ synthesized sound, displayed to the ears, including directional sound and simulated sound fields;
❖ display of synthesized forces and other haptic sensations to the kinesthetic senses;
❖ devices, such as tracked gloves with pushbuttons, by which the user specifies interactions with virtual objects; and
❖ interaction techniques that substitute for the real interactions possible with the physical world (1999).

With current technology, we can have a computer-generated world that mimics reality in quite dynamic ways. Often when we speak of VR currently we envision VR headsets such as HTC Vive, released in 2016 and costing around $800 (not inclusive of the computer system to run it). Vive has a natural 110° field of view, a feeling of natural movement, and almost flawless visuals. Vive also has a forward-facing camera system that is intended to alert you of real obstacles surrounding you as you move around. Tech Radar describes how it works:

HTC Vive has two base stations, which sit on the wall attached to the included wall mounts or a high shelf and help map track your movements as you walk around in the 3D world. What the stations track are small divots on the top of the two controllers and on the headset itself. There are 72 of these dots speckling the controllers and helmet that help accurately track the Vive (Pino, 2016).

Discrepancies in the meaning of VR exist just as they do for artificial intelligence. We debate about what senses need to be imitated and to what levels to qualify, for example. We also debate about the relationships and differences between physical- and virtual-world personalities and embodiments. These two areas demonstrate that not all technologies are clear cut or agreed upon, although technological advancements they are. It is not even agreed if some of these technologies should be perceived as a good thing or negative thing for societies generally speaking. What we can say, however, is that they are playing an active role, for better or worse.

The Science Behind the Technology

In order to track position and orientation, mediated and VR systems use varying technologies such as digital cameras, optical sensors, accelerometers, global positioning systems (GPS), gyroscopes, solid state compasses, radio frequency identification (RFID), and wireless sensors.

Computer Science

Computer science is the science of information processes and their interactions with the world. It studies the principles and use of computer-based systems from programming to computations used in correspondence with computer systems. Most students in computer science degree programs will learn a language such as Java, C++, or PHP. Generally speaking, they will have advanced math skills and will be comfortable working with algorithms. In most cases, they will also have experience working directly with databases. Their skills are not just in programming but in analysis, reasoning, and logic necessary for programming games and applications in virtual environments.

Object-oriented Programming

Object-oriented programming uses predefined programming stored in modular units such as objects or classes in order to make programming faster. Object-oriented languages include Java, C++, C#, Python, PHP, Ruby, and Perl, almost all of which are taught in computer science programs.

Of any field that would have its graduate working in mediated or VR, this would likely be it. Some computer science programs, such as at the University of Illinois, even offer courses on VR development via their computer science program (Moone, 2015).

Earth Science

Earth science deals with the physical constitution of the earth and its atmosphere. It includes, among other things, geology, oceanography, and meteorology. Let's briefly describe each:

- ❖ Geology: Science of the earth. Geology is the study of the solid earth, the rocks that compose it, and the processes by which they change
- ❖ Oceanography: Science of the oceans. Oceanography is the study of the ocean and its ecosystem dynamics.
- ❖ Meteorology: Science of the atmosphere. Meteorology is the study of the atmosphere and how processes in the atmosphere determine Earth's weather and climate.

These are relevant because often knowledge of one or more areas is necessary to make a believable, accurate, and functional depiction of our world inside a virtual or mediated reality space. Added to that, with augmented or mediated reality, programmers must understand the mapping and tracking technologies used by each along with understanding reasons for variation in measurement in order to be able to successfully add augmented features to what already exists. Refer to the location and tracking technologies chapter to better understand some of the technologies currently in use in earth science.

Career Connections

The field of VR is growing and job demand is increasing. The website Road to VR completed a study in May 2015 and noted, "While its full impact may be years away, there are a growing number of companies catching on and hiring in the virtual reality market. According to our WANTED Analytics hiring demand and talent supply data, there were about 200 employers advertising for candidates with virtual reality knowledge in March. Demand for this skill set was up about 37% year-over-year" (Zito Rowe, 2015). They also note that information, professional scientific and technical services, educational services, retail, and manufacturing were among the sectors with the greatest need for VR talent.

At issue is the fact that not many people have experience with VR. Thus prior experience and skillsets are not commonplace. This is good for job seekers generally speaking. Monster.com notes, "If you're interested in the software side of the industry, you need experience designing and developing with 3D modeling software, programming experience with C/C++, game development or graphics programming" (Zaayer Kaufman, 2015). Beyond that, you need to be able to learn quickly and be forward thinking.

Steve Santamaria, Chief Operating Officer (COO) of Envelop VR in Seattle lists a few more areas of consideration: "A newer skill set is optics, which are critical to great visualization. Hardware engineers that can help to continue to shrink the form factor, increase performance and optics fidelity, and improve battery and connectivity capability will be in high demand" (Zaayer Kaufman, 2015).

If you are considering augmented reality then CW Jobs has a few recommendations, "Most augmented reality applications rely on superimposing either 3D-generated computer imagery or some form of descriptive knowledge over the real-time images obtained through a camera, webcam or phone. This requires a good understanding of image processing and computer vision techniques, mainly for tracking either markers or the natural features on which this imagery is superimposed. Computer-generated imagery has to look realistic and be properly aligned with the real environment in order to create an authentic impression. Most of the applications are designed for the general public so a good understanding of intuitive user interfaces is also required to provide a seamless experience." (Chippindale, Van Rijswijk, & Kirkland, n.d.).

Here are some types of jobs titles you might find: research engineer, software engineer, creative director, machine learning specialist, VR engineer, applications engineer, game developer, software developer, software architect, VR programmer, user interface designer, applied engineer

Modular Activities

Discussions

❖ Post an original discussion in the online discussion board on the following topic: Some people believe mediated and VR will end up causing us to lose touch with what is actually real. What are your thoughts and why? Next, comment on the post(s) of a minimum of one other student in a thoughtful and academic way that enhances the conversation. See rubric for grading and assessment measures.

❖ Post an original discussion in the online discussion board on the following topic: How likely do you think you would be to watch television, movies and video in VR? Why? Next, comment on the post(s) of a minimum of one other student in a thoughtful and academic way that enhances the conversation. See rubric for grading and assessment measures.

❖ Post an original discussion in the online discussion board on the following topic: Some say that VR is meant for the younger generations. However, some VR systems are aimed at the elderly, such as written about in http://www.npr.org/sections/health-shots/2016/06/29/483790504/virtual-reality-aimed-at-the-elderly-finds-new-fans. What are your thoughts? Where do you see the most relevant markets and why? Next, comment on the post(s) of a minimum of one other student in a thoughtful and academic way that enhances the conversation. See rubric for grading and assessment measures.

Tests

❖ Online graded quiz on overall chapter content, written in multiple choice format. When submitted, we recommend giving the correct answer along with the page number in which it is found for questions students did not answer correctly.

❖ For students who have Steam installed (free to install via http://store.steampowered. com/), install and run the Steam VR Performance Test, found at http://store. steampowered.com/app/323910/, to see if your system is VR ready. Submit your results to your instructor.

Research

❖ Mediated or VR technology report: Choose a mediated or VR technology of your choice and write a two-page informational report about it. This report must use proper citation and use a minimum of two sources.

❖ Increasingly, VR is being used for broadcast of events, movies, and the like. One example is the 2015 Rio Olympics. Write a three-page academic research report about the use of VR in this context. This report must use proper citation and use a minimum of three academic sources.

❖ Read the article "The effect of balance training on postural control in people with multiple sclerosis using the CAREN VR system: a pilot randomized controlled trial," which was written by Alon Kalron, Ilia Fonkatz, Lior Frid, Hani Baransi, and Anat Achiron and can be found at https://jneuroengrehab .biomedcentral.com/articles/10.1186/s12984-016-0124-y and write up a review of the article.

Note: You need to be willing to look up terms you do not understand to be successful with this assignment

❖ Include the following:
  ➤ A brief summary of the article in your own words (copying and pasting is not acceptable), describing the main points (100 word minimum)
  ➤ Give a brief summary of the research methods used (150 word minimum)
  ➤ Give a brief summary of the results of the research (100 word minimum)
  ➤ What is your personal analysis of this article from an academic standpoint? (100 word minimum)

Design and/or Build Projects

❖ VR Video: Create a mini video (around 2 minutes or so) that explains what VR is (using the Internet and other sources as necessary to find information) and either post it on YouTube, Vimeo, or some other source and provide a link. You must create your own original content; simply using video content created by another person will not be accepted, although you can use other videos and websites for research. Cite where you obtained your information at the end of the video clip.

❖ You work for an earthquake damage assessment and repair company that deals primarily with buildings, bridges, and infrastructure. Your company decides it would like to create a VR room after reading about Virginia Tech's giant VR room as described at http://www.theverge.com/2015/3/13/8204193/virginia-tech-icat-vr-research-oculus-rift and in the 6-minute video at https://www.youtube.com/watch?v=2viFCKPl7CE. Your job is to design such a room with technology that actually exists today (i.e., it has to be possible with current technology that you can name). Create a mockup of such a room and explain the technological resources you would use specifically (remember, they have to actually exist).

Assessment Tasks

❖ Watch Games and Education Scholar James Paul Gee at https://www.youtube.com/watch?v=LNfPdaKYOPI (5:50 minute) What do you think of the viewpoints of this video? What are your thoughts and why? (200 word minimum)

❖ Watch the eDrawings video at https://www.youtube.com/watch?v=rVcIaBAQSE4 (2 minute). What are your thoughts on the usefulness of this technology and why? (200 word minimum) Give a specific example of how this technology could be used in industry (25 word minimum).

❖ Volvo is using Google's cardboard VR headsets (standard cost of about $15—see https://vr.google.com/cardboard/index.html) to advertise their latest cars via what they refer to as the Volvo Reality App as described at http://www.volvocars.com/us/about/our-points-of-pride/google-cardboard. What are your thoughts on this method of marketing? (200 word minimum).

Terms

Augmented reality (AR)—a live direct or indirect view of a physical, real-world environment whose elements are augmented (or supplemented) by computer-generated sensory input such as sound, video, graphics, or GPS data.

Cave automatic virtual environment (CAVE)—an environment where projectors are directed at (or projected from) three to six walls of a room-sized cube. In this environment, the participant then wears special glasses that are synchronized with the projectors so they can actually seem to walk around an image and study it from all angles. It is also meant to reference The Allegory of the Cave as found in Plato's The Republic in which the philosopher plays with ideas of perception, reality, and illusion.

Collective intelligence (CI)—intelligence that comes from shared or group participation. Via collaboration, collective effort, and competition there comes consensus or other definitive decision making.

Computational linguistics—an interdisciplinary field concerned with the statistical or rule-based modeling of natural language from a computational perspective.

Haptics—relating to the sense of touch

Immersive experience—generally seen as a system that seems to (or actually does) surround the player or audience so that they feel completely involved and encompassed by something.

Multimodal—multisensory

Olfaction—of or relating to the sense of smell

Proprioception—the unconscious perception of movement and spatial orientation arising from stimuli within the body itself.

Speech recognition—the ability of a machine or program to identify words and phrases in spoken language and convert them to a machine-readable format.

Stereo projection systems—the extraction of 3D information from two distinct digital images. By comparing information about a scene from two vantage points, 3D information can be extracted by examination of the relative positions of objects in the two panels. This is similar to the biological process stereopsis.

Telepresence—involves the use of VR technology, especially for remote control of machinery or for apparent participation in distant events. It is essentially robotic remote control from potentially vast distances

Virtual reality—immersive and generated worlds that are so convincing that users are inclined to react the way they would in real life.

References

Booch, G. (1999). Frederick (Fred) Brooks. A.M. Turing Award. Association for Computing Machinery. Retrieved from http://amturing.acm.org/award_winners/brooks_1002187.cfm

Brooks, F. (1999). What's real about virtual reality? Special report. Chapel Hill, NC: University of North Carolina. Retrieved from http://www.cs.unc.edu/~brooks/WhatsReal.pdf

Burch, A. (2015, November 11). The VR (virtual reality) consumer sentiment report—infographic. *Touchstone Research.* Retrieved from https://touchstoneresearch.com/the-vr-virtual-reality-consumer-sentiment-report-infographic/

Chippindale, J., Van Rijswijk, F., & Kirkland, K. (n.d.). The 10 things you need to know about augmented reality. *CW Jobs.* Retrieved from http://www.cwjobs.co.uk/careers-advice/it-glossary/the-10-things-you-need-to-know-about-augmented-reality

Dickerson, K. (2013, October 19). The 11 best astronomy apps for amateur star gazers. *Business Insider.* Retrieved from http://www.businessinsider.com/11-best-astronomy-apps-for-amateurs-2013-10

ecoMUVE. (n.d.). ecoMUVE overview. Retrieved from http://ecolearn.gse.harvard.edu/ecoMUVE/overview.php

Entertainment Software Association (ESA). (2015). Essential facts about the computer and video game industry. *2015 sales, demographic, and usage data.* Retrieved from http://www.theesa.com/wp-content/uploads/2015/04/ESA-Essential-Facts-2015.pdf

Grant, D. (2016). Observer. The News Big Thing in Television: Virtual Reality Gets Ready for Its Close-up. Retrieved from http://observer.com/2016/01/the-next-big-thing-in-television-virtual-reality-gets-ready-for-its-close-up/

Grimes, B. (2013, January 30). University of Illinois at Chicago: Virtual reality's CAVE pioneer. *Ed. Tech.* Retrieved from http://www.edtechmagazine.com/higher/article/2013/01/university-illinois-chicago-virtual-realitys-cave-pioneer

Lateef, F. (2010). Simulation-based learning: Just like the real thing. *Journal of Emergencies, Trauma and Shock, 3*(4), 348–352. doi:10.4103/0974-2700.70743

Lowood, H. (n.d.) Virtual reality. *Encyclopedia Britannica.* Retrieved from http://www.britannica.com/technology/virtual-reality

Malone, T., & Bernstein, M. (n.d.). The collective intelligence handbook [tentative title]. *MIT Center for Collective Intelligence.* Retrieved from http://cci.mit.edu/CIchapterlinks.html

Matthews, C. (2015, September 25). Now you can watch Netflix using virtual reality. *Fortune.* Retrieved from http://fortune.com/2015/09/25/netflix-virtual-reality/

McGonigal, J. (2007). *Why I love bees: A case study in collective intelligence gaming.* Retrieved from http://www.avantgame.com/McGonigal_WhyILoveBees_Feb2007.pdf

Merriam-Webster. (n.d.). Simulation. Retrieved from http://www.merriam-webster.com/dictionary/simulation

MIT Institute Archives & Special Collections. (2009). Project whirlwind. Retrieved from https://libraries.mit.edu/archives/exhibits/project-whirlwind/

Moone, T. (2015). Virtual reality course brings students to the forefront of technology. *Engineering at Illinois.* Retrieved from https://cs.illinois.edu/news/virtual-reality-course-brings-students-forefront-technology

Noor, A. (Ed.). (1995). *The CAVEÔ automatic virtual environment: Characteristics and applications.* Retrieved from https://www.cs.uic.edu/~kenyon/Conferences/NASA/Workshop_Noor.html

Park, A. (2015, September 24). John Carmack on developing the Netflix app for oculus. *Netflix Tech Blog.* Retrieved from http://techblog.netflix.com/2015/09/john-carmack-on-developing-netflix-app.html

Pino, N. (2016, August 25). HTC vive. *Techradar.* Retrieved from http://www.techradar.com/us/reviews/wearables/htc-vive-1286775/review

Rouse, M. (2011, April). Thermal imaging. *Tech Target*. Retrieved from http://whatis.techtarget.com/definition/thermal-imaging

Rouse, M. (2015). Telepresence. Whatis.com. Retrieved from http://whatis.techtarget.com/definition/telepresence

Statista. (2016). Virtual reality software and hardware market size worldwide from 2016 to 2020 (in billion U.S. dollars). Retrieved from http://www.statista.com/statistics/528779/virtual-reality-market-size-worldwide/

Walsh, B. (2014, November 25). Virtual reality, real science. *Usable Knowledge*. Retrieved from https://www.gse.harvard.edu/news/uk/14/11/virtual-reality-real-science

Zaayer Kaufman, C. (n.d.). How to land a job in virtual reality tech. Monster.com. Retrieved from https://www.monster.com/career-advice/article/virtual-reality-tech-land-a-job

Zito Rowe, A. (2015, May 29). 200 companies now hiring—A look at the growing virtual reality jobs market. *RoadtoVR*. Retrieved from http://www.roadtovr.com/200-companies-now-hiring-a-look-at-the-growing-virtual-reality-jobs-market/

Chapter 10

Medical Technology

(Continued)

(*Continued*)

- Ribonucleic acid (RNA)
- Genes
- Chromosomes
- Proteins
- Genetic mutations
- Recombinant DNA (rDNA)
- Genetic testing
- Genetics and personalized medicine
- Pharmacogenomics
- Biotechnology in medicine
  - Bioengineered parts
  - Foxo gene
- Nanotech medical innovations
- Digital health technologies
  - Electronic medical records
  - Radio frequency identification (RFID)
  - Information technology in health care
  - Telehealth tools
  - Patient identification
- The science behind the technology
  - Kinetics
  - Nuclear Chemistry
  - Organic Chemistry
  - Chemical Bonds
- Career connections
  - Biomedical Engineers
  - Radiologic and MRI Technologist
  - Surgical Technologist
- Modular activities
  - Discussions
  - Tests
  - Research
  - Design and/or Build Projects
  - Assessment Tasks
- Terms
- References
- Further reading

Chapter Learning Objectives

- ❖ Define bioengineering
- ❖ Identify the information that can be found in an electronic health record
- ❖ Explain the process of tissue and organ regeneration
- ❖ Examine the role of genetic engineering in medicine
- ❖ Describe how personalized medicine is changing health care
- ❖ Distinguish the difference between an MRI and a CT scan
- ❖ Summarize the findings of the Human Genome Project and its importance
- ❖ Compare imaging technologies and their use in medicine
- ❖ Differentiate between the different types and use of anesthesia; local, regional, and general

Overview

Today we live in a society where modern health care is readily available. Hospital and physicians are easily accessed. In fact there are 2.5 physicians for every 1000 people in the United States and there are 2.9 hospital beds for every 1000 US residents (The World Bank, 2016a, b). There has been an increase in the average life expectancy in the last century, which in large part is the result of developments in science, medical technologies, and techniques. Many of these improvements in health care can be attributed directly to technology-driven innovation (Spekowius & Wendler, 2006). Science and technological developments are critical components in health care today and they have greatly contributed to the increased life expectancy that we now have.

There is a vast array of medical technologies that are integral elements in modern day health care. Technologies such as those used in diagnostic imaging have made it possible to view organs and bones inside of our bodies without the need to cut into tissue. Surgical techniques have helped to extend our life span and lessened pain and time in recovery. Pharmaceutical technologies can help heal diseases, or make them manageable and can help to improve the quality of our life. Information technologies are also important in staying healthy and in providing high quality health care. Medicine is now becoming personalized and treatments for disease will be customized to the patient seeking treatment. All these technologies are used to help a person maintain a healthy life and to treat illness and disease.

There has been a convergence of technologies in physical and life sciences, electronics, information technology, and bioinformatics. Big data is being used in conjunction with genetic engineering to develop customized treatment plans for people with diseases such as cancer. Many of the technologies in use today were brought about by innovations to existing technologies and new technologies are being developed every day. Innovation in medical technology is a complex process involving many stakeholders, including industry, academia, regulatory bodies, and clinical institutions.

This chapter will give a brief overview of the history of medical technologies. The importance of the development of the microscope and its role in science and medicine will be discussed. Several noninvasive imaging technologies will be addressed including X-ray imaging, magnetic resonance imaging (MRI), mammograms, computed tomography (CT) scans, positron emission tomography (PET) scans, and optical imaging. An exploration of some of the technologies used in surgery will take place, moving from a historical use of surgical technologies to robotic surgery and the use of pulse oximeters.

The human genome project (HGP) and the science behind advances in our understanding of genetic-based medicine will be included. There will be a discussion of biotechnology applications in medicine that incorporates tissue and organ regeneration as well as the Foxo gene and its role in longevity. There will also be a discussion of pharmaceutical technologies. Genetic engineering and personalized medicine will also be addressed in this chapter. The increasing role that nanotechnology has in medicine is also investigated.

Information technology and its continually increasing presence in health care will be delved into. The role that electronic medical records will have in improving health outcomes, as well of Health Insurance Portability and Accountability Act of 1996 (HIPAA) guidelines and security issues will be studied. The use of radio frequency identification (RFID) tags and other wearable devices and monitoring technologies will also be explored in this chapter.

Rationale

Quality health care is one of the most important issues in today's society. "Innovation in medical technology, which has been estimated to account for about half of health care cost increases over the past 50 years" (National Research Council, 2010), has helped us to increase our life spans but has also led to some of the high costs we are seeing in health care delivery today.

In 2014, US health care spending was over $3 trillion dollars or about $9523 per person per year. A full 17.5% of the gross domestic product is spent on health care. Hospital care ($971.8 billion) and physician ($603.7 billion) and professional services ($84.4 billion) accounted for the majority of expenditures with prescription drug spending coming in at $297.7 billion dollars and durable medical equipment at $46.4 billion, other areas included dental services, residential and personal care services, home health care, and nursing care facility in addition to nondurable medical product spending (Centers for Medicare and Medicaid Services, 2014). According to the Organization for Economic Co-operation and Development (OECD) (2011) "the United States spends two-and-a-half times more than the OECDs average health expenditure per person," yet as Gossink and Souquet (2006) state that there are few "quantitative studies on the net effect of new medical technology on health care costs."

There are several areas that impact the growth in medical technologies, including the aging of society and the increase in chronic disease. Health care is changing very quickly, and many of the changes are associated with advances in electronics and molecular biology. One of the most recent developments in health care is the use of deoxyribonucleic acid (DNA) sequencing in the diagnosis and treatment of diseases. Our ability to understand our genome is revolutionizing health care.

Brief History of Medical Technology

In ancient times physicians would examine a patient with their ears and eyes as there were not many medical devices or technologies. They would also examine human bile, urine, and phlegm. Ancient Greeks thought all diseases were associated with disorders of bodily fluids that they called humors, which were black bile, yellow bile, phlegm, and blood. Hippocrates (300 BC) believed that a physician should listen to a patient's lungs, observe the color of their skin and any other outward signs, as well as taste the patient's urine when diagnosing patients. Christians in medieval Europe thought that disease was either a punishment for sinning or a result of witchcraft among other things (Berger, 1999).

By the end of the 1600s drugs were injected intravenously. There were transfusions of blood performed and temperature and pulse rates were used to determine the health status of patients (Berger, 1999). Hospitals in the eighteenth century were unlike hospitals today. They were filthy and disease-ridden places, and people often died from surgeries because of septicemia or blood poisoning from a bacterial infection.

During the industrial age of the nineteenth century diagnostic techniques and laboratories started to gain importance, though there was skepticism of science in

America at that time. Younger physicians often went abroad to study medicine in France and Germany at higher educational institutions. The introduction of antiseptic and new principles of cleanliness in surgery in the late nineteenth century by Joseph Lister greatly reduced deaths from surgery and injury (Science Museum, 2016). By the 1850s new medical technologies such as the stethoscope and laryngoscope allowed physicians greater diagnostic powers. X-rays, chemical and bacteriological tests, and the electrocardiographs allowed for measurement and quantification of data so that classification of individuals and evaluation of deviations could occur (Berger, 1999).

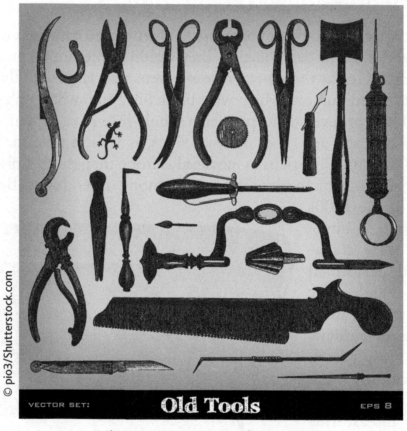

Vintage surgery tool set, 1851

There has been unprecedented progress in medical science since the late twentieth century which has contributed significantly to the increase in the average life expectancy. The leading causes of death have shifted from infectious disease and parasitic diseases to chronic conditions heart disease and noncommunicable diseases (National Institute on Aging, 2011). Biomedical technologies will again revolutionize medicine in the near future and life expectancy is expected to increase even further. Information on health care and disease is accessible to many people through the Internet. Medical records are now stored online and accessible 24 hours a day. Diagnostic techniques continue to improve as well as treatment options.

The average life expectancy at birth has increased dramatically in the twentieth century.

Year	Average Life Expectancy in the United States	Average Global Life Expectancy
1900	47.3	31
1950	68.3	48
2001	77	67
2015	<u>79.3</u>	71.4

Source: Life expectancy in the USA 1900–1998 (1998); World Health Organization (2016)

Microscope

The development of the microscope was critically important to the progression of scientific knowledge and medicine. The use of microscopes helped to fuel the scientific revolution and the use of the scientific method. People were able to observe things first hand and they could verify the same observations that others had. Doctors, researchers, and clinicians still rely on the ability of the microscope to see objects that are smaller than the naked eye can see. They are even able to view objects that are a fraction of the size of a nanometer.

Microscopes are used to see objects that are smaller than the naked eye can see. The origin of the microscope goes back many centuries, a Roman philosopher named Seneca actually described the concept of magnification in the first century AD, when he described a glass globe filled with water making images appear larger than they were. The concept of optical lenses came about around the end of the thirteenth century, when Italian glassmakers started making the first form of eyeglasses or spectacles, which were magnifying lenses set in bone, metal, or leather that were used for reading. The invention of the first compound microscope occurred in the late sixteenth century, and is largely attributed to Hans and Zacharias Janssen. It consisted of two lenses in a tube. This allowed people to investigate what had previously been invisible and allowed the beginning of investigation into causes of diseases (Berger, 1999).

In 1675, Anton van Leeuwenhoek made improvements to the microscope and was the first person to observe things at the cellular level. He saw things nobody else had ever seen before such as bacteria, sperm, and muscle fibers (Logan, 2016; History of the Microscope, 2016).

Microscopes use either a single lens or multiple lenses to make objects appear larger than they are. A lens is a curved piece of glass that focuses light; it can be either concave or convex. Modern day optical microscopes have two lenses and use light. One is the objective lens that collects the light to magnify the image and sits closest to the object being observed. The other lens is an eyepiece lens and sits closer to the eye and magnifies an image a second time (Logan, 2016). Both the lenses help in magnifying an object. Each lens may have different magnifying powers. For example, the eyepiece lens will usually magnify something 10× and the objective lens will usually magnify something 40×. 10× means that that the image will appear to be 10 times larger than it actually is. To calculate total magnification you will need to multiple the magnification of the objective lens by the magnification of the eyepiece lens. So if viewing an object through a microscope that had a 10× eyepiece lens and a 40× objective lens you would multiple 10× by 40× which equals 400× so the object would appear to be 400 times larger than it actually is.

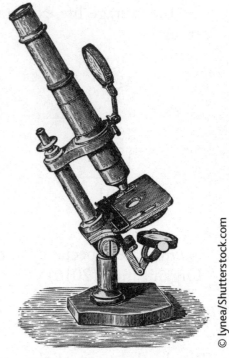

© lynea/Shutterstock.com

There are several different types of microscopes in use today. Three types of electron microscopes exist: the scanning electron microscope (SEM), transmission electron microscope (TEM), and hybrid models that combine both scanning and transmission electron microscope properties. They allow one to see objects that are smaller than those that can be seen by optical or light microscope. The scanning electron microscope scans the surface of an object with a focused electron beam. Signals are produced when the electrons interact with atoms in the sample, and they can achieve a very high resolution even at the nano level.

A transmission electron microscope (TEMrequires that samples be prepared carefully, as electrons are passed through an ultra-thin sample. The interaction of the electrons being transmitted through the specimen creates an image. The image is magnified and focused on a layer of photographic film or the image can be captured on a high quality camera that has charge-coupled device for digital imaging. There are also hybrid microscopes that combine both the SEM and TEM technologies.

In addition to the optical and electron microscopes there are different types of scanning probe microscopes, such as atomic-force microscopy and scanning-force microscopy. These have extremely high resolution and one can see fractions of a nanometer. They are developed for very specific uses. For example, magnetic resonance force microscopy captures magnetic resonance images (MRI images at the nano level.

See Chapter 12: Nanotechnology for more information about nanotechnology.

Imaging Technologies

Imaging technologies are noninvasive ways to view organs and other physical structures inside the body. Before these technologies were available, a physician would have to cut into a body to see what was wrong. This could be dangerous to the patient and not always accurate. Medical professionals now have many options to view, diagnose, and treat illness and injury. They can determine what is happening inside a patient with imaging technology and decide whether surgery is even needed. Most imaging procedures are painless and have been used successfully for years.

There are several medical imaging technologies available including ultrasound imaging, MRI-Magnetic Resonance Imaging, Computed Tomography (CT, fluoroscopy, mammography, and X-rays or radiography. Each imaging technology provides different information that can be used to study or treat different diseases and injuries or to determine a medical treatment's effectiveness. Hybrid imaging or the combination of two or more imaging modalities is now occurring because of the improved diagnostic power. There is also a convergence of imaging technologies and information technologies. The merging of capabilities allows images to be digitized as well as stored and transmitted online. This allows physicians to easily compare images to determine if there are changes occurring, which may indicate a progression of a disease (U.S. Food and Drug Administration, 2016a,).

X-rays

You are probably most familiar with X-ray imaging. Very few people get through life without breaking a bone or two. In 1895, Wilhelm Konard Roentgen was the first person to discover electromagnetic radiation in a wavelength range commonly known as the X-ray (Wilhelm Konard Roentgen, 2016). He also produced the first X-ray image ever, which was of his wife's hand. An X-ray uses electromagnetic waves to produce an image. Electromagnetic waves pass through solid materials such as tissue, but denser materials such as bones will absorb some or all of the waves resulting in the film behind the bones not being exposed completely, and these areas will appear white on an X-ray. X-rays are one of the most common diagnostic tools in use today. An X-ray can use photographic plates as a detector. An X-ray detector is a device that measures

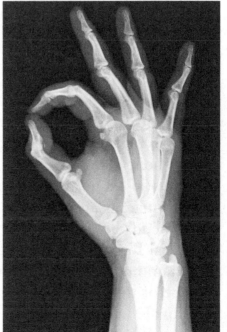

the properties of X-rays, and an X-ray image that uses a detector to produce digital images are called radiographs. X-ray radiography is used to detect broken bones and pneumonia, calcifications, foreign objects, and certain tumors. (National Institute of Biomedical Imaging and Bioengineering, 2016 b,)

X-rays produce ionizing radiation that have the potential to cause harm to living tissue and there is a small risk that a person who is exposed to X-rays will develop cancer later on in life. The risk of developing cancer from exposure to radiation is small and "depends on at least three factors: the amount of radiation dose, the age at exposure, and the sex of the person exposed." Women have a somewhat higher lifetime risk of developing radiation-associated cancer (National Institute of Biomedical Imaging and Bioengineering, 2016b) b).

Magnetic Resonance Imaging (MRI)

Magnetic resonance imaging or MRI is a noninvasive painless procedure that is used to examine organs and tissues inside the body and does not use radiation. In 1977, the first full body MR image was created (Lam, 2016) and in 1980 the first MR image of the brain was performed (Busse, 2006). Since then their use has become quite common. An MRI uses magnetic fields from two powerful magnets, radio waves and a computer to create detailed images of tissues and organs (Lam, 2016). MRI scans are used to determine if abnormalities such as tumors or cysts are in the brain, spinal cord, or other parts of the body (U.S. Food and Drug Administration, 2016a). MRIs can also be used to monitor blood flow in the brain of people who have migraine headaches or aneurysms and can help surgeons plan microsurgical breast reconstruction for breast cancer patients.

Mammography

A mammogram uses a low-dose X-ray to screen or diagnose cancer and other diseases or abnormalities in the breast tissue. Any women over 40 have likely had a mammogram to screen for breast cancer. While a mammogram cannot necessarily distinguish among noncancerous benign cysts, complex cysts, or cancerous tumors, it can identify unusual tissue. If an area of concern is found, the radiologist may recommend further imaging tests such as an ultrasound or an MRI of the breast. They also may recommend that a biopsy be performed on the tissue to determine if there is cancer (National Institute of Biomedical Imaging and Bioengineering, 2016a, b, c).

Mammograms can be done on film, but digital mammography is increasing in use because of the ability to increase the contrast of images. This may limit the need for additional images thus decreasing radiation exposure. It also gives medical professionals the ability to use computer-aided diagnostics and the ability to transmit images for second opinions. The National Cancer Institute found that digital mammography was superior to film mammography for several populations, including women under 50, women who have not gone through menopause, and women with dense breast tissue (National Institute of Biomedical Imaging and Bioengineering, 2016a, b, c).

Computed Tomography (CT)

CT or computed tomography scanning was invented by Sir Godfrey Hounsfield and Allan Cormack in 1972 (Busse, 2006) and is in widespread use today. The CT scanner rotates around the patient and takes a large number of X-ray images often referred to as slices from different angles and converts them into pictures that physicians then use to help in the diagnosis of diseases such as cancer or kidney. The images can be reformatted for viewing and even three-dimensional images can be produced. Contrasting agents can be injected into the bloodstream to further aid in diagnosis. A major advantage of the CT scan is that it can image soft tissue, bone, and blood vessels all at the same time. As with X-rays, a small amount of radiation is delivered during a CT scan. It is also noninvasive and painless. Some common uses of a CT scan include the following:

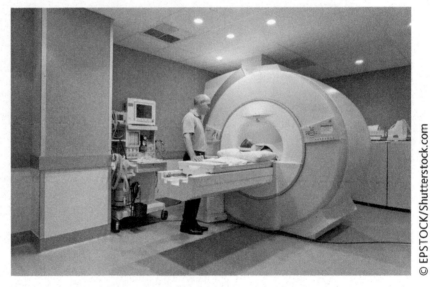

© EPSTOCK/Shutterstock.com

- ❖ Determining the extent of injuries in trauma patients
- ❖ Detecting different cancers, such as lymphoma, and tumors in the lungs, kidneys, ovaries, and pancreas
- ❖ Diagnosing and treatment of vascular illnesses such as heart disease and strokes (RadiologyInfo.org 2016)

Positron Emission Tomography (PET)

Positron Emission Tomography (PET) was developed at the University of Washington in 1974 (Busse, 2006). PET scans use a radioactive substance called a tracer to look for disease in the body and show how organs and tissues are working (U.S. National Library of Medicine, 2016e). PET scans can be used to detect cancer or to see if a cancer treatment plan is working. They can also be used to view the functioning of the brain and heart. Combination PET–CT hybrid systems are now common.

Ultrasound

In 1953, John Julian Wild and John Reid built the first real-time ultrasound device (Busse, 2006). Ultrasound or medical sonography uses high frequency sound waves to view images inside the body in real time. Ultrasound imaging can show movement in real time such as blood flow through blood vessels. An image is produced from the reflection of sound waves off the body's structures. "The strength (amplitude) of the sound signal and the time it takes for the wave to travel through the body provide the information necessary to produce an image" (U.S. Food and Drug Administration, 2016).

Ultrasound imagining is used in a wide variety of medical disciplines including but not limited to cardiology, surgery, critical care, pediatrics, sports medicine, and emergency care. Probably one of the most exciting ultrasound images people see is the first images of their unborn baby while it is still in the womb. Ultrasound imagining can be used to visualize abdominal tissue and breast tissue, an echocardiogram allows viewing of the heart, and Doppler ultrasound allows visualization of blood flow through a blood vessel, organ, or other structure (U.S. Food and Drug Administration, 2016b). Telesonography allows real-time video transmission of ultrasound images (Tsung, 2016).

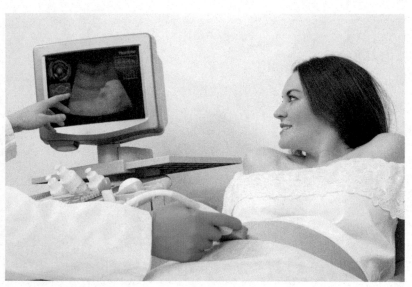

© Khakimullin Aleksandr/Shutterstock.com

Woman viewing an ultrasound image of her unborn child

Cutting Edge Imaging

Optical imaging is at the forefront of imaging technologies. The retina is the only place in the body where both nerves and blood vessels are visible. Machine visioning is used to collect and save data on eye movement at a rate of 320 frames per second to test for double vision or eye misalignment (diplopia), which is a common neuro-ophthalmologic symptom that leads to nearly one million ambulatory and emergency room visits annually. It is hoped that soon physicians will be able to scan a person's eyes to determine if they have had a closed head injury, if there is the beginning stage of dementia, or if they have experienced a stroke or aneurysm (Drake, 2016).

Surgical Technologies

There is evidence that surgery has been practiced for thousands of years, though definitely not in the manner that it is practiced today. Developing an understanding of human anatomy, methods to stop bleeding, antiseptic principals, and the use of anesthesia makes modern day surgical procedures possible. The word surgery itself is from ancient Geek meaning "hand work." Surgery is used to treat disease or an injury, and it can help improve bodily functions or appearances (think of cosmetic surgery) or can be used to repair an area such as a broken bone or a ruptured tendon.

Hippocrates of Kos is often considered the "father of western medicine" (Encyclopedia Britannica, 2016). The Hippocratic Oath that was written around fifth century BC still informs many modern day physicians on ethical practices and professional conduct that doctors should abide by when treating patients. The Hippocratic corpus is composed of multiple volumes on topics such as joints, fractures, injuries of the head, surgery, ulcers, and even hemorrhoids.

Many of the early surgical techniques were used to treat injuries or trauma that people experienced from war. There were tools such as bone forceps that were used to remove pieces of bone from the skull if there was a skull fracture. There was a wide variety of scalpels in use.

In addition there are examples of bloodletting cups, which were used in procedures where blood was removed to cure or prevent disease and illness. This was very popular technique during ancient times and Greeks, Egyptians, Mayans, and even the Aztecs used this technique. Bloodletting was still a relatively common practice in the United States through the nineteenth century and is still performed in parts of the world today, despite "centuries of harm" (Engelhaupt, 2015).

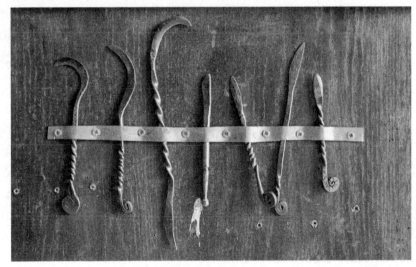

© Sergio Foto/Shutterstock.com

Medieval surgeons used a variety of medical tools to perform surgery. There were different types of scalpels developed that were used in surgeries as well as surgical saws for amputations. Some of these instruments were very ornate. An example of medieval scalpels with decorative handles is seen below.

Today many technologies are required in modern surgical procedures. Your health history will be taken and checked against your electronic health record (EHR) usually by a nurse using a computer. You will be asked if you are allergic to any medications. Your blood pressure and temperature will be taken to help ensure that you are healthy enough to undergo surgery. You will be asked to bathe prior to surgery using an antibacterial soap and to dress in a clean surgical gown, and you may even be given a sterile hair net and socks. It is very important to keep things clean and great lengths are taken in hospitals and surgical centers to ensure that there is an antiseptic environment to perform the surgery. There are even infection detection technologies which are the tools that are used to detect sepsis earlier, thus making it easier for hospitals to diagnose and manage infections more successfully, and less expensively.

Anesthesia

Anesthesia is used to minimize the amount of pain and sensation that a person feels during surgery or certain medical procedures. Different types of anesthesia and doses of medication are used depending on the type of procedure being done. Anesthesia can be considered local, regional, or general. When a regional or general anesthetic is used, an anesthesiologist, who is a doctor practicing anesthesia, will administer and manage the anesthesia. They will also make sure that the patient's breathing, heart rate, and blood pressure are functioning properly. Anesthesiologists are trained to diagnose and treat medical problems during and right after surgery.

Local Anesthesia

A local anesthetic agent will stop a sense of pain for a short period of time in a small or localized area. Usually, the patient will remain conscious and the anesthetic agent will be either applied directly to the area or injected into the area; sometimes both are done. For example, if you need a filling done by the dentist they may rub a small amount of a numbing agent on the gum area prior to injecting Novocain, which causes a loss of feeling or numbness of the skin and mucous membranes (Hopkins Medicine, 2016).

Regional Anesthesia

Only the portion of the body that is going to be worked is numbed with a regional anesthesia and oftentimes the patient remains awake during the procedure. A spinal anesthetic is used in surgeries of the lower limbs, the abdomen, and rectal area. One dose of the anesthetic is injected into the subarachnoid space, which surrounds the spinal cord in the lower back. An epidural anesthetic is also used for surgery of the lower extremities and it is used in labor and delivery. One of the differences between a spinal and an epidural anesthetic is that a catheter is used in the epidural which will provide a continuous infusion of anesthetic medication. The catheter is put in the space outside of the subarachnoid space in the lower back, which can also be placed higher along the spine for chest and abdominal surgeries (Hopkins Medicine, 2016). Oftentimes, an epidural will be used during a cesarean section (C-section) birth so that the mother can be awake during the birthing process but not feel the pain of the incisions.

General Anesthesia

A general anesthetic will cause a person to lose consciousness during surgery. The general anesthetic can be given either through an intravenous line (IV) or a patient can inhale it through a tube or breathing mask. After the surgery is done the anesthesiologist will stop administering the anesthetic and the patient will be taken to a recovery area to wake up (Hopkins Medicine, 2016).

Robotics in Surgery

The use of robots in surgery has been occurring for over 20 years. Minimally invasive surgical techniques began in 1987 with laparoscopic surgeries. A laparoscopy is where a fiber–optic instrument is inserted through a very small incision, typically a half inch or so, into the abdominal wall so that organs can be viewed or a surgical procedure, such as removing a gallbladder, can be performed. Laparoscopic surgeries have improved surgical outcomes by reducing the degree of trauma a patient experiences because surgeons are able to make smaller incisions. There are many advantages with the use of smaller incisions: they result in lowering the risk of infection, decreasing the length of recovery time, there are better cosmetic outcomes, and there is decreased pain.

In the mid-to-late 1980s the use of robots in surgery started to occur. There was the PROBOT, a robot designed to perform surgery on enlarged prostates to relieve urinary symptoms and the ROBODOC, which did precision machining of the femur in hip replacement surgeries. Researchers at National Air and Space Administration's (NASA) Ames Research Center were interested in developing telepresence surgery. By the early 1990s they were working with researchers at Stanford Research Institute to develop a telemanipulator for hand surgery, with one of the goals being to give the surgeon a direct sense of operating on the patient. Personnel at the US Army were very interested in the concept of telemedicine and envisioned its use in wartime applications. They even designed a vehicle with robotic surgical equipment. Some of these engineers and surgeons went on to form commercial surgical robotic companies; the Da Vinci surgical system is an outcome of one of these ventures (Lanfranco, Castellanos, Desai, & Meyers, 2004).

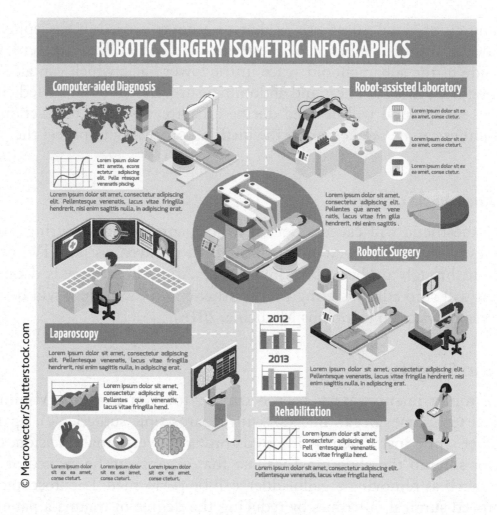

© Macrovector/Shutterstock.com

The Food and Drug Administration (FDA) has approved the use of several surgical robotic systems including the AESOP (Automated Endoscopic System for Optimal Positioning) system which is a voice-activated endoscope. The Da Vinci and Zeus surgical robotic systems are considered comprehensive master–slave robots that have multiple arms and are operated from a remote console that has computer enhancement and video-assisted visualization.

You can view a short 5 minute video on robotic surgery here: https://www.youtube.com/watch?v=vb79-_hGLkc

Augmented reality has resulted in new training methods for surgery and diagnosis. Robotic-assisted surgery is now taking place and "one of the most promising advances in surgical technology is the introduction of robotic-assisted MIS, which allows procedures to take place that would otherwise be prohibited by the confines of the operating environment" (Imperial National Institute for Health Research, 2016).

Pulse Oximeter

One of the most important medical technologies to have during surgical procedures that requires a patient to be under sedation or general anesthesia is the pulse oximeter. These medical devices monitor oxygen levels in the blood indirectly. Pulse oximeters can sound an alarm if oxygen levels become too low during surgery, making this technology a favorite of anesthesiologists. Besides being used in the operating room, pulse oximeters are also used in the recovery room after the surgery is completed, in critical care units of hospitals in ambulances and you can even buy a pulse oximeter to use in your home from local drug store.

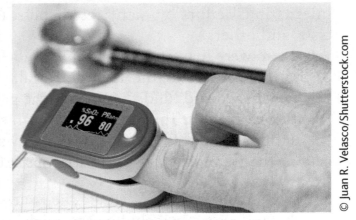

A pulse oximeter used to measure pulse rate and oxygen levels

A pulse oximeter is a small piece of equipment that is usually attached to the end of a finger. They are relatively inexpensive and fairly easy to use. A pulse oximeter uses light to measure oxygen levels or oxygen saturation in the blood. There is a light source on one side and a light detector on the other side of the pulse oximeter. Oxygen saturation is calculated by measuring how much of the available hemoglobin in the blood is carrying oxygen and is referred to as a percentage. Hemoglobin absorbs light, so when light passes through the finger the part of the light that is not absorbed by the hemoglobin will reach the light detector. An oxygen saturation level of 95% or greater is usually considered normal, but if oxygen saturation levels fall below 92% oxygen supplementation may be necessary (How Equipment Works, 2016).

The Human Genome Project

The Human Genome Project (HGP) was a collaborative international research program with the goal of mapping all of the genes in humans. The mapping of the human genome was completed in 2003 and was considered one the greatest achievements in history. The goal of the HGP was to develop an understanding of genetic factors in human disease, with hopes that new ways of diagnosing, treating, and preventing diseases would be developed. There were over three billion letters or base pairs that needed to be sequenced in the human genome, and this required a massive

international effort by many researchers, technologists, computer scientists, ethicists, and others.

There were many ethical, legal, and social issues to consider with the sequencing of the HGP. In 1990, the Ethical, Legal, and Social Implications (ELSI) Program was established and funded as part of the genomics research program at the National Institute of Health. This research program supports studies on the ethical, legal, and social implications of genomics, with many studies helping to inform policy and education. The ELSI research priorities consist of four main areas:

❖ Genomic research: What are the issues that may come from designing and conducting research, including the production and analysis and sharing of individual information, especially with regard to electronic health records?
❖ Genomic health care: How does information from genomic research impact the way health care is provided and how does it affect individuals, families, and community health?
❖ Broader societal issues: What are societies' beliefs, practices, and policies regarding genomic information and what implications are there for understanding diseases and a person's responsibility for their health?
❖ Legal, regulatory, and public policy: What are the effects of existing policies, regulations, and genomic research, and how does that inform the development of new policies and regulations? (McEwen et al., 2014)

The Human Genome Project was critical to the development of the biotechnology industry. As of 2013, more than 1800 genes that are associated with diseases have been found, and there are over 2000 genetic tests for different humane conditions. There are also over 350 biotechnology-based products in clinical trials that have resulted from the HGP. Treatments for diseases, including some types of cancer have already been developed and are in use (National Institute of Health, 2013). For example, Herceptin also called trastuzumab, is a drug used to treat an aggressive form of breast cancer tumors, has been developed and is in use today.

An interactive timeline of the human genome can be accessed here: https://unlockinglifescode.org/timeline?tid=4.

The Biology behind the Human Genome Project

In order to understand how the diagnosis of diseases that are hereditary in nature and the development of treatments that are targeted and personalized are made, it is important to have an understanding of the basic structures that make up the human genome. This section will review cells, DNA, mitochondrial DNA, ribonucleic acid (RNA), genes, chromosomes, proteins, genetic mutation, and recombinant DNA (rDNA).

Cells

Human bodies are composed of trillions of cells which are the basic units or building blocks of all living things. Each cell has a nucleus which is the command center of the cell and genetic material (DNA) resides in the nucleus. Cells also contain cytoplasm which is a fluid with other structures that surrounds the nucleus. Cells contain organelles, which are structures that carry out specific tasks within the cell, the nucleus is an organelle as are ribosomes and mitochondria. The nucleus, cytoplasm and organelles are surrounded by a cell membrane. (National Institute of Health, 2016).

Cells perform several functions: they provide the structure of the body, bring in nutrients from food, which they convert to energy, and they carry out specialized functions. Cells can also make copies of themselves.

Deoxyribonucleic Acid

DNA is short for **d**eoxyribo**n**ucleic **a**cid which is the hereditary material in most organisms including humans and it is self-replicating. DNA has the instructions for our genes. Almost all of a person's DNA is located in the cell's nucleus but some can be found in the mitochondrial. DNA code has four chemical bases: adenine (A), guanine (G), cytosine (C), and thymine (T). A human has approximately 3 billion bases and 99% of these bases are the same in all people. We are much more alike than different from each other.

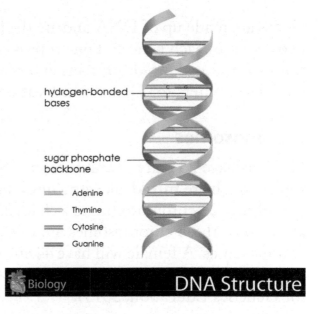

hydrogen-bonded bases

sugar phosphate backbone

— Adenine
— Thymine
— Cytosine
— Guanine

Biology   DNA Structure

Each base pairs up with another base, **A** and **T** pair up and **C** pairs up with **G**; these units are called base pairs. Each base is also attached to a sugar molecule and a phosphate molecule and together they are called nucleotide. A double helix is formed from nucleotides arranged along two long strands, with the backbone being made of the sugar and phosphate molecules and the rungs from the base pairs (U.S. National Library of Medicine, 2016a).

Mitochondrial DNA

Each cell has many mitochondria and they are located in the cytoplasm which is the fluid that surrounds the nucleus and is the structure within a cell that converts energy from food into forms that cells use. Mitochondrial DNA is the DNA that is located in the mitochondria, and contains 37 genes which are required for normal functioning. Mitochondria have several other functions including the production of cholesterol and part of the hemoglobin called heme, and it also helps with the regulation of the self-destruction of cells. Some of the genes in mitochondrial DNA are instructions for making cousins of DNA called transfer RNA (tRNA) and ribosomal RNA (rRNA). These chemical cousins assist the amino acids to become functioning proteins (U.S. National Library of Medicine, 2016b).

Ribonucleic Acid (RNA)

RNA is short for ribonucleic acid that is present in living cells. RNA plays an important role and carries instructions from DNA for controlling protein synthesis. It also carries virus's genetic information.

Genes

Genes are made up of DNA and are the basic units of heredity. Each person will have two copies of each gene and one copy comes from each parent. Genes are the instructions for the making of proteins and most genes are positioned on chromosomes in humans (Centers for Genetics Education, 2015).

Chromosomes

Humans have 23 pairs of chromosomes and most genes are arranged on the chromosomes which are thread-like structures housed in the nucleus of each cell. Each parent contributes 23 chromosomes to a child, 22 of these chromosomes are called autosomes and two of these chromosomes are called X and Y chromosomes, which are the sex chromosomes. A female will have 44 autosomes plus two copies of the X chromosome and a male will have 44 autosomes plus a copy of the X and Y chromosome (Centers for Genetics Education, 2015).

Proteins

Proteins are molecules that do most of the work in cells and are made up of 20 different types of amino acids which determine the proteins specific function. Some examples of protein functions include enzymes, antibodies, messengers, structural components, and transport and storage of atoms and small molecules (U.S. National Library of Medicine, 2016c).

Genetic Mutations

Gene mutations are alterations in the DNA sequences that compose the gene and can range in size and affect a single base pair or a large segment of a chromosome that has multiple genes. Gene mutations can be hereditary, which means that they are inherited from a parent or they can be acquired from the environment or when a mistake is made during cell division.

Recombinant DNA (rDNA)

When a two or more gene sequences are formed together in a lab the result is called rDNA; these are usually engineered for very specific purposes. DNA is extracted from a donor organism and is cut up into fragments that contain one or more genes. These gene fragments are allowed to insert themselves individually into small autonomously replicating DNA molecules such as bacterial plasmids or a virus. These small molecules act as carriers and are also referred to as vectors for the donated DNA. The vector molecules with their inserted DNA are called rDNA because they have DNA from the donor and DNA from the molecule they were inserted into (Griffiths, Miller, Suzuki, Lewontin, Gelbart, 2000).

Genetic Testing

Each person has a unique DNA sequence, sometimes referred to as their *DNA fingerprint* with the exception of identical twins. In forensic genetic testing, blood, urine, saliva, hair, semen, cheek cells, tissues, and bones can be used for DNA analysis. This type of analysis can be used to help determine if someone committed a crime, or to identify catastrophe victims, and to determine biological relationships such as establishing paternity (U.S. National Library of Medicine, 2016d).

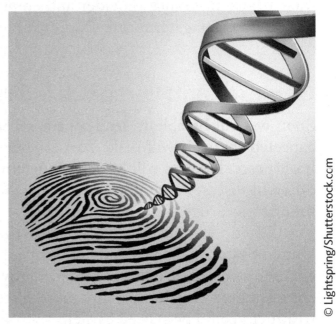

© Lightspring/Shutterstock.com

People can have genetic testing which is a type of medical test that detects changes in genes, chromosomes, or proteins. Genetic tests can provide information about a person's genes and chromosomes. A person can have diagnostic tests on tissue biopsies for example, that can help aid in treatment planning. They can have their newborn tested to see if there are genetic disorders that can be treated early, and they can be tested to see if they are carriers of certain genetic disorders. In addition, genetic testing can be done on a fetus's genes or chromosomes before birth and there are even preimplantation tests which are used to detect genetic changes in embryos. Predictive and presymptomatic tests can help determine if a person is at risk of developing a genetic disorder such as Huntington's disease.

People should think carefully about genetic testing before consenting to have it done. There are many benefits associated with genetic testing but there are also some risks and drawbacks that are associated with genetic testing. Some of the benefits associated with testing are that some of the uncertainty may be removed regarding a health issue, and sometimes treatment plans can be tailored to the individual being tested. If people know that they are at an increased risk of developing a disease later in life, they may get screenings earlier and make changes to their diets and exercise regimes in order to minimize those risks. Testing positive for something does not always mean that you are going to develop a disease. It is important that if you are thinking about having genetic testing then you should meet with a genetic counselor who can go over what a particular test can tell you.

Some of the drawbacks associated with genetic testing are that learning that you may be at an increased risk of developing a disease can be very emotional, some people get scared and depressed upon finding out the results of genetic testing. There may also be costs associated with testing, though some of these are covered by insurance plans. One of the biggest areas of concern about genetic testing is discrimination by your insurance company or employer. The Genetic Nondiscrimination Act was passed in 2008 by Congress and it protects people from discrimination by their health insurance provider and employer but it does not apply to disability, life insurance, or long-term care (National Human Genome Research Institute, 2015).

Genetics and Personalized Medicine

Soon we will be able to read a person's entire genome for less than $1000 and it is expected that this will make having your genome analyzed as a routine part of your health care. Doctors will be able to determine whether you have an increased risk for different diseases and they will also be able to tell what your likely response, if any, is to different drugs used in treatment.

Genetic engineering has enabled the advent of personalized medicine. In the United States, all babies that are born are tested for between 29 and 50 diseases that are inherited, severe and most importantly treatable. If whole genome sequencing was completed on newborns many more diseases and conditions could be detected and treatment could be started earlier.

Pharmacogenomics

There are over 150 FDA approved drugs that involve the use of an individual's genome in determining effectiveness of a particular therapy (National Institute of General Medical Sciences, 2016). The study of this field is called pharmacogenomics and anti-viral drugs, cardiovascular drugs, analgesics, and anticancer therapeutics are already developed; research will definitely continue in this area.

Almost all pharmaceutical companies and biomedical research labs now use genetic information in the development of diagnostic techniques and in the development of new drugs to treat disease. Pharmaceutical manufacturers have seen increasing pressure from regulatory health care reforms, increased competition, shifts in patient behavior, and emerging technologies. We are now seeing increased collaboration between the life sciences industry, academic partners, and information technology.

Biotechnology in Medicine

The convergence of medicine, nutrition, informatics, and agriculture is enabling new technologies and approaches to improve global health and food security. The ability to rapidly diagnose disease using microarrays, genomic sequencing, and bioinformatics is allowing medical professionals to develop personalized therapies and diets which are based on a person's genotype are now available. Regenerative medicine using adult and embryonic stem cells is a reality.

Bioengineered Parts

The National Institute of Health funded much of the basic research necessary for genetic engineering. This research was the starting point for what is now the more than $40 billion biotech industry. This year in the United States more than 120,000 people are on the transplant list waiting for life-saving organs (U.S. Department of Health and Human Services, 2016a). People who receive transplanted organs have to take immunosuppressant drugs for the rest of their lives so that the organ is not rejected by the body. The ability to regenerate organs and tissue using the patient's own cells reduces the occurrence of rejection because the immune systems recognize the tissue. According to the National Institute of Biomedical Imaging and Bioengineering (2016c), "tissue engineering refers to the practice of combining scaffolding, cells, and biologically active molecules in functional tissues." The goal of tissue engineering is to assemble functional constructs that restore, maintain, or improve damaged tissues or whole organs. The FDA has already approved engineered artificial skin and cartilage.

The use of a multidisciplinary approach by surgeons, scientists, and engineers has allowed for the development of tissue engineering which results in tissue and organs being developed that can now replace damaged or diseased tissue and organs. Source cells from the patient, a cadaver, or stem cells are cultured and seeded in an extracellular matrix scaffold which grow into the desired tissue or organ and are then implanted into the patient. The extracellular matrix can be a naturally occurring substance such as collagen or a synthetic degradable polymer such as polyglycolide that can assimilate into the body (Mungadi, 2012). These tissue-engineered implants do not require antirejection drugs and can grow, remodel, and respond to injury. Some tissues and organs that are being developed include skin, bladders, hearts, livers, kidneys, tracheas, back discs, and intestines (Danigelis, 2013). Experiments to implant tissue-engineered skin grafts, cartilage, and small arteries have already occurred, but they are very costly and are still considered experimental. There have also been successful lab creations of complex organs such as liver, heart, and lung tissue, but they are long way from human patient trials National Institute of Biomedical Imaging and Bioengineering (2016a, b, c).

To view a video on tissue engineering click on the link below:

http://www.pbslearningmedia.org/resource/biot09.biotech.app.tissueeng/tissue-engineering/.

Foxo Gene

Since the beginning of time people have sought the elusive fountain of youth. We all want to live healthy long lives and FOXO proteins may be the key. The functional deterioration of physiological mechanisms from the passage of time can be defined as aging (Martins, Lithgow, & Link, 2016). Cancer, cardiovascular diseases, and neuro-degenerative disorders are all associated with aging. Many environmental factors such as physical activity, diet, health habits, and psychosocial factors and genetic factors influence how long a human will live and if they will age in good health. Genetic factors appear to be more important in people who live extremely long lives (living longer than 100 years) than environmental factors (Martins et al., 2016). In fact several studies have indicated that the FOXO1 and FOXO3 genes are associated with longevity. "FOXOs are master regulators that translate environmental stimuli arising from insulin, growth factors, nutrients and oxidative stress into specific gene expression programmes. The role of FOXO3 in longevity may involve upregulation of target genes involved in stress resistance, metabolism, cell cycle arrest and apoptosis (normal death of cells)" (Morris, Willcox, Donlon, & Willcox, 2015).

Evidence suggests that FOXO factors are important in tissue homeostasis (maintaining balance, internal balance or equilibrium) and stem cell biology though it is not yet understood how FOXO factors influence human aging. It is thought that the control of the FOX3 in response to environmental factors will be vital in preventing age-related diseases such cancer, type 2 diabetes, cardiovascular disease, and neuro-degenerative diseases (Morris et al., 2015). Research will continue on in this very important area.

Nanotech Medical Innovations

Nanotechnology is the ability to manipulate properties and structures at the nanoscale, which is very very small, in fact a nanometer is one billionth of a meter. There is a great deal of potential to use nanotechnology in medicine for things such as gene therapy, diagnostics, and drug delivery (Paddock, 2012). Genetic therapies may involve manipulation of individual genes or their molecular pathways. This would require extremely small tools which could be created by nanotechnologies. In fact Harvard Medical School researchers made a programmable "origami nanorobot" out of DNA that carried molecules with instructions to trigger cell suicide in leukemia and lymphoma cells (Paddock, 2012).

Nanotechnology in the medical field will continue to grow, research is being done on nanobots and nanostars for drug delivery, and the building of nanofactories to produce drugs in situ (within the body) are being researched. The development of nanofibers may help people with spinal cord injuries and may be used in tissue engineering and the development of artificial organs (Paddock, 2012).

You can view a short 7 minute video on nanotechnology and medicine here: https://www.youtube.com/watch?v=_GjqbUPmcWQ

See Chapter 12: Nanotechnology for more information on nanotechnology.

Digital Health Technologies

Electronic Health Records

Electronic health records and genetic data are being used to improve health care outcomes and lower costs, as physicians and insurance companies can better track patient outcomes over the long term. (Industry Week: Advancing the Business of Manufacturing, 2013). Data services can assist in the management and organization of data from multiple organizations such as R&D, research labs, and academic institution which is leading to improved understanding and new solutions related to drug and medical device development. Social media has also started to play a role in the pharmaceutical industry by helping manufacturers to better understand their customers, for example, by creating a Facebook page for patients to discuss symptoms and treatment options.

A broad and growing area of digital health technologies exists today and includes health information technology, mobile and telehealth, wearable devices, telemedicine, and personalized medicine. These technologies are used to improve access to health care and to improve quality of health care by making medicine more personalized for patients, while at the same time reducing costs and inefficiencies.

The increasing prevalence of smart phones, social networks, and mobile medical apps has changed the way we communicate with each other and is also changing the way we monitor our health. We now have unprecedented convergence of health care information, connectivity, and technology which is leading the way to improved health outcomes. Medical devices, such as wearable heart monitors, can connect and communicate with other systems and devices. They can send information in real time to doctors, who can monitor at a distance but respond quickly when needed.

Digital health care has a variety of stakeholders that includes doctors and their patients, other health care practitioners, biomedical technicians, the traditional medical device industry, application developers, and information security professionals.

Electronic Health Records (EHRs) allow medical professionals to share your health information with other health care practitioners in an effort to improve the quality of care. Your EHR may include any or all of the following:

❖ Immunizations
❖ Lab results
❖ Medications
❖ Diagnoses
❖ Vital signs
❖ X-rays
❖ Surgeries
❖ Doctor notes

The Health Insurance Portability and Accountability Act (HIPAA) of 1996 Privacy Rule ensures that you have rights over your own health information. There is also a HIPAA Security Rule that requires organizations to put safeguards in place such as encryption and access control, in order to protect your information for security breaches (Department of Health and Human Services, 2016).

Radio Frequency Identification (RFID)

Radio frequency identification tags commonly known as RFID have many applications in the health care industry, including the ability to keep track of hospital assets, to monitor patients, and to automate payments. Surgical instruments are tracked in hospitals to ensure that none of them are misplaced, before, during, or after surgery. Medical equipment and assets such as beds, pumps, monitoring devices, and electronic medical equipment used to be tracked by serial number and bar codes in hospitals, and employees would have to visually inspect the items, which was very time-consuming and costly. RFID readers can now be installed in strategic locations in the hospital and users can track items according to the last zone that the item was detected in. Handheld readers can also locate assets in specific zones and help aid in inventory management (RFID Journal, 2016). Many medical devices such as electronic heart monitors have to be periodically serviced according to strict FDA guidelines, and RFID tags can assist biomedical technicians in locating and servicing the equipment. Another use of RFID tags is the tracking of urine and blood samples in the laboratory; many of the tests require controlled temperature environments and the samples frequently have to be moved from one piece of equipment to another.

Real-time location systems can also be employed in maternity wards, where transponders are attached to both the baby and the mother, as well as the badges worn by doctors and nurses. If a baby is moved by an unauthorized person, an alarm will sound. RFID tags can also be used to track patients and to share that information with family members as well as hospital administrators, for example, family members can be notified that a patient is out of surgery and is in the recovery room. (http://www. rfidjournal.com/articles/view?14462).

Information Technology in Health Care

There are many information technologies that can be considered patient friendly. A centralized scheduling area would make scheduling of appointments and tests easier for patients, though many departments want to keep control of scheduling. Having a master patient index, which is a database with unique patient identifiers that is HIPAA compliant, helps to eliminate patients from having to complete multiple forms.

Telehealth Tools

We are also starting to see health care come into the home. Patients can now communicate easily with their doctors using bidirectional video and audio formats. Applications such as Skype allow this mode of communication and it is especially useful in rural areas or in areas where accessing medical care is not easy. For example, if a solider is wounded in warfare, a paramedic could Skype into the hospital using a smart phone and show an image of the wound to a doctor and receive information on how to care for the wound immediately. This type of communication is also useful for referrals and consultations with doctors. This helps to bring down the cost of health care while providing immediate assistance to the patient.

Patient Identification

Usually when a patient is admitted to a hospital or goes into a surgical center for an outpatient procedure a plastic bracelet is put on. This bracelet will have information such as the patient's name, birthdate, and medical record number. There also may be barcode and/or a quick response (QR) code on it. The bracelet can be scanned and used as a bedside medication verification system that helps to ensure that the right patient is getting the correct medication. It can also be scanned to ensure that the correct patient is getting the correct procedure, which is another safety measure.

The Science behind the Technology

In the above sections, we addressed varying areas of biology that are involved in the HGP among other areas of medical technology. We also touched on various imaging and surgical technologies. Now, let us address a few areas of chemistry that play a vital role. Chemistry is the study of matter. More accurately, it is a branch of physical science that studies the composition, structure, properties, and change of matter. As we look at altering chemicals or atoms and molecules in human bodies, we move into the

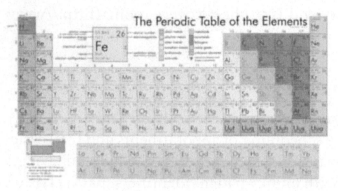

Periodic table of elements

realm of chemistry. For many of us, this may bring up images of the periodic table of elements, but they are, after all, what make all matter as we know it.

Kinetics

The study of reaction rates is known as kinetics. It helps us understand how fast certain reactions will occur and what we can do to control that rate.

Nuclear Chemistry

Nuclear chemistry considers reactions that change the composition of an atom's nucleus. It also addresses nuclear stability and decay.

Organic Chemistry

Organic chemistry studies the structure, properties, and reactions of carbon-based structures (organic compounds and materials usually, but can include man-made substances).

Chemical Bonds

Chemical bonds are the glue that hold molecules together. The type of chemical bonding that occurs affects the properties of the material.

In a number of cases, it is the manipulation and modification of cell structures that permits medical technology innovations to occur. This is the case, for example, in genetic modification. Often medical technology and the clinical laboratory sciences necessitate both biology and chemistry-based knowledge.

Career Connections

Biomedical Engineers

Biomedical engineers combine engineering with biology to create medical devices and procedures that are used in the diagnosis and treatment of diseases and health-related problems.

Graduates of bioengineering programs are able to work in "in management,

© A and N photography/Shutterstock.com

production or research and development in a variety of industries – such as medical devices, diagnostics, genetics, health care industry support, pharmaceutical manufacture, drug discovery, environmental remediation, or agricultural advancement – as well as in nonprofit and academic research" (University of California, 2015).

Radiologic and MRI Technologist

A radiologic technologist or radiographer operates diagnostic imaging equipment such as an X-ray machine. They help the patient get into the correct position and operate the machinery. Many of radiographers work in hospitals and other health care facilities. Most states require that the radiologic technologists be licensed or certified, whereas most states do not license MRI technologists though employers still prefer applicants to have certification. An associate degree is typically required to enter this field (U.S. Department of Labor, Bureau of Labor Statistics, 2015a).

Surgical Technologist

Surgical technologists work in operating rooms and are sometimes referred to as operating room technicians. They help to assist with surgical procedures and do activities such as arranging the equipment, preparing the operating room, and helping the surgeons during the procedures. A surgical technologist generally has some type of education beyond high school, which can be obtained from a community college or university. They can be trained in setting up technical or robotic equipment, sterilization techniques, and care and safety of patients (U.S. Department of Labor, Bureau of Labor Statistics, 2015b).

Modular Activities

Discussions

❖ Post an original discussion in the online discussion board on the following topic: Choose an area of interest to you from world health statistics found at http://www.who.int/gho/publications/world_health_statistics/en/ and post an entry about it. What are your thoughts and why? (200 words minimum). Next, comment on the post(s) of a minimum of one other student in a thoughtful and academic way that enhances the conversation. See rubric for grading and assessment measures.

❖ Post an original discussion in the online discussion board on the following topic: Some people say we are taking genetic coding, tracking, modification, and manipulation too far. What are your thoughts and why? (200 words minimum).

Next, comment on the post(s) of a minimum of one other student in a thoughtful and academic way that enhances the conversation. See rubric for grading and assessment measures.

❖ Post an original discussion in the online discussion board on the following topic: There is some belief that in the near future bioengineered body parts will be commonplace. What are your thoughts and why? (200 words minimum). Next, comment on the post(s) of a minimum of one other student in a thoughtful and academic way that enhances the conversation. See rubric for grading and assessment measures.

❖ Listen to the one minute audio file about a rare genetic mutation that makes people very short but also protects them from cancer and diabetes found at http://sciencenetlinks.com/science-news/science-updates/cancer-diabetes-resistance/. What role might medical understanding and related medical technology advancements such as this play in society? (200 words minimum). Next, comment on the post(s) of a minimum of one other student in a thoughtful and academic way that enhances the conversation. See rubric for grading and assessment measures.

❖ Listen to the one minute audio file about reversing cell division found at http://sciencenetlinks.com/science-news/science-updates/cells-in-reverse/. What role could this play in medical technology? What are some things that should be considered with this type of technique? (200 words minimum). Next, comment on the post(s) of a minimum of one other student in a thoughtful and academic way that enhances the conversation. See rubric for grading and assessment measures.

Tests

❖ Online graded quiz on overall chapter content, written in multiple choice format. When submitted, we recommend giving the correct answer along with the page number in which it is found for questions students did not answer correctly.

❖ Use the website Socrative found at http://www.socrative.com/ (free of charge) to set up a live interactive quiz where students can instantly see the overall results for the class. Ask the following questions:
  ➢ What do you feel are the most important areas of medical technology development? (short answer)
  ➢ I believe medical advances are, overall, positive (true/false)
  ➢ I believe there should be equal access to medical resources nationwide (true/false)
  ➢ I believe there should be equal access to medical resources worldwide (true/false)
  ➢ I am comfortable with the idea of purposeful genetic modification (true/false)
❖ Discuss the overall results in the classroom.

Research

* Medical technology: Choose a current or upcoming (being developed for the future or developed within the last 5 years) medical technology of interest to you and write a 3-page informative paper about it.
* Interview a person who works in a field related to medical technology. Ask the following questions at a minimum and then come up with four (4) questions of your own:
  - Which technologies play the greatest role in your job or field and why?
  - What are challenging aspects of your job?
  - What are your favorite parts of your job?
  - What would you recommend to someone going in to this field?
  Submit both the questions and the answers.
* Research one of the following higher education academic areas:
  - Chemistry
  - Biochemistry
  - Pharmacology
  - Pathology
  - Medical technology

This might include review of offerings or degrees at your institution or at another location, contacting a professional/faculty in the field, assessing higher education level academic events or educational resources, and so on. Submit a 200-word minimum summary of your findings. Make sure to include your source(s).

Design and/or Build Projects

* You have been asked to modify and improve the "peering inside the body" site found at http://www.exploratorium.edu/exploring/bodies_mag/peering.html. What would you change, how, and why? Create a simple mock-up of your recommendations with rationale for changes made.
* You have been asked by a local family practice clinic to make a poster that describes cancer risks (factors that increase your change of developing cancer) and what can be done to mitigate that risk. The idea came to them when they came across the site http://sciencenetlinks.com/lessons/cancer-risks/ although other sources are likely useful as well. Create a poster for the family practice clinic. You will be graded on the depth, quality, accuracy, and professionalism of your information.
* You have been told that your educational institution is obtaining some high quality three-dimensional (3D) printers and are interested in potentially printing items, on a limited basis, that could be useful in medicine as a type of service to the community. You know that some web resources offer free 3D print models or stereolithography (STL) files used by 3D systems. Find three resources that you believe would be helpful and explain how they could be used.

Assessment Tasks

❖ Consider the medical field of cosmetic surgeries. Choose a cosmetic surgery such as facial fillers, tummy tucks, eyelid surgery, body liposuction, face lifts, or similar and assess the benefits and drawbacks of this particular technology. Write a two-page assessment of your findings using a minimum of two sources.

❖ Historically, there was a woman named Henrietta Lacks whose cells were taken without her knowledge and then bought and sold to millions as an immortal cell line known as HeLa. Research this situation and write up your assessment of what occurred. Items to include are as follows: (a) a summary, as you understand it, of the Henrietta Lacks and HeLa; (b) a personal assessment of the events in history related to HeLa; (c) a personal assessment of the medical appropriateness of HeLa; and (d) a list of sources used.

Terms

Bioengineering—"The biological or medical application of engineering principles (as the theory of control systems in models of the nervous system) or engineering equipment (as in the construction of artificial organs)—called also biomedical engineering; the application of biological techniques (as genetic recombination) to create modified versions of organisms (as crops); especially: genetic engineering" (Meriam-webster.com/dictionary/bioengineering).

DNA (deoxyribonucleic acid)—"The molecule that encodes genetic information. DNA is a double-stranded molecule made of two twisting, paired strands held together by weak bonds between base pairs of nucleotides" (U.S. Department of Health and Human Services, National Human Genome Research Institute, 2003).

Gene—"The fundamental physical and functional unit of heredity. A gene is an ordered sequence of nucleotides located in a particular potion within the genome that encodes a specific functional product (i.e., a protein or RNA molecule)" (U.S. Department of Health and Human Services, National Human Genome Research Institute, 2003).

Genetic engineering—"The group of applied techniques of genetics and biotechnology used to cut up and join together genetic material and especially DNA from one or more species of organism and to introduce the result into an organism in order to change one or more of its characteristics" (http://www.merriam-webster.com/dictionary/genetic%20engineering).

Genome—"All of the genetic material of a particular organism; its size is generally given as its total number of base pairs, or as its total number of genes" (U.S. Department of Health and Human Services, National Human Genome Research Institute, 2003).

"An organism's complete set of DNA, including its genes" (https://ghr.nlm.nih.gov/primer/hgp/genome).

Genomic medicine—The National Human Genome Research Institute defines genomic medicine as "an emerging medical discipline that involves using genomic information about an individual as part of their clinical care (e.g., for diagnostic or therapeutic decision-making) and the health outcomes and policy implications of that clinical use" (https://www.genome.gov/27552451/what-is-genomic-medicine/).

Nucleotide bases—"The basic subunits of DNA or RNA. Thousands of nucleotides are linked to forma DNA or RNA molecule. The four nucleotides in DNA contain the bases adenine (A), guanine (G), cytosine (C), and thymine (T). In nature, base pairs form only between A and T and between G and C; thus the base sequence of each single strand can be deduced from that of its partner" (U.S. Department of Health and Human Services, National Human Genome Research Institute, 2003).

Protein—"A large molecule composed of one or more chains of amino acids in a specific order; the order is determined by the base sequence of nucleotides in the gene that codes for the protein. Proteins are required for the structure, function and regulation of the body's cells, tissues and organs, and each protein has unique function. Examples are hormones, enzymes and antibodies" (U.S. Department of Health and Human Services, National Human Genome Research Institute, 2003).

Sequencing—"Determination of the order of nucleotides (base sequences) in a DNA or RNA molecule" (U.S. Department of Health and Human Services, National Human Genome Research Institute, 2003).

References

Berger, D. (1999). A brief history of medical diagnosis and birth of clinical laboratory: Part 1 – Ancient times through the 19th century. *Medical Laboratory Observer*. Retrieved from http://www.academia.dk/Blog/wp-content/uploads/KlinLab-Hist/LabHistory1.pdf

Busse, F. (2006). Diagnostic imaging. In G. Spekowius & T. Wendler (Eds.), *Advances and trends in healthcare technology* (pp. 15–34). The Netherlands: Springer.

Centers for Genetics Education. (2015). *Fact sheet 1: An introduction to DNA, genes and chromosomes.* Retrieved from http://www.genetics.edu.au/Publications-and-Resources/Genetics-Fact-Sheets/FactSheetDNAGenesChromosomes

Centers for Medicare and Medicaid Services. (2014). *National health expenditures 2014 highlights.* Retrieved from https://www.cms.gov/research-statistics-data-and-systems/statistics-trends-and-reports/nationalhealthexpenddata/downloads/highlights.pdf

Danigelis, A. (2013).10 bioengineered body parts that could change medicine. Retrieved from http://mashable.com/2013/07/23/bioengineered-body-parts/#oEN5tRv3wPqI

Department of Health and Human Services. Privacy, security and electronic health records. Retrieved from http://www.hhs.gov/sites/default/files/ocr/privacy/hipaa/understanding/consumers/privacy-security-electronic-records.pdf

Drake, Charles, (2016). Contributing author.

Encyclopedia Britannica. (2016). *Hippocrates*. Retrieved from http://www.britannica.com/biography/Hippocrates

Engelhaupt, E. (2015), Bloodletting is still happening, despite centuries of harm. *National Geographic*. Retrieved from http://phenomena.nationalgeographic.com/2015/10/27/bloodletting-is-still-happening-despite-centuries-of-harm/

Gossink, R., & Souquet, J. (2006). Stopping diseases before they start. In G. Spekowius & T. Wendler (Eds.), *Advances and trends in healthcare technology* (pp. 1–14). The Netherlands: Springer.

Griffiths, A. J. F., Miller, J. H., Suzuki, D. T., Lewontin, R. C., Gelbart, W. M. (2000). *An introduction to genetic analysis: Making recombinant DNA 7th Edition*. New York, NY: W. H. Freeman. Retrieved from http://www.ncbi.nlm.nih.gov/books/NBK21881/

History of the Microscope. (2016). Anton van Leeuwenhoek. Retrieved from http://www.history-of-the-microscope.org/anton-van-leeuwenhoek-microscope-history.php

Hopkins Medicine. (2016). Types of anesthesia and your anesthesiologist. Retrieved from http://www.hopkinsmedicine.org/healthlibrary/conditions/surgical_care/types_of_anesthesia_and_your_anesthesiologist_85,P01391/

How equipment works, pulse oximeter. Retrieved from http://www.howequipmentworks.com/pulse_oximeter/

Imperial National Institute for Health Research. (2016). Surgery and surgical technologies. Retrieved from http://imperialbrc.org/our-research/research-themes/surgery-and-surgical-technology

Industry Week: Advancing the Business of Manufacturing. (2013). *Six trends that will shape the pharmaceutical industry*. Retrieved from http://www.industryweek.com/emerging-technologies/six-tech-trends-will-shape-pharmaceutical-industry-2013?page=2

Lam, P., (2016). MRI scans: How do they work? *Medical News Today*. Retrieved from http://www.medicalnewstoday.com/articles/146309.php#background_of_the_MRI_scanner

Lanfranco, A. R., Castellanos, A. E., Desai, J. P., & Meyers, W. C. (2004). Robotic surgery: A current perspective. *Annals of Surgery, 239*(1), 14–21. Retrieved from http://doi.org/10.1097/01.sla.0000103020.19595.7d

Life expectancy in the USA 1900–1998. (1998). Retrieved from http://u.demog.berkeley.edu/~andrew/1918/figure2.html

Logan, L. (2016). Early microscopes revealed a new world of tiny living things. *Smithsonian*. Retrieved from http://www.smithsonianmag.com/science-nature/early-microscopes-revealed-new-world-tiny-living-things-180958912/?no-ist

Martins, R., Lithgow, G. J., & Link, W. (2016). Long live FOXO: Unraveling the role of FOXO proteins in aging and longevity. *Aging Cell, 15*(2), 196–207. Retrieved from http://onlinelibrary.wiley.com/enhanced/doi/10.1111/acel.12427/

McEwen, J. E., Boyer, J. T., Sun, K. Y., Rothenberg, K. H., Lockhart, N. C., & Guyer, M. S. (2014). The Ethical, Legal, and Social Implications Program of the National Human Genome Research Institute: Reflections on an Ongoing Experiment. *Annual Review of Genomics and Human Genetics, 15*, 481–505. doi:10.1146/annurev-genom-090413-025327. Retrieved from http://www.annualreviews.org/eprint/eDSR5xjQy7XjwMQ9VDXs/full/10.1146/annurev-genom-090413-025327

Merriam-Webster (n.d.). Bioengineering. Retrieved from http://www.merriam-webster.com/dictionary/bioengineering

Morris, B. J., Willcox, D. C., Donlon, T. A., & Willcox, B. J. (2015). FOXO3A major gene for human longevity—A mini-review. *Gerontology, 61*, 515–525. Retrieved from http://www.karger.com/Article/FullText/375235#

Mungadi, I. A. (2012). Bioengineering tissue for organ repair, regeneration, and renewal. *Journal of Surgical Technique and Case Report, 4*(2), 77–78. doi:10.4103/2006-8808.110247

National Human Genome Institute. (2015). Frequently asked questions about genetic testing. Retrieved from https://www.genome.gov/19516567/faq-about-genetic-testing/#al-3

National Institute on Aging, National Institute of Health. (2011). Global health and aging. NIH Publication no. 11-7737. Retrieved from https://www.nia.nih.gov/research/publication/global-health-and-aging/living-longer

National Institute of Biomedical Imaging and Bioengineering. (2016a). Mammography. Retrieved from https://www.nibib.nih.gov/science-education/science-topics/mammography

National Institute of Biomedical Imaging and Bioengineering, (2016b). X-rays. Retrieved from nibib.nih.gov/science-education/science-topics/x-rays

National Institute of Biomedical Imaging and Bioengineering (2016c). *Tissue engineering and regenerative medicine.* Retrieved from https://www.nibib.nih.gov/science-education/science-topics/tissue-engineering-and-regenerative-medicine

National Institute of General Medical Sciences. (2016). Pharmacogenomics fact sheet. Retrieved from https://www.nigms.nih.gov/education/pages/factsheet-pharmacogenomics.aspx

National Institute of Health. (2013). Human genome fact sheet. Retrieved from https://report.nih.gov/NIHfactsheets/Pdfs/HumanGenomeProject(NHGRI).pdf

National Institute of Health. (2016). Genetics home reference. What is a cell? Retrieved on December 13, 2016 from https://ghr.nlm.nih.gov/primer/basics/cell

National Research Council (US) Committee on National Statistics. (2010). Improving health care cost projections for the Medicare population: Summary of a workshop. In *Modeling medical technology* (p. 3). Washington, DC: National Academies Press. Retrieved from http://www.ncbi.nlm.nih.gov/books/NBK52816/

Paddock, C. (2012). Nanotechnology in medicine: Huge potential, but what are the risks? Retrieved from http://www.medicalnewstoday.com/articles/244972.php

RadiologyInfo.org (2016). Computed tomography (CT) - Body. Retrieved from http://www.radiologyinfo.org/en/info.cfm?pg=bodyct#common-uses

RFID Journal. (2016). Health care news. Retrieved from http://www.rfidjournal.com/healthcare

Science Museum. *Brought to life: Exploring the history of medicine.* Joseph Lister (1827–1912). Retrieved from http://www.sciencemuseum.org.uk/broughttolife/people/josephlister

Spekowius, G., & Wendler, T. (2006). *Advances in healthcare technology: Shaping the future of medical care.* The Netherlands: Springer.

Tsung, J. 2016. History of ultrasound and technological advances. Retrieved from http://www.wcume.org/wp-content/uploads/2011/05/Tsung.pdf

The World Bank. (2016a). Physicians (per 1000 people). Retrieved from http://data.worldbank.org/indicator/SH.MED.PHYS.ZS

The World Bank. (2016b). Hospital beds (per 1000 people). Retrieved from http://data.worldbank.org/indicator/SH.MED.BEDS.ZS

U.S. Department of Health and Human Services, National Institutes of Health, The National Human Genome Research Institute. (2003). Glossary of terms. Retrieved from https://www .genome.gov/pages/education/modules/blueprinttoyou/blueprintinsideback.pdf

University of California, Berkeley. (2015). What is bioengineering? Retrieved from http://bioeng. berkeley.edu/about-us/what-is-bioengineering on 5-20-2016

U.S. Department of Health and Human Services. (2016a). Organ procurement and transplantation network. Retrieved from https://optn.transplant.hrsa.gov/

U.S. Department of Health and Human Services. (2016b). Privacy, security and electronic health records. Retrieved from http://www.hhs.gov/sites/default/files/ocr/privacy/hipaa/ understanding/consumers/privacy-security-electronic-records.pdf

U.S. Department of Labor, Bureau of Labor Statistics. (2015a). *Radiologic and MRI technologists*. Retrieved from http://www.bls.gov/ooh/healthcare/radiologic-technologists.htm

U.S. Department of Labor, Bureau of Labor Statistics. (2015b). Surgical technologist. Retrieved from http://www.bls.gov/ooh/healthcare/surgical-technologists.htm#tab-4

U.S. Food and Drug Administration (FDA). (2016a). MRI (magnetic resonance imaging). Retrieved from http://www.fda.gov/Radiation-EmittingProducts/RadiationEmittingProductsandProcedures/ MedicalImaging/MRI/default.htm

U.S. Food and Drug Administration (FDA). (2016b). Ultrasound imaging. Retrieved from http:// www.fda.gov/Radiation-EmittingProducts/RadiationEmittingProductsandProcedures/ MedicalImaging/ucm115357.htm

U.S. National Library of Medicine. (2016a). *Genetics Home Reference.* What is DNA? Retrieved from https://ghr.nlm.nih.gov/primer/basics/dna

U.S. National Library of Medicine. (2016b) *Genetics Home Reference.* What is mitochondrial DNA? Retrieved from https://ghr.nlm.nih.gov/primer/basics/mtdna

U.S. National Library of Medicine. (2016c). *Genetics home reference.* What are proteins and what do they do? Retrieved from https://ghr.nlm.nih.gov/primer/howgeneswork/protein

U.S. National Library of Medicine. (2016d). *Genetics home reference.* What are the types of genetic tests? Retrieved from https://ghr.nlm.nih.gov/primer/testing/uses

U.S. National Library of Medicine (2016e). *Medline Plus.* PET scan. Retrieved from https://www .nlm.nih.gov/medlineplus/ency/article/003827.htm

National Institute on Health, National Institute on Aging. (2011). Retrieved from https://www.nia. nih.gov/research/publication/global-health-and-aging/living-longer

Wilhelm Konard Roentgen. (2016). Retrieved from https://explorable.com/wilhelm-conrad-roentgen

World Health Organization. (2016). Life expectancy at birth, 2015. Retrieved from http://www .who.int/gho/mortality_burden_disease/life_tables/situation_trends/en/

Further Reading

Bureau of Labor Statistics. (2016). *Occupational outlook handbook. Surgical Technologisits*. Retrieved from http://www.bls.gov/ooh/healthcare/surgical-technologists.htm#tab-4

Cordis Europa. (2013). Nanotechnology documentary—Revolutionizing medicine and healthcare. [video]Retrieved from https://www.youtube.com/watch?v=_GjqbUPmcWQ https://www. youtube.com/watch?v=_GjqbUPmcWQ

Encyclopedia.com. (2001). *Life expectancy.* Retrieved from http://www.encyclopedia.com/topic/ Life_expectancy.aspx

Food and Drug Administration. *FDA.gov.* Retrieved from http://www.fda.gov/Radiation-Emitting Products/RadiationEmittingProductsandProcedures/MedicalImaging/ucm2005914.htm

Chapter 11

Biotechnology

(Continued)

(*Continued*)

* ❖ Modular activities
    * ➤ Discussions
    * ➤ Tests
    * ➤ Research
    * ➤ Design and/or build projects
    * ➤ Assessment tasks
* ❖ Terms
* ❖ References
* ❖ Further reading

Chapter Learning Objectives

* ❖ Define agriculture biotechnology
* ❖ Identify traits that can be genetically engineered into crops
* ❖ Explain how bioremediation works
* ❖ Examine biodiversity and its importance
* ❖ Describe the use of biotechnology in crop production
* ❖ Distinguish the difference between the cloning techniques of artificial twinning and somatic cell nuclear transfer
* ❖ Summarize biotechnology use in fuel production
* ❖ Compare the benefits and concerns of genetically engineered food crops
* ❖ Differentiate the roles of the FDA, EPA, and USDA in regulating the genetic modification of crops in the United States

Overview

Biotechnology has many applications in many different industries, including but not limited to, agriculture and food production, pharmaceutical production, genetic engineering and gene therapy, immunology and virology, fuel production, and bioremediation of waste products. It refers to the manipulation of living organisms or their components to produce products or it can also be considered as the application of biological science to produce products (Merriam-Webster, 2016). Humans have used biotechnology for thousands of years although the term *biotechnology* was only first used in the beginning of the twentieth century. The making of bread, cheese, and beer requires the use of biotechnological processes, though we often do not consider these processes to be biotechnology processes when in fact they are.

The modern era of the biotechnology industry started in the mid-1970s and today many people associate biotechnology with genetic engineering. Genetically modified organisms referred to as GMOs are sometimes thought to be the only biotechnological organisms but biotechnology is much broader than just GMOs. Biotechnology is

very much of an interdisciplinary field that integrates aspects of biology with biochemistry, information technology, molecular biology, nanotechnology, pathology, engineering, and other fields (Newton, 2003).

This chapter focuses on several biotechnology applications including crop and livestock production, biotechnology developed fuels, mining applications, bioremediation, and biopharming applications. The influence that biotechnology can have on maintaining biodiversity will be addressed. We will also review regulations that impact the biotechnology industry.

The use of biotechnology has many positive aspects: food can be made more nutritious, waste can be converted into fuel, drugs can be produced more efficiently, and resources could be used more effectively. There is even a term called bioeconomy, which "refers to the set of economic activities relating to the invention, development, production and use of biological products and processes" (OECD, 2009).

Cultural, social, economic, and industrial areas will be impacted by developments in biotechnology. Regulation and policy formulation in agriculture, medicine, and industrial applications will be necessary in order to optimize the potential benefits of biotechnology while at the same time minimizing the potential harm. We are seeing a coevolution of biotechnology applications in the aforementioned industries that coincides with an evolution in social and regulatory areas that takes place within local and global economies. Basically biotechnology is changing the way we live, eat, stay healthy, and produce goods almost everywhere.

As the world population continues to expand and the per capita income continues to grow there will be increasing demands on energy and food resources as well as increased medical demands due to aging populations. Some of these demands can be met with biotechnological solutions. There is room for caution however in the development and use of some biotechnological processes and products. We still do not know the full impact of some of these biotechnologies on the human body and the environment.

Brief History of Biotechnology

The use of biotechnology processes is not new in agricultural practices; in fact, it is thousands of years old. During the hunting and gathering period, food used to be harvested from plants in the surrounding areas and there was no domestication of animals for breeding and food production. Overtime though, people instead of just gathering food began to prorogate crops, and they started saving the seeds from the "best" plants. By planting the seeds from the plants that had the traits that were desirable, such as larger seeds, bigger fruits or a shorter time to maturity, different cultivars or crop varieties began to appear. These practices are still in use today in agriculture, different varieties of tomatoes for example, are being produced. Some people today want heirloom tomato varieties, such as Brandywine that is noted for its "tomatoey" flavor and takes 90–100 days to fruit, instead of the "Early Girl" tomato that takes only 63 days to reach maturity, but is not as flavorful as the Brandywine tomato.

Modern agricultural practices have changed the way we live and work, in 1870 about 53% of the US population were involved in some form of farming and there were under 39 million people living in the United States. Today with over 275 million

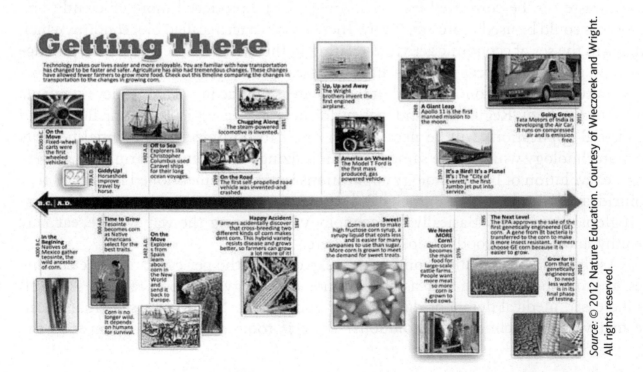

people living and working in the United States, only about 1.8% of the population are involved in farming (Wieczorek & Wright, 2012).

Animal biotechnology has also been around for thousands of years. People started to domesticate animals as early as 5000 BC and would cross different strains of animals to introduce greater genetic variety. This process is known as hybridizing, and hybridizing was also used in crop production. A good example, that has been around for about 3000 years, is when a female horse is bred with a male donkey their offspring is called a mule, and mules are very good work animals (North Carolina Association for Biomedical Research, 2016).

Agriculture-Related Biotechnologies

According to the U.S. Department of Agriculture (USDA), "Agricultural biotechnology is a range of tools, including traditional breeding techniques that alter living organisms, or parts of organisms, to make or modify products; improve plants or animals; or develop microorganisms for specific agricultural uses" (U.S. Department of Agriculture, 2016). The technologies used in genetic engineering are considered biotechnology tools today.

Farmers use biotechnology tools to make the growing of crops and animals less expensive and more manageable, and to produce products that are more appealing to the consumers, such as redder tomatoes. Crops can be genetically modified to be tolerant or resistant to herbicides, they can also have plant-incorporated protectants that will protect the plant from insect damage. For example, crops can be genetically engineered to carry a gene from the soil called bacterium *Bacillus thuringiensis* (Bt). This "bacterium produces proteins that are toxic to some pests but are non-toxic to humans and other mammals. Crops containing the Bt gene are able to produce this toxin thereby providing protection for the plant" (U.S. Department of Agriculture, 2013). In the United States, there is widespread planting of corn and cotton that has Bt genetically engineered into the seed and is commercially available.

Biotechnology in Crop Production

The use of genetically engineered crops has been occurring for more than 30 years. An antibiotic resistant tobacco placed was the first genetically modified plant produced in 1983, and in the early 1990s China commercialized tobacco that was virus resistant (Bawa & Anilakumar, 2013). By the mid-1990s, the U.S. Food and Drug Administration (FDA) had approved the "Flavour Saver tomato" which was genetically engineered to ripen slower after picking. Today there are many foods and crops produced that are genetically modified including but not limited to the following: carrots,

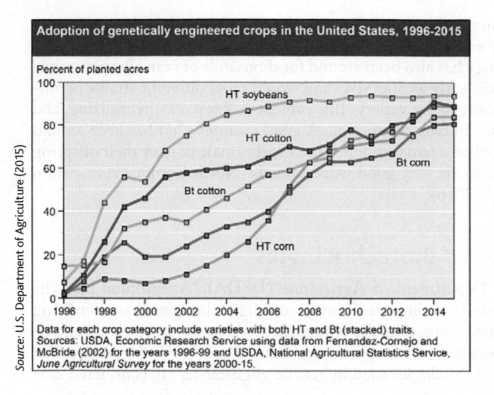

Adoption of genetically engineered crops in the United States, 1996-2015

Percent of planted acres

Source: U.S. Department of Agriculture (2015)

Data for each crop category include varieties with both HT and Bt (stacked) traits.
Sources: USDA, Economic Research Service using data from Fernandez-Cornejo and
McBride (2002) for the years 1996-99 and USDA, National Agricultural Statistics Service,
June Agricultural Survey for the years 2000-15.

cantaloupe, cotton, corn, eggplant, lettuce, papayas, potatoes, soybeans, strawberries, sugar beets, and sweet potatoes.

Biotechnology in crop production is big business and growers in the United States have embraced the technology, especially for some of the major crops such as corn, soybeans, cotton, and sugar beets. One of the biggest players in this field is Monsanto Corporation; they were the first company to introduce glyphosate-resistant soybeans in 1996 and glyphosate-resistant corn in 1998. Some of the other producers of genetically modified crops include Dow, Du Pont, Bayer/Genective, Pioneer, Stine Seed, AgrEvo, Northrup King, Plant Genetic systems, DeKalb, and Syngenta (Johnson & O'Connor, 2015).

Multiple traits can be incorporated into plants, such as incorporating an insecticide to prevent damage from insects and incorporating a Roundup ready gene that will prevent the plant from being killed by an herbicide. Some crops have been engineered to be resistant/tolerant to certain herbicides, such as Roundup that uses glyphosate to kill flora, this makes weed control much easier and it is far more efficient for the farmer to plant. For example, a farmer can spray herbicide while at the same time planting a field of corn that was genetically engineered to be resistant of the herbicide Roundup.

The use of herbicide-tolerant crops and insect-resistant crops has increased dramatically since their inception in the United States. In the United States, there are up to 33 different varieties of genetically modified corn, including varieties that are moth and butterfly resistant, European Corn Borer Resistant, rootworm resistant, drought resistant, male sterile, and herbicide tolerant to name a few (Johnson and O'Connor, 2015).

An important measure of research and development activity in genetically engineered crops is the number of field releases given by the USDA's Animal and Plant Health Inspection Services to test genetically modified crops. "As of September 2013, about 7,800 releases were approved for GE [genetically engineered] corn, more than 2,200 for GE soybeans, more than 1,100 for GE cotton, and about 900 for GE potatoes. Releases were approved for GE varieties with herbicide tolerance (6,772 releases), insect resistance (4,809), product quality such as flavor or nutrition (4,896), agronomic properties like drought resistance (5,190), and virus/fungal resistance (2,616)"

© Marcin Balcerzak/Shutterstock.com

(Fernandez-Cornejo, Wechsler, Livingston, & Mitchell, 2014). In 2013, as much as 85% of the acreage used to grow corn in the United States was genetically engineered to be herbicide tolerant and 82% of the acreage devoted to growing cotton had herbicide-resistant properties (Fernandez-Cornejo et al., 2014).

Benefits of biotechnology in crop production

There are many benefits of biotechnology use in the agricultural industry to farmers, producers, and consumers. Weeds can be controlled more easily with reduced-risk herbicides that break down easily and are not toxic to humans and wildlife. Crops can be protected from destructive diseases, for example, the Hawaiian papaya industry was saved from the papaya ring spot virus which threatened to wipe out the papaya production and now the papaya industry is thriving.

Farmers can produce higher yields with some genetically engineered crops and they can also increase the quality of crops. Farming can be made safer with less pesticide use which can cause harm to the environment and ground water. Crops can also be engineered to be more nutritious, for example, rice can be genetically modified to have increased levels of beta-carotene which could reduce vitamin A deficiency. Even the makeup of oil in soybeans and canola can be structured to be healthier. Crops have also been modified to be drought tolerant, which is extremely important in areas where drought has become a problem such as California.

Safety and concerns of genetically engineered crops

There are concerns and controversies surrounding the use of biotechnology techniques in agricultural crop production. The most pressing concerns are about human safety and environmental safety of crops used for food and feed. There are also issues regarding labeling and consumer choice of genetically modified food products. Environmental conservation and biodiversity are also important topics that people want to see addressed. In addition there are issues regarding food security, poverty reduction, and ethics (Bawa & Anilakumar, 2013).

There are, however, tests that can detect and quantify genetically modified organisms (GMOs) in food sources; these tests would assist in the ability to develop labeling of genetically engineered food. Manufacturers are opposed to labeling because they fear it will have a negative impact on sales if consumers know that a crop used in production of the food was genetically modified in some manner. The European Union (EU) has far stricter requirements for labeling of genetically modified food. The EU has adopted a mandatory labeling requirement for even the nonintentional presence of genetically engineered material in food or feed, using a traceability concept. Traceability as defined by the EU General Food Law Regulation 178/2002/EC refers to the "ability to trace and follow food, feed, food producing animals and other substances intended to or expected to be incorporated into food or feed, through all stages of production, processing and distribution."

Some evidence suggests that there has been an increase in herbicide (glyphosate) resistant weeds as a result of planting glyphosate crops (Fernandez-Cornejo et al., 2014). This would result in an increase in the use of herbicides. Most of the scientific research that has been conducted to date indicates that there is not any significant risk associated with the growing and use of genetically engineered crops (Nicolia, Manzo, Veronesi, & Rosellini, 2014).

Biotechnology in Livestock Production

Though traditional biotechnology practices in animals have occurred for thousands of years, genetically modifying animals has lagged behind the genetic modification of plants (Shmaefsky, 2013). The three main areas in animal biotechnology are animal genomics, animal cloning, and genetic engineering of animals. Animal biotechnology can be used to improve human and animal health, to develop more nutritious food, and in the conservation of animals and the environment (Bio: Biotechnology Industry Organization, n.d.).

Animal genomics

Unlocking the genetic code of animals using DNA sequencing techniques is very important to society. Using sequencing can benefit agricultural practices, medical and health for both humans and animals and it enhances our basic understanding of science.

In agriculture studying the genomes of domestic livestock can help to improve production and reduce disease which can ultimately have economic benefits to producers and consumers. We have bred animals for different traits for thousands of years. Some of the traits that are looked at in domestic livestock include: how fast the animal grows during gestation and after birth, what are the traits of the carcass and how is fat distributed between muscles, how tender and tasteful is the meat, how many eggs are produced, and what is the fat content and volume of milk produced. Traditional methods of breeding can take many years, multiple tries, and multiple generations to develop desired traits. Developing an understanding of the animals' genome can lead to much greater precision in the developing desired traits with the use of genetic engineering (Pool & Waddell, 2002).

Developing a basic scientific understanding of the genomic sequencing of animals can help us understand the evolution of different species and their relationship to each other. Millions of years of natural selection and mutation that were necessary to adapt to the environment have led to the genomes of domesticated animals that are sequenced today. By sequencing genomes of different species we can find out how they diverged into different forms and how they are related to each other (Pool & Waddell, 2002). For example, scientists through DNA analysis have established that wolves are the ancestor of dogs. Consider how our beloved dogs are related to the wolf (Public Broadcasting Station, n.d.).

Medical research for both humans and animals can benefit from the genetic sequencing of animals. There are many similarities and, of course, differences between the genomes of animals and human.

Biodiversity Conservation

The conservation of both domestic animals (livestock) and wild species is important for environment, economy, history, and culture. There are many advantages associated with maintaining biodiversity in livestock production including the preservation of nutritional and nutraceutical properties. There are also social, economic, and cultural reasons for preserving the biodiversity of livestock. With the worldwide population expecting to be over 9 billion people by 2050 (United Nations, 2015), the ability to maintain adequate food sources is going to be increasingly important.

© Protasov AN/Shutterstock.com

According to the Animal Genetic Resources, it is necessary to have genetic diversity so that animals can adapt to their environments and for improvement. Some of the main causes for loss of biodiversity in livestock can be attributed to livestock industry practices, disasters and emergencies, and animal disease epidemics and control measure (Pizzi et al., 2013)

There are several ways of conserving biodiversity including in situ and ex situ conservation strategies. In situ biodiversity conservation is when the conditions exist to perpetuate the breed within natural habitats or ecosystems or the area in which the breed was developed. Cryoconservation is one of the primary methods of ex situ conservation and it involves the freezing and storage of semen, ova, embryos, or tissues. There has also been the development of bio-banks which are similar to a seed bank, genetic information is stored and cataloged in bio-banks. Digital analysis and preservation is also important.

The cryopreservation of genetic materials can be used for several things including the following:

- Reconstruction of a breed in cases of extinction or loss of a large number of animals
- Develop new breeds
- As a backup plan or to modify different selective breeding programs
- In conjunction with in vivo bioconservation plans
- Resource for research (Pizzi et al., 2013)

Farming and livestock production has become increasingly industrialized since the 1950s. The larger agricultural establishments focused on the development of a few very specialized breeds. For example, there are many different breeds of cattle, some are bred for meat production and some are bred for milk production. In the United States the most popular milk producing breed is Holstein breed. This breed is a reliable cow that produces high volumes of milk. Different breeds of cattle are used in meat production and some cattle have been bred for dual use such as milk and meat production. The increasing industry specialization has led to the decreasing biodiversity of livestock.

A good example of livestock biodiversity preservation comes for the conservation of the Reggiana cattle breed from Italy. Many of you may enjoy Parmigiano Reggiano cheese on your favorite Italian dish. This cheese is produced from the Reggiana cattle breed. Farmers started to use the Holstein cattle breed because of its superior milk production capabilities and the Reggiana cattle became far less popular until the brand "Parmigiano Reggiano" was created in the 1980s. Parmigiano Reggiano uses only milk for the Reggiana cattle, which is a taste many consumers appreciate. The farmers also appreciate this breed because of the premium price they receive from the products produced using the Reggiana cattle. Another example is from the Guernsey breed of cattle, this breed produces milk with a high butterfat and high protein levels and is known for the delicious dairy products produced from the milk.

There are several reasons for the decline in biodiversity amongst wild animal populations. Much of the decline is due to degradation and loss of habitat. Things such as changes in moisture and temperature and atmospheric conditions can lead to loss of optimal habitats for species of animals, as well as toxins and pesticides use, acid rain and increased CO_2 emissions. Human population and expansion also cause loss of habitat. In the United States, in many areas, you will see growth of subdivisions and development, the same is true in many other parts of the world. The development and use of natural resources along with the poaching of animals can also lead to loss of population in species.

There are several strategies that can be employed to stem the loss of biodiversity in wild animal populations. Obtaining information on wildlife populations and putting in place a plan for the monitoring of the species population and their habitat is necessary and is part of a population vulnerability analysis. It is also important that the ecosystem in which the species lives be preserved. This takes widespread efforts by multiple stakeholders including governments at the local and national levels, conservation groups and concerned citizens living in areas where endangered species live. For example the European Union (EU) has a network of areas that are protected and the US Fish and Wildlife Services publish information about critical habitats and information on protected species. In addition, biotechnological applications such as cloning of animals on the verge of extinction are being developed. A program for the cloning of Pandas is being developed in China (Matheson, 2015).

Cloning

A clone is an exact genetic copy of an organism, with the entire DNA being the same. Identical twins are clones and cloning occurs naturally in nature; cloning can also occur in the lab. Dolly the sheep was the first mammal to be cloned from an adult cell in 1997 and she helped to usher in the modern era of biotechnology in animal production and made people aware of cloning. There are two ways to clone an organism in a lab: one is called somatic cell nuclear transfer and the other artificial embryo twinning (Genetic Science Learning Center, 2014).

Scientists have experimented with cloning over the last 50 years and have cloned different types of animals. The first successful cloning of identical mice occurred in 1979, where mouse embryos were spilt and implanted into the womb of an adult mouse. Cats, dogs, deer, horses, sheep, mules, oxen, rabbits, rats, and even monkeys have been cloned using the artificial twinning or embryo splitting techniques (National Institute of Health, National Genome Research Institute, 2016). Cloning from embryonic cells is actually much easier to do than cloning from adult cells.

Artificial Twinning

Artificial twinning is very similar to natural twinning except that the twinning will take place in a petri dish instead of in the mother's body. Identical twins are formed shortly after an egg and sperm join forming an embryo. When an embryo splits into two and each half continues developing they are considered identical twins because each one carries the same DNA and they both came from the same fertilized egg (Genetic Science Learning Center, 2014).

Somatic Cell Nuclear Transfer

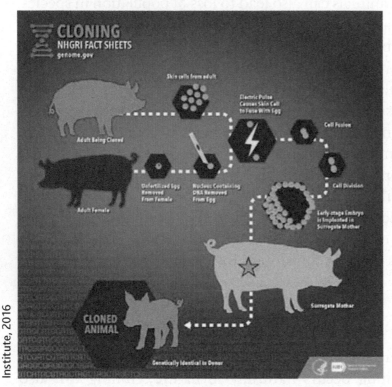

Source: National Institute of Health, National Genome Research Institute, 2016

With somatic cell nuclear transfer a somatic cell (any cell that is not a sperm or egg cell) is isolated, in the case of Dolly the somatic cell was an udder cell. Once the DNA in the somatic cell is isolated, then the nucleus, including all of its DNA is removed from an unfertilized egg cell. The next step is transferring of the DNA from the somatic cell into the egg cell that has had its nucleus removed. Scientists can apply electrical current to fuse the membranes of the somatic cell with the egg cell, which causes the egg cell with the new nucleus to act like a freshly fertilized egg. The fertilized egg develops through cell division, into an embryo. The resulting embryo is then implanted into a surrogate mother that carries it to term. The new organism will be an exact genetic copy of the organism that the somatic cell was taken from. The primary difference between a naturally fertilized embryo and somatic cell nuclear transfer is that in natural fertilization an egg and sperm join creating the two sets of chromosomes, and with somatic cell nuclear transfer the complete set of chromosomes comes from the somatic cell (Genetic Science Learning Center, 2014).

Cloning of animals has some potential applications that could be beneficial. Scientists, for example, have cloned sheep that have been genetically modified to produce milk that has a human protein required for blood clotting. Cloning could also help in the rebuilding of endangered animals' populations. Some people have even wanted to clone their pets.

There are several drawbacks associated with the cloning of animals, though, one of which is the expense. It is also very inefficient, for example it took 277 attempts to clone Dolly the sheep. There are also safety concerns and adverse health effects associated with cloning of animals. Problems associated with cloning include, defects in organs such as the brain, liver, and heart, increased birth size, premature ageing, and problems with the immune system of cloned animals.

Biotechnology in Fuel Production

It has been well known for many years that fossil fuels such as oil and natural gas have finite limits and that using fossil fuels has an effect on atmospheric CO_2 levels and the climate. Even with conservation measures in many countries energy

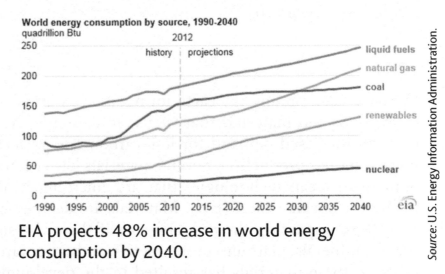

EIA projects 48% increase in world energy consumption by 2040.

Source: U.S. Energy Information Administration.

use is climbing. The International Energy Outlook 2016 projects that the energy consumption in the world will grow by 48% between 2012 and 2040. The growth in the economies in China and India will account for more than half of the increase in energy consumption during this time period. More than 75% of our energy consumption comes from fossil fuels, though nonfossil fuel growth is expected to be faster than fossil fuel growth during this time period (U.S. Energy Information Administration, 2016).

Renewable energy resources such as solar power, wind, geothermal heat, biomass, and water can replace fossil fuel use and they are abundant. Some of the primary reasons that renewable energy resources have not been widely adapted include the cost, the technological infrastructure, and distribution networks.

Biomass is the main renewable source for production of liquid fuels, including biodiesel and ethanol and they are suitable for use in the transportation sector. Biodiesel is made from vegetable oils, recycled restaurant oils, and animal fats. It produces less pollution than diesel made from fossil fuels (U.S. Department of Energy, n.d.).

Ethanol is an alcohol fuel that is made for plants such as corn and grasses and sugar cane. It is often mixed with gasoline. Almost all of the gasoline sold in the United States has up to 10% ethanol in it. Flex fuel, or E85 is a blend of gasoline and ethanol that has between 51% and 83% ethanol and it can be used in flex fuel vehicles (U.S. Department of Energy, n.d.a). In 2008, a researcher, Mariam Sticklen of Michigan State University, was able to genetically modify corn to produce cellulose-degrading enzymes in the plant's stems and leaves. The genetic modification of the corn was done with the goal of making the process of conversion into ethanol more affordable (Goho, 2008).

The use of ethanol is thought to reduce greenhouse gas emissions and our dependence on foreign oil (U.S. Department of Energy, n.d.b). There are concerns however regarding the use of biofuels on society and the environment. The land used to produce crops used for ethanol productions could also be used for food production. There are also concerns about water scarcity, loss of biodiversity, and nitrogen pollution from fertilizers (Stecker, 2014).

Mining and Biotechnology Applications

We have been mining the earth for natural resources for thousands of years to produce goods and materials that we use in daily life. As society progresses there has been an increased demand for low-carbon energy technologies which use certain metals in their production. Several materials have been deemed critical to the development of clean technologies that are emerging in the United States as well as in the European Union (EU). Materials are deemed critical if there is a high risk of adverse impact on the economy and if there is a supply shortage. These can be metals, minerals, platinum group metals, and other rare earth elements. The importance of these materials has resulted in the development of technologies that are used to extract, recover, separate, and purify metals is becoming increasingly crucial (Zhuang et al., 2015).

Biotechnologies can be used for the extraction and recovery of metals from ores and from waste material. The term biomining is sometimes used to describe processes

Courtesy of Dr. Michael Bunds

and technologies that use biological systems to assist in the recovery and extraction of metals from waste materials and ore. Biomining is a process that involves the interactions between metals or metal bearing materials and microorganisms such as *Acidithiobacillus*, a bacterium. Biomining technologies are used to generate approximately 15% of copper and 5% of gold as well as small amounts of other metals (Johnson, 2014).

Utah Brigham Canyon Mine (also known as Kennecott Copper Mine)

Biomining started in the 1960 in Utah's Bingham Canyon mine operated by the Kennecott Copper Corporation. They used it to extract copper from the waste rock which was dumped into large mounds at the mine. They would irrigate the mounds with a diluted sulfuric acid which would stimulate mineral-oxidizing bacteria. The copper rich liquid would then trickle down and would be collected in ponds. This process of bioleaching has evolved; some operations now grind the ore to a small grain size and the mounds are placed on impermeable membranes or liners. They aerate the mounds of ore providing carbon dioxide and oxygen to the bioleaching microorganisms, and also irrigate them and some are inoculated with acidophilic microorganisms. These mounds are sometimes called bio-heaps and they can be stacked on one another. Bioleaching is a relatively slow process but is considered more environment friendly than smelting processes (Johnson, 2014).

There are also stirred tank bioreactors that use bio-oxidation and there is in situ biomining. In situ mining involves the flooding of the mine and the metal enriched liquids are pumped to the surface where the metal is extracted, this method has been in use since medieval times. One of the problems associated with in situ mining is that it is difficult to get carbon dioxide and oxygen, which promote the microbial growth, delivered to the area. Genetic manipulation of the biomining microorganisms has not been pursued as much as other biotechnologies because they cannot prevent the microorganisms used in the processes from being released into the environment (Johnson, 2014).

We can also mine metals through recovery methods such as recycling. There are three areas where recovery of metals could occur, scrap, or residue generated during manufacturing production processes, recycling of products, and landfill mining. These areas, sometimes referred to as "waste streams" have received attention because of the potentially high levels of platinum group metals and rare earth elements which can be found in nickel hybrid batteries, permanent magnets, car catalysts, and electronic waste. There are also liquid waste streams that could be used, such as waste water from pharmaceutical, chemical and glass industries (Zhuang et al., 2015).

Bioremediation

Humans have been polluting earth since the dawn of time, though with the advent of the industrial age, pollution was taken to a whole new level. Oil spills on land and sea require actions to clean them up. We are even seeing genetically engineered plants being developed to help detoxify pollutants in the soil and this process is known as phytoremediation (U.S. Department of Agriculture, 2016). The use of bacteria, fungi, green plants, or other biological agents to neutralize or remove contaminants from polluted water or soil is called bioremediation (The American Heritage Science Dictionary, n.d.).

With bioremediation microbes are used to clean up soil and ground water that is contaminated from things such as oil and petroleum products, solvents, and pesticides. Microbes are small organisms such as bacteria that occur naturally in the environment and certain microbes will use these contaminants as sources of food and energy. During the digestion process the contaminants are usually changed into harmless gases such as ethane and carbon dioxide and water. Sometimes bioaugmentation is necessary and it requires the addition of the right microbes needed for bioremediation.

Several things are needed in order for bioremediation to be effective; you must have the right temperature, food and nutrients so that the microbes will grow and multiply thus eating more of the contaminants. Amendments can improve the conditions, and things such as molasses and vegetable oil, air, and chemicals can be added to improve the conditions necessary for effective bioremediation. In situ or in place bioremediation occurs if these amendments are put into the contaminated area, for example pumping the amendments into the groundwater through wells. But sometimes the conditions may not be right and the soil will have to be dug up before bioremediation can occur and this is considered ex situ or not in place bioremediation. Some bioremediation processes require oxygen to work and this is called an aerobic environment, while others can only be bioremediated if there is no oxygen present, this would be called an anaerobic environment (U.S. Environmental Protection Agency, 2016).

Microorganisms that bioremediate the environment can occur naturally or they can be created through the use of biotechnology. Some microorganisms seek out nutrients and energy from organic substances and will digest them until all of the food source is gone (Center for Public Environmental Oversight, n.d.).

Biopharming

Biopharming can be considered as the production and use of pharmaceutical substances from trans-genetically engineered plants and animals for use in humans and animals (Goven, 2014). Biopharming is heavily dependent upon genetic technologies and information technologies. We will continue to see increased use of these technologies in the development of pharmaceutical products.

Transgenic Animal Use in Medicine

The use of transgenic animals in medicine has been occurring for several decades. The decoding on the human genome and the mouse and rat genomes was important to the development of medicines using transgenic animals. A transgenic animal is one that has been altered by the introduction of recombinant DNA through human intervention (U.S. Food and Drug Admiration, 1995). Recombinant DNA is when a specific gene is taken from one organism and is inserted into another organism using a process similar to cutting and pasting.

Transgenic animals are used for drug research and development, as well as production of drugs and possibly xenotransplantation which is the transplantation of cells or organs from one species to another. Transgenic animals for use in medicine have been genetically engineered to have traits that simulate certain human pathologies. The drug ATryn, an anticoagulant, was the first drug to receive US FDA approval for use in humans (Bagle, Kunkulol, Baig, & More, 2013).

Today transgenic mice and rats are widely used in medical research and investigation of human diseases. They have been developed specifically for various diseases such as human immunodeficiency virus/acquired immunodeficiency syndrome (HIV/AIDS), Alzheimer's disease, cardiovascular disease, diabetes mellitus, angiogenesis, and cancer. The first transgenic animal to receive a patent was the Oncomouse, which was genetically mod-

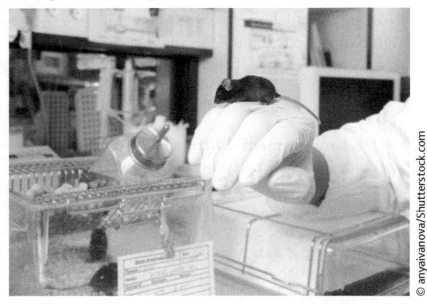

Work with transgenic mouse in modern laboratory.

© anyaivanova/Shutterstock.com

ified to develop cancer easily and is used to research and develop drugs that are used in cancer treatment (Brown, 2000).

Bioengineered Crops in Medicine

Pharmaceuticals such as edible vaccines and antibodies can be produced in plant crops and animals now. Scientists have been able to genetically modify rice to increase the provitamin A content, which is important in the prevention of blindness in children, especially in developing countries. Today about one-third of all deaths globally are due to infectious diseases, such as HIV, hepatitis C, and SARS, even some cancers may be caused from infectious agents. Vaccinations can prevent someone from developing these diseases and the use of plant-based vaccines has been gaining an increasing amount of market share (Ahmad et al., 2012).

Plants have been genetically engineered to produce therapeutic recombinant proteins, with the most important being plant-based vaccines (Goven, 2014). These plant-based systems can be scaled up for production at the industrial level, and can be cost-effective for production of recombinant proteins. Some of the crops used for the production of biopharmaceutical proteins include leafy crops, legumes, fruits, and vegetables, and vaccines have been developed in tomatoes, carrots, and potatoes. Certain plant tissues can also be used for the administration of vaccines which reduces the costs and labor of administering an injectable vaccine, plus it does not hurt.

Regulatory Agencies in the Biotechnology Industry

There are several agencies within the United States that regulate and oversee the biotechnology applications. The three agencies that have oversight of genetically engineered crops are the US Environmental Protection Agency (EPA), the US Food and Drug Administration (FDA, and the U.S. Department of Agriculture (USDA. Each of these agencies has specific responsibilities and powers.

❖ **US Environmental Protection Agency (EPA)**: Regulates toxins and biopesticides in genetically engineered crops under the Federal Insecticide, Fungicide, and Rodenticide Act. They regulate the use, sale, distribution, and testing of pesticidal substances. Plants that are engineered to be insecticide resistant or that carry the gene for a Bt toxin must go through rigorous testing to determine if it is safe for the environment and safe for consumption, and steps are also taken to assess whether the resulting plant is allergenic using food safety analysis (USDA, 2013).

❖ **US Department of Agriculture (USDA)**: In the United States, commercial production of genetically engineered crops requires that biotechnology companies apply for a "deregulated status" from the USDA to plant and distribute the crops without any type of restrictions (Johnson & O'connor, 2015). Out of the thousands of tests on genetically modified crops that the Animal and Plant Health Inspection Services received, only 145 petitions for deregulation were made by 2013 and of those 145 petitions only 96 were approved with the majority being for corn, followed by cotton, soybeans, and tomatoes (Fernandez-Cornejo et al., 2014). In 2015, in the United States, there are up to 33 different varieties of genetically modified corn that have deregulated status, including varieties that are moth and butterfly resistant, European Corn Borer Resistant, rootworm resistant, drought resistant, male sterile, and herbicide tolerant to name a few (Johnson & O'Connor, 2015).

  ➤ *USDA's Animal and Plant Health Inspection Services* protect agriculture from pests and diseases. They regulate organisms that are known to be or that might be suspected of being a pest or of posing a risk to plants and this includes plants that have been genetically engineered or modified (USDA, 2013).

❖ **US Food and Drug Administration (FDA):** The FDA ensures the safety of plant-derived food and feed as well as ensuring that proper labeling procedures are followed. "Under the Federal Food, Drug, and Cosmetic Act, it is the responsibility of food and feed manufacturers to ensure that the products they market are safe and properly labeled" (USDA, 2013). They encourage the food and feed manufacturers to participate in a voluntary consultation process before commercialization. This voluntary process is supposed to ensure that any type of human food or animal feed safety issues are resolved before the products go to market. The FDA also regulates food additives which are substances added to food that are not pesticides and are not "generally" recognized as safe by expert scientists (USDA, 2013). The FDA designates most genetically modified crops as "generally recognized as safe" and so most genetically modified crops do not have to go through an approval process. If there is an insertion of genes that causes expression of proteins that are significantly different in their function, structure, or the quality of the natural plant proteins and that are potentially harmful to human health more rigorous provisions of the Federal Food, Drug, and Cosmetic Act may apply (Federation of American Scientists, 2011). In 2008, the US FDA deemed that the milk from cloned animals was safe to drink (National Institute of Health, National Genome Research Institute, 2016).

Many people are concerned about the safety and use of transgenic crops and animals which have been developed through biotechnology and genetic engineering in particular. The National Institutes of Health have created very strict rules and regulations regarding the use and disposal of genetically engineered plants. The European Union has much stricter regulations regarding the use, labeling, and selling of genetically *engineered* products.

The Science behind the Technology

Throughout this chapter we have addressed the sciences behind various technologies. Perhaps the most visible is that which relates to chemistry and biology as used in bioengineering of virtually every sort. The use of transgenic animals in medicine is one such example. Biofuel creation is another. Bioengineered crops are still another. In each case, molecular structures are modified or controlled to fit our needs. Synergies and the sharing of expertise exist between chemistry, biology, and biotechnology in order to successfully produce various chemical commodities and biomaterials. Now we have new areas of study developing based on these synergies, such as pharmaceutical chemistry, combinatorial chemistry, or agrochemistry. In each case it is the use of scientific methods with organisms to produce new products or new forms of organisms.

Biotechnology is truly multidisciplinary. It involves science from all areas—physical, life, and social; as well as mathematics, and applied science such as computer science, agriculture, instrumentation, and technology management. At its base though is often chemistry and biology in the form of biochemistry, microbiology, cell biology, or molecular biology. Technology working with recombinant DNA, for example, as addressed earlier in this chapter, involves removing, modifying, and inserting genes of interest. It requires working with plasmids (a genetic structure in a cell that can replicate independently of the chromosomes), vectors (a DNA molecule used as a vehicle to artificially carry foreign genetic material into another cell), restriction enzymes, sequencing, and, of course, living organisms.

Career Connections

The career field of biotechnology is vast and varied. It includes, among more, genetic engineering, bioinformatics, immunology, plant and animal biotechnology, environmental biotechnology, nano-biotechnology, pharmacology, and marine biotechnology. Biotechnology is such a new and growing field and there are almost innumerable career options. It has a role in medicine, diagnostics, agriculture, pharmaceuticals, energy, waste management, and many other areas. Within each you could pursue careers in research, quality control, manufacturing and production, regulation, administration, information technology systems, or marketing and sales.

Biological Technician

Biological technicians work in laboratories with medical and biological scientists whom they help in the conducting laboratory experiments and tests. They prepare materials for testing and conduct tests and analyze data. Generally you will need a four-year college degree in biology or a related field to obtain employment, and it is also very important that you work in a lab during your time in school. Job growth is expected to be about 5% and demand is expected to increase as the biotechnology industry grows. The median annual wage for a biological technician was $41,650 in 2015 (U.S. Department of Labor, Bureau of Labor Statistics, 2016).

Biochemists and Biophysicists

Biochemists and biophysicists study the chemical and physical principles of living things and biological processes such as cell development, growth, heredity, and diseases. Individuals in this field can expect to make about $82,150 a year if they have a doctoral or professional degree. The job outlook from 2014 to 2024 is an 8% growth rate (Bureau of Labor Statistics, 2015).

Modular Activities

Discussions

❖ Post an original discussion in the online discussion board on the following topic: Some say there are situations where biotechnology goes too far based on how it is used and by whom. What are your thoughts and why? (200 words minimum). Next, comment on the post(s) of a minimum of one other student in a thoughtful and academic way that enhances the conversation. See rubric for grading and assessment measures.

Tests

❖ Online graded quiz on overall chapter content, written in multiple choice format. When submitted, we recommend giving the correct answer along with the page number in which it is found for questions students did not answer correctly.

Research

❖ Genetic food modification: Write a three-page report explaining what some specific type of genetic food modification is and its effects. This report must use proper citation and use a minimum of three sources.
❖ Interview: Find someone who is a farmer or works in an agricultural field and interview them about biotechnology—ultimately creating a two-page report (absolute minimum 600 words, no maximum) about the interview. Please include the questions you create in the paper.
❖ Research one of the following higher education academic areas:
  ➤ Biotechnology
  ➤ Bioinformatics

This might include review of offerings or degrees at your institution or at another location, contacting a professional/faculty in the field, assessing higher education level academic events or educational resources, and so on. Submit a 200-word minimum summary of your findings. Make sure to include your source(s).

Design and/or Build Projects

❖ Create a visual diagram that expresses the many aspects of biotechnology. You will be assessed on the depth, accuracy, and detail of your diagram.
❖ Create an informative flyer (handbill) that advertises either the advantages or disadvantages (choose one) of genetic modification of food (or a specific type of food). You will be assessed on the depth, accuracy, and detail of your flyer.

Assessment Tasks

❖ Choose one of the following biotechnology areas and assess the benefits and drawbacks of the use of biotechnology in that chosen area: pharmaceuticals, genetic engineering, cell and molecular biology, immunology, plant biotechnology, pesticides/fertilizers, chemical engineering, diagnostics, DNA fingerprinting, gene profiling, or biotechnology for energy. Write a two-page assessment of your findings. This report must use proper citation and use a minimum of two sources.

❖ Peruse the Coordinated Framework for Regulation of Biotechnology, which spells out the basic federal policy for regulating the development and introduction of products derived from biotechnology at https://www.aphis.usda.gov/brs/fedregister/coordinated_framework.pdf. Write a two-page assessment of what caught your attention most and why.

Terms

Agricultural biotechnology—"A range of tools, including traditional breeding techniques, that alter living organisms, or parts of organisms, to make or modify products; improve plants or animals; or develop microorganisms for specific agricultural uses. Modern biotechnology today includes the tools of genetic engineering" (USDA, 2013a).

Bioinformatics—"the collection, classification, storage, and analysis of biochemical and biological information using computers especially as applied to molecular genetics and genomics." (Merriam-Webster, 2016a).

Biotechnology—"the manipulation (as through genetic engineering) of living organisms or their components to produce useful usually commercial products (as pest resistant crops, new bacterial strains, or novel pharmaceuticals); also: any of various applications of biological science used in such manipulation" (Merriam-Webster Dictionary, 2016b).

Bacillus thuringiensis (Bt)—"A soil bacterium that produces toxins that are deadly to some pests. The ability to produce Bt toxins has been engineered into some crops." (USDA, 2013a).

Biopharming—"Biopharming is the production and use of transgenic plants and animals genetically engineered to produce pharmaceutical substances for use in humans or animals" (Goven, 2014).

Bt crops—"Crops that are genetically engineered to carry a gene from the soil bacterium *Bacillus thuringiensis* (Bt). The bacterium produces proteins that are toxic to some pests but nontoxic to humans and other mammals. Crops containing the Bt gene are able to produce this toxin thereby providing protection for the plants. Bt corn and Bt cotton are examples of commercially available Bt crops" (USDA, 2013a).

Clone—"A genetic replica of an organism created without sexual reproduction" (USDA, 2013a).

Gene mapping—"Determining the relative physical locations of genes on a chromosome. Useful for plant and animal breeding" (USDA, 2013a).

Genetic engineering—"Manipulation of an organism's genes by introducing, eliminating, or rearranging specific genes using the methods of modern molecular biology, particularly those techniques referred to as recombinant DNA techniques" (USDA, 2013a).

Genetically engineered organism (GEO)—"An organism produced through genetic engineering" (USDA, 2013a).

Genetic modification—"The production of heritable improvements in plants or animals for specific uses, via either genetic engineering or other more traditional methods. Some countries other than the United States use this term to refer specifically to genetic engineering" (USDA, 2013a).

Genetically modified organism (GMO)—"An organism produced through genetic modification" (USDA, 2013a).

Genetics—"The study of the patterns of inheritance of specific traits" (USDA, 2013a).

Genomics—"The mapping and sequencing of genetic material in the DNA of a particular organism as well as the use of that information to better understand what genes do, how they are controlled, how they work together, and what their physical locations are on the chromosome" (USDA, 2013a).

Genomic library—"A collection of biomolecules made from DNA fragments of a genome that represent the genetic information of an organism that can be propagated and then systematically screened for particular properties. The DNA may be derived from the genomic DNA of an organism or from DNA copies made from messenger RNA molecules. A computer-based collection of genetic information from these biomolecules can be a "virtual genomic library" (USDA, 2013a).

Genotype—"The genetic identity of an individual. Genotype often is evident by outward characteristics, but may also be reflected in more subtle biochemical ways not visually evident" (USDA, 2013a).

Herbicide-tolerant crops—"Crops that have been developed to survive application(s) of particular herbicides by the incorporation of certain gene(s) either through genetic engineering or traditional breeding methods. The genes allow the herbicides to be applied to the crop to provide effective weed control without damaging the crop itself" (USDA, 2013a).

Insecticide resistance—"The development or selection of heritable traits (genes) in an insect population that allow individuals expressing the trait to survive in the presence of levels of an insecticide (biological or chemical control agent) that would otherwise debilitate or kill this species of insect. The presence of such resistant insects makes the insecticide less useful for managing pest populations" (USDA, 2013a).

Insect-resistant crops—"Plants with the ability to withstand, deter, or repel insects and thereby preventing them from feeding on the plant. The traits (genes) determining resistance may be selected by plant breeders through cross-pollination with other varieties of this crop or through the introduction of novel genes such as Bt genes through genetic engineering" (USDA, 2013a).

Pest-resistant crops—"Plants with the ability to withstand, deter, or repel pests and thereby preventing them from damaging the plants. Plant pests may include insects, nematodes, fungi, viruses, bacteria, weeds, and other "(USDA, 2013a).

Pesticide resistance—"The development or selection of heritable traits (genes) in a pest population that allow individuals expressing the trait to survive in the presence of levels of a pesticide (biological or chemical control agent) that would otherwise debilitate or kill this pest. The presence of such resistant pests makes the pesticide less useful for managing pest populations" (USDA, 2013a).

Protein—"A molecule composed of one or more chains of amino acids in a specific order. Proteins are required for the structure, function, and regulation of the body's cells, tissues, and organs, and each protein has a unique function" (USDA, 2013a).

Recombinant DNA or rDNA—"Genetically engineered DNA incorporating DNA from more than one species of organisms" (Merriam-Webster Dictionary, http://www.merriam-webster.com/dictionary/recombinant%20DNA)

Recombinant DNA technology—"Procedures used to join together DNA segments in a cell-free system (e.g., in a test tube outside living cells or organisms). Under appropriate conditions, a recombinant DNA molecule can be introduced into a cell and copy itself (replicate), either as an independent entity (autonomously) or as an integral part of a cellular chromosome" (USDA, 2013a).

Ribonucleic Acid (RNA)—"A chemical substance made up of nucleotides compound of sugars, phosphates, and derivatives of the four bases adenine (A), guanine (G), cytosine (C), and uracil (U). RNAs function in cells as messengers of information from DNA that are translated into protein or as molecules that have certain structural or catalytic functions in the synthesis of proteins. RNA is also the carrier of genetic information for certain viruses. RNAs may be single or double stranded" (USDA, 2013a).

Selectable marker—"A gene, often encoding resistance to an antibiotic or an herbicide, introduced into a group of cells to allow identification of those cells that contain the gene of interest from the cells that do not. Selectable markers are used in genetic engineering to facilitate identification of cells that have incorporated another desirable trait that is not easy to identify in individual cells" (USDA, 2013a).

Transgene—"A gene from one organism inserted into another organism by recombinant DNA techniques" (USDA, 2013a) .

Transgenic organism—An organism resulting from the insertion of genetic material from another organism using recombinant DNA techniques" (USDA, 2013a).

References

Ahmad, P., Ashraf, M., Younis, M., Hu, X., Kumar, A., Akram, N. A., & Al-Qurainy, F. (2012). Role of transgenic plants in agriculture and biopharming. *Biotechnology Advances, 30*(3), 524–540. doi:10.1016/j.biotechadv.2011.09.006

Bagle, T., Kunkulol, R., Baig, M., & More, S. (2013). Transgenic animals and their application in medicine. *International Journal of Medical Research & Health Sciences, 1*(2), 107–116.

Bawa, A. S., & Anilakumar, K. R. (2013). Genetically modified foods: Safety, risks and public concerns—A review. *Journal of Food Science and Technology, 50*(6), 1035–1046. doi:10.1007/s13197-012-0899-1

Bio: Biotechnology Industry Organization. (n.d.). What is animal biotechnology? Retrieved from https://www.bio.org/sites/default/files/files/Animal_onepager.pdf

The American heritage® science dictionary. (n.d.). Bioremediation. Retrieved from http://www.dictionary.com/browse/bioremediation

Brown, C. (2000). Patenting life: Genetically altered mice an invention, court declares. Retrieved from http://www.ncbi.nlm.nih.gov/pmc/articles/PMC80518/

Bureau of Labor Statistics. (2015). Biochemists and biophysicists. In *Occupational outlook handbook* (2016–2017 ed.), U.S. Department of Labor. Retrieved from http://www.bls.gov/ooh/life-physical-and-social-science/biochemists-and-biophysicists.htm

Center for Public Environmental Oversight. (n.d.). Bioremediation and enhancement. Retrieved from http://www.cpeo.org/techtree/ttdescript/ensolmx.htm

Federation of American Scientists. (2011). *U.S. regulation of genetically modified crops.* Retrieved from http://fas.org/biosecurity/education/dualuse-agriculture/2.-agricultural-biotechnology/us-regulation-of-genetically-engineered-crops.html

Fernandez-Cornejo, J., Wechsler, S. J., & Livingston, M. (2014). Adoption of genetically engineered crops by U.S. farmers has increased steadily for over 15 years. *Amber Waves, ,* 1A.

Fernandez-Cornejo, J., Wechsler, S., Livingston, M., & Mitchell, L. (2014). *Genetically engineered crops in the United States, ERR-162 U.S. Department of Agriculture, Economic Research Service.* Retrieved from https://www.ers.usda.gov/webdocs/publications/err162/43667_err162_summary.pdf

Genetic Science Learning Center. (2014, June 22). What is cloning? *Learn. Genetics.* Retrieved from http://learn.genetics.utah.edu/content/cloning/whatiscloning/

Goho, A. (2008). Corn primed for making biofuel. *Technology Review.* Retrieved from https://www.technologyreview.com/s/409913/corn-primed-for-making-biofuel/

Goven, J. (2014). Encyclopedia of food and agricultural ethics. In P. B. Thompson & D. M. Kaplan (Eds.), *Biopharming.* Dordrecht: Springer Science-Business Media.

Griffiths, A. J. F., Miller, J. H., Suzuki, D. T., Lewontin, R. C., & Gelbart, W. M. (2000). *An introduction to genetic analysis.* New York, NY: W. H. Freeman. Retrieved from http://www.ncbi.nlm.nih.gov/books/NBK21881/

Johnson, D., & O'Connor, S. (2015). These Charts show every genetically modified food people already eat in the U.S. *Time.* Retrieved from http://time.com/3840073/gmo-food-charts/

Johnson, D. B. (2014). Biomining-biotechnologies for extracting and recovering metals from ores and waste materials. *Current Opinion in Biotechnology, 30*, 24–31. doi:10.1016/j.copbio.2014.04.008

Matheson, S. (2015). Cloning could save endangered giant pandas, researchers say. *Nature World News*. Retrieved from http://www.natureworldnews.com/articles/17303/20151006/cloning-save-endangered-giant-pandas-researchers.htm

Merriam-Webster Dictionary (2016a). *Bioinformatics*. Retrieved from https://www.merriam-webster.com/dictionary/bioinformatics

Merriam-Webster Dictionary (2016b). *Biotechnology*. Retrieved from http://www.merriam-webster.com/dictionary/biotechnology

National Institute of Health, National Genome Research Institute. (2016). *Advancing human health through genomics research, cloning*. Retrieved from https://www.genome.gov/25020028/cloning-fact-sheet/#al-6

Newton, D. E. (2003) *Environmental encyclopedia, biotechnology*. Retrieved form http://www.encyclopedia.com/article-1G2-3404800201/biotechnology.html

Nicolia, A., Manzo, A., Veronesi, F., & Rosellini, D. (2014). An overview of the last 10 years of genetically engineered crop safety research. *Critical Reviews in Biotechnology, 34*(1), 77–88. doi: 10.3109/07388551.2013.823595

North Carolina Association for Biomedical Research. (2016). *About bioscience: Animal biotechnology*. Retrieved from http://www.aboutbioscience.org/topics/animalbiotechnology

Organization for Economic Co-operation and Development (OECD), (2009): *The bioeconomy to 2030: Designing a policy agenda*. Retrieved from http://www.oecd.org/futures/long-termtechnologicalsocietalchallenges/thebioeconomyto2030designingapolicyagenda.htm

Public Broadcasting Station (n.d.). Evolution of the dog. *Evolution library.*. Retrieved from http://www.pbs.org/wgbh/evolution/library/01/5/l_015_02.html

Pizzi, F., Caroli, A. M., Landini, M., Galluccio, N., Mezzelani, A., & Milanesi, L. (2013). Conservation of endangered animals: From biotechnologies to digital preservation. *Natural Science, 5*(8), 903–913. doi:10.4236/ns.2013.58109

Pool R, Waddell K; National Research Council (US). (2002). Exploring horizons for domestic animal genomics: Workshop summary. In *The value of sequencing domestic animal genomes*. Washington, DC: National Academies Press (US). Retrieved from http://www.ncbi.nlm.nih.gov/books/NBK207584/

Shmaefsky, B. R. (2013). *Agricultural biotechnology: History, science, and society (October 2013): Animal biotechnology*. Retrieved from http://pustaka2.upsi.edu.my/eprints/515/1/Agricultural%20Biotechnology%20History,%20Science,%20and%20Society.pdf

Stecker. (2014), *Scientific America, reprinted from climate wire*. Retrieved from http://www.scientificamerican.com/article/biofuels-might-hold-back-progress-combating-climate-change/

United Nations. (2015). *Department of economic and social affairs, population division, world population, 2015*. Retrieved from https://esa.un.org/unpd/wpp/Publications/Files/World_Population_2015_Wallchart.pdf

U.S. Department of Agriculture. (2013). *Glossary of agricultural biotechnology terms*. Retrieved from http://www.usda.gov/wps/portal/usda/usdahome?navid=BIOTECH_GLOSS&navtype=RT&parentnav=BIOTECH

U.S. Department of Agriculture. (2015). *Adoption of genetically engineered crops in the United States, 1996-2015, figure*. Retrieved from https://www.ers.usda.gov/data-products/adoption-of-genetically-engineered-crops-in-the-us/recent-trends-in-ge-adoption.aspx

U.S. Department of Agriculture. (2016). *Biotechnology frequently asked questions*. Retrieved from http://www.usda.gov/wps/portal/usda/usdahome?navid=BIOTECH_FAQ&navtype= RT&parentnav=BIOTECH

U.S. Department of Agriculture Economic Research Service. (2014). Retrieved from http://www .ers.usda.gov/amber-waves/2014-march/adoption-of-genetically-engineered-crops-by-us- farmers-has-increased-steadily-for-over-15-years.aspx#.V2m5brgrI2w

U.S. Department of Energy. (n.d.a). Biodiesel. *Energy, Efficiency & Renewable Energy*. Retrieved from http://www.fueleconomy.gov/feg/biodiesel.shtml

U.S. Department of Energy. (n.d.b). Ethanol. *Energy, Efficiency & Renewable Energy*. Retrieved from http://www.fueleconomy.gov/feg/ethanol.shtml

U.S. Department of Labor, Bureau of Labor Statistics. (2016). *Biological technicians*. Retrieved from http://www.bls.gov/ooh/life-physical-and-social-science/biological-technicians.htm

U.S. Energy Information Administration (2016). *Today in energy*. Retrieved from http://www.eia .gov/todayinenergy/detail.cfm?id=26212

U.S. Environmental Protection Agency. (2016). *A citizen's guide to bioremediation*. Retrieved from https://www.epa.gov/sites/production/files/2015-04/documents/a_citizens_guide_to_ bioremediation.pdf

U.S. Food and Drug Administration, Centers for Biologics Evaluation and Research. (1995). *Points to consider in the manufacture and testing of therapeutic products for human use derived from transgenic animals*. Retrieved from http://www.fda.gov/downloads/biologicsbloodvaccines/guidanc ecomplianceregulatoryinformation/otherrecommendationsformanufacturers/ucm153306.pdf

Wieczorek, A. M., & Wright, M. G. (2012). History of agricultural biotechnology: How crop development has evolved. *Nature Education Knowledge, 3*(10), 9. Retrieved from http://www.nature.com/scitable/ knowledge/library/history-of-agricultural-biotechnology-how-crop-development-25885295

Zhuang, W., Fitts, J. P., Ajo-Franklin, C. M., Maes, S., Alvarez-Cohen, L., & Hennebel, T. (2015). Recovery of critical metals using biometallurgy. *Current Opinion in Biotechnology, 33*, 327–335. doi:10.1016/j.copbio.2015.03.019

Further Reading

Barker, D. (2014). Genetically engineered (GE) crops: A misguided strategy for the twenty-first century? *Development, 57*(2), 192–200. doi:10.1057/dev.2014.68

Beyond Pesticides. (n.d.) Herbicide tolerant crops. Retrieved from http://www.beyondpesticides .org/programs/genetic-engineering/herbicide-tolerance

Biotechnology Innovation Organization. (n.d.). History of biotechnology. Retrieved from https:// www.bio.org/articles/history-biotechnology

Camacho, A., Van Deynze, A., Chi-Ham, C., & Bennett, A. B. (2014). Genetically engineered crops that fly under the US regulatory radar. *Nature Biotechnology, 32*(11), 1087–1091. doi:10.1038/nbt.3057

Dragan, Š., Radmila, M., Jelena, N., Branko, P., Stamen, R., & Svetlana, G. (2015). The application of biotechnology in animal nutrition. *Veterinarski Glasnik, 69*(1), 127–137. doi: 10.2298/ VETGL1502127S

Fulekar, M. H. (2010). *Bioremediation technology: Recent advances*. Dordrecht: Springer.

Gillespie, I. M. M., & Philp, J. C. (2013). Bioremediation, an environmental remediation technology for the bioeconomy. *Trends in Biotechnology, 31*(6), 329–332. doi:10.1016/j.tibtech.2013.01.015

Global Animal Biotechnology Technologies, Markets and Companies Report 2015. (2015). *Veterinary Week*. p. 433.

Howard, H., Jones, K., & Rudenko, L. (2012). Agency perspectives on food safety for the products of animal biotechnology. *Reproduction in Domestic Animals, 47*(S4), 127–133. doi:10.1111/j.1439-0531.2012.02066.x

Maes, S., Boon, N., Hennebel, T., & Lenz, M. (2015). Biotechnologies for critical raw material recovery from primary and secondary sources: R&D priorities and future perspectives. *New Biotechnology, 32*(1), 121–127. doi:10.1016/j.nbt.2013.08.004

Maksimenko, O. G., Deykin, A. V., Khodarovich, Y. M., & Georgiev, P. G. (2013). Use of transgenic animals in biotechnology: Prospects and problems. *Acta Naturae, 5*(1), 33–46.

Mary Beth Griggs (August 29, 2011). How microbes clean up our environmental messes. *Popularmechanics*. Retrieved from http://www.popularmechanics.com/science/environment/a7176/how-microbes-will-clean-up-our-messes/

Milne, R. (2012). Pharmaceutical prospects: Biopharming and the geography of technological expectations. *Social Studies of Science, 42*(2), 290–306.

Mohee, R., Mudhoo, A., & Ebrary, I. (2012). *Bioremediation and sustainability: Research and applications*. Hoboken, NJ: Wiley.

Ormandy, E. H., Dale, J., & Griffin, G. (2011). Genetic engineering of animals: Ethical issues, including welfare concerns. *Canadian Veterinary Journal, 52*(5), 544–550.

Petetin, L. (2012). The revival of modern agricultural biotechnology by the UK government: What role for animal cloning? *European Food and Feed Law Review: EFFL, 7*(6), 296.

Singh, O. V., & Harvey, S. P. (2010). (Editors) *Sustainable biotechnology: Sources of renewable energy*.

Tizard, M., Hallerman, E., Fahrenkrug, S., Newell-McGloughlin, M., Gibson, J., de Loos, F., … Doran, T. (2016). Strategies to enable the adoption of animal biotechnology to sustainably improve global food safety and security. *Transgenic Research, 25*(5), 575–595. doi:10.1007/s11248-016-9965-1

U.S. Department of Agriculture, (2013b). *How the Federal Government regulates biotech plants*. Retrieved from http://www.usda.gov/wps/portal/usda/usdahome?contentidonly=true&contentid=biotech-plants.xml

Zhao, J., Xu, J., Wang, J., & Li, N. (2013). Nutritional composition analysis of meat from human lactoferrin transgenic bulls. *Animal Biotechnology, 24*(1), 44.

Chapter 12

Nanotechnology

Outline

- ❖ Chapter learning objectives
- ❖ Overview
- ❖ Introduction
- ❖ Nanofabrication techniques
  - ➢ Evaporation
  - ➢ Sputtering
  - ➢ Chemical vapor deposition
- ❖ Nanomaterials
  - ➢ Fullerenes and carbon
  - ➢ Nanowires
  - ➢ Metal nanoparticles and quantum dots
  - ➢ Self-assembling monolayers
  - ➢ Smart materials
- ❖ Tools of nanoscience
  - ➢ Electron microscopy
  - ➢ Scanning probe microscopy
- ❖ Uses of nanotechnology
- ❖ The science behind the technology
  - ➢ Atoms and bonds
  - ➢ Scaling of forces
  - ➢ Quantum mechanics
- ❖ Career connections
- ❖ Modular activities
  - ➢ Discussions
  - ➢ Tests

(Continued)

(*Continued*)

> ➤ Research
> ➤ Design and/or Build Projects
> ❖ Terms
> ❖ Bibliographys

Chapter Learning Objectives

❖ Identify varying types of nanomaterials
❖ Explain nanofabrication techniques
❖ Examine atoms and how they bond
❖ Describe nanomaterials and their potential uses
❖ Distinguish between tools used in nanotechnology

Overview

In the 2015 superhero movie *Ant-Man* based on Marvel Comics books, a super-suit was invented and developed. Armed with the suit, a normal-size person can shrink smaller than an ant and be powered with the ant's strength. We can say that the idea of nanotechnologies is rooted in the creation of the super-suit. In real life, nanotechnologies are around us everywhere. The thumb drive you are using to back up your pictures and files, the pregnant test strips sold on the shelf of Walgreens, the tennis rack with the carbon fiber string, and the nanofabric sportswear that can keep you dry by letting the sweat out are everyday examples of nanotechnologies.

In this chapter, we focus on learning about nanotechnology. In essence, nanotechnologies are microscopic building components, much smaller than a human hair. We look at various manufacturing methods of nanotechnologies and some of the current uses.

Introduction

Technology is a term that describes inventions which apply scientific principles to create practical devices to solve problems or do useful things.[i] Nanotechnology innovations have components with at least one dimension which is smaller than 100 nanometer. A nanometer is one billionth of a meter (1/1,000,000,000). If you took

---

i   A combination of definitions taken from Merriam-Webster.com and yourdictionary.com.

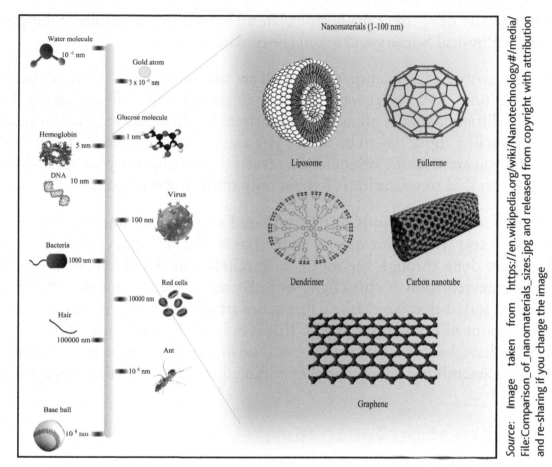

Nanomaterial sizes

a human hair and divided it into 1000 equal strands, the strands would be approximately 50–100 nm thick. It is commonly considered that hair is about the smallest thing that can be seen by the human eye. Therefore, these components need special equipment just to see them and characterize them completely.

Having made its way into popular print many times, nanotechnology is understood to refer to the industries developed by companies that utilize nano-sized components in their products. These include products used in computers, communications, sporting goods, pharmaceuticals, medical diagnostics, cosmetics, aerospace, cars, clothing, and apparel; just about every facet of our modern lives is currently impacted by nanotechnologies.

Working with devices this small is extremely difficult, often compared to picking up a pin while wearing boxing gloves. In order to manufacture, assemble, and use pieces this small, engineers and chemists have had to devise some advanced fabrication processes. In 1959, Richard Feynman, a legendary physicist, gave a talk that revolutionized the approach we would take to create technologies. In his speech entitled

"There is plenty of room at the bottom," he addressed the assembled colleagues of the American Physical Society and asked (Feynman, 1992):

> Why can't we manufacture these small computers somewhat like we manufacture the big ones? Why can't we drill holes, cut things, solder things, stamp things out, mold different shapes all at an infinitesimal level? What are the limitations as to how small a thing has to be before you can no longer mold it? How many times when you are working on something frustratingly tiny like your wife's wrist watch, have you said to yourself, "If I could only train an ant to do this!"

We have developed methods to fabricate and assemble nanotechnologies into usable systems and it does not involve trained ants. The drilling, cutting, stamping, and molding are done through the means of carefully controlled chemical processes and using unexpected tools such as light and plasma. In fact, although difficult to verify, it has been stated that we manufacture more transistors every year in the world than we grow grains of rice (Fraser, 2005). This is only possible because of how small those transistors have become. But we are even moving beyond these manufacturing techniques and assembling nanotechnologies directly from the smallest pieces of matter.

Nanofabrication Techniques

Many of the pioneering efforts in the manufacture of nanotechnologies were made in the computer chip industry. Early efforts led to a manufacturing process often called top–down manufacturing. Top–down manufacturing can be compared to an artist sculpting in clay. The artist starts with a lump of clay and shapes it into the final

Figure 1.   Electronic integrated circuit chip.

figurine by carefully adding, shaping selecting, and removing clay. The clay with which the nanotechnologist works is commonly the element silicon. Highly pure and crystalline silicon is made into round wafers which are anywhere from 100 to 450 millimeters in diameter and less than a millimeter thick. Although the processes can be complicated and contain many dozens of steps they can be classified as steps which either add material, select material by patterning, or remove the previously selected material.

The additive steps are often called deposition. Materials can be deposited on the wafer to provide the working surface for the design of nano-sized devices. There are many ways to deposit materials but we will consider only three.

Evaporation

The first is evaporation, wherein the metals are heated to a high enough temperature so that the atoms begin to evaporate inside a closed container called a chamber. Similar to steam coming off hot water, these metal atoms float upwards to wafers that are attached to the top of the container. These hot atoms collide and stick to the cool wafer surfaces and create a metal coating. The thickness of the coating can be monitored by sensors and is primarily dictated by the amount of time the wafers are exposed to the evaporated metal.

Sputtering

Sputtering of metals is accomplished by the collision of gas ions with a metal surface. Special sputtering chambers are used to enclose this deposition process. Large non-reactive ions can be produced by putting high voltage on argon gas to form what is called plasma. This plasma is a collection of positively charged argon ions and negatively charged electrons. The positive argon ions can be accelerated to a metal surface by applying a relative negative charge to the metal. The argons then impact the metal like a comet colliding with a planet's surface and the energy of the impact causes the surface atoms of the metal to break off the surface and descend toward the bottom of the chamber where they land on the silicon wafer and attach to form a layer. Again the thickness is generally controlled by the amount of time the sputtering occurs.

Chemical Vapor Deposition

Chemical vapor deposition (CVD) is used to apply coatings to the wafer inside a chamber wherein the temperature and chemical composition of the atmosphere are tightly controlled. By controlling the chemical precursors that enter the chamber the atoms gain energy due to the heat and react to form a surface layer on the wafer. All three of these techniques coat the entire exposed wafer surface. But generally the device we are producing only needs the freshly coated materials in very specific areas.

So now the other two steps of the process, a patterning step followed by a subtractive step, can be used to hopefully wind up with the material in the necessary places.

Etching is generally accomplished using wet or dry chemical processes. Wet chemical etching involves dunking or washing the surface of the wafer in an acid, base, or other chemical which will chemically react with the surface and remove the metal or other materials we do not want on the surface. An example of dry etching involves using a technique called reactive ion etching (RIE). This technique involves chemical gas plasma similar to sputtering but in this technique the ions of the plasma are turned and accelerated toward the wafer. Removal of the materials on the wafer surface happens due to the impact of ions and their reaction with surface atoms. Both wet and dry chemical etching will attack the entire surface of the wafer and will remove all the material unless we block that attack by strategically placing a protective covering on certain parts of the surface. We accomplish this by using a patterning technique called photolithography.

Photolithography means writing with light. This is not writing with lasers in the surface but transferring a pattern to the surface using a sudden exposure of light similar to making shadow puppets on the wall with a flashlight shining behind you. The first thing we need to do is to coat the surface with a material that will record the images projected on the surface by the light. This coating is made with a photoactive polymer called photoresist. Photoresist often looks like nail polish, red in color, smelly, and about the same thickness as polish. This liquid is spread on the surface by dropping it on the wafer while it is spinning very fast. Because the wafer is spinning the photoresist spreads all over the surface in a flat and even coating. Next we bake or heat up the photoresist to dry it and create a strong chemical resistant coating. Now the special thing about photoresist is that it changes when certain types of light are shined on it, specifically, when ultraviolet light hits the surface, it changes the molecules in the photoresist, weakening them. Because of this special ability we can write patterns onto the surface. The way we do this is by using a mask. A mask is a glass plate where the parts have been covered by a metal pattern; this metal pattern is what we want transferred to the photoresist. When we place the mask over the photoresist and turn on an ultraviolet light above the mask then shadows are formed on the photoresist directly under the metal mask features; therefore, in those areas the photoresist stays strong. Conversely, ultraviolet light saturates the photoresist surface between the shadows weakening the photoresist. Finally we dunk the wafer in a developing solution which removes the weakened photoresist and leaves the protective coating only in the areas where we want it. After this, we can do an etching treatment of the wafer with full confidence that the materials we deposited in the first step will be safely protected under the areas that are still covered by photoresist. When all the unwanted materials are removed and the photoresist has done its job, we remove the photoresist using a solvent like acetone (which coincidentally also does a good job removing fingernail polish), and we are ready for the next processing step. Micro and nano-sized technologies can be made by repeating similar steps over and over again, up to dozens of times. Just remember that it is extremely important that each and every cycle is lined up as perfectly as possible with the preceding patterns, or your device will not work as you expected.

These processes were pioneered by the microelectronics (now nanoelectronics) industry for manufacturing computer chips such as the central processor and random access memory. Engineers and designers have expanded beyond these initial applications into sensors, and even devices that move. Micro electromechanical system (MEMS) devices use similar top–down processing to create very small accelerometers for deploying your airbags, pressure sensors for monitoring the tire pressure in your car, gyroscopes to keep track of twisting or turning motions in your cell phone or video game controller; the list of applications goes on and on. Nanotechnology is spreading beyond MEMS and microelectronics into the medical and consumer product applications as well. This shift has been helped by a new fabrication strategy called bottom–up manufacturing.

Biologists and chemists have studied the way that nature creates and sustains natural organisms without man-made intervention. When I look out my window, I see a beautiful tree that someone planted but the trunk, branches, and leaves are all self-organized. On the cellular level we see how plants and animals have a complex network of processes that take the basic building blocks of life and assemble the necessary machinery to keep them alive. Bottom–up fabrication embraces similar principles where small molecules and nano-sized materials are used to fabricate technologies. Many times this process has been compared to building with molecular Legos which can assemble themselves according to nature's rules of thermodynamics. Simply put these rules cause certain features of molecules to be either attracted to or repelled by other molecules. The exciting possibility with bottom–up fabrication is that manufacturers can presumably control every aspect of the technology.

In studying nature, it has been observed that self-assembly requires three basic principles, structured building blocks (molecules or particles), binding forces (the cement between building blocks), and driving forces (a reason for the building blocks to move around, arrange, and assemble themselves). These three aspects of self-assembly represent the factors that designers can control to create new nanotechnologies. There are certain challenges faced when analyzing and designing self-assembling systems. When you look at biological structures and try to understand how nature assembles the DNA, proteins, cells, tissue, organs, and so on, you are often looking backwards from the final product trying to piece together how they were developed. When you are trying to create something new you are often left with the task of trying to predict what will happen when you combine your building blocks in a way that they can respond to the driving forces and binding forces to form structures. Even if you identify a possible process there is always the question of yield. If all of these ingredients are combined according to plan, how many will actually form the right product? Will it be 1% or 100% of what was predicted? This is one of the largest challenges that the bottom–up manufacturing is facing (Pelesko, 2007).

Nanomaterials

Fullerenes and Carbon

The discovery and engineering of our molecular Legos is the job of chemists, materials scientists, and engineers. Just as the major epochs of human technological development have been characterized by the materials we used, that is, Stone Age, Bronze Age, Iron Age, so the discovery and utilization of nanomaterials is enabling new advances in nanotechnology. In 1985, Richard Smalley and Harold Kroto were looking for forms of carbon that might explain certain astronomical observations in space. It so happens that they found some very intriguing but puzzling results when they noticed that carbon structures with exactly 60 carbon atoms were very common when they vaporized carbon with powerful lasers. But they had no idea what this carbon was, it was unlike diamond the crystalline form of carbon, or graphite which is the stuff in pencil leads. It turns out that they had found a new form of carbon that they named a buckyball after the architect Buckminster Fuller who was a designer of geodesic domes which are similar in structure. In fact, the 60 carbon atoms formed a cage structure similar to a soccer ball with hexagons and linking pentagons on the sides which were only a few nanometers in size. This discovery garnered them and their collaborator Robert F. Curl, Jr. the Nobel Prize in 1996. (This was not the end of new nano-sized discoveries for these new carbon fullerenes other structures such

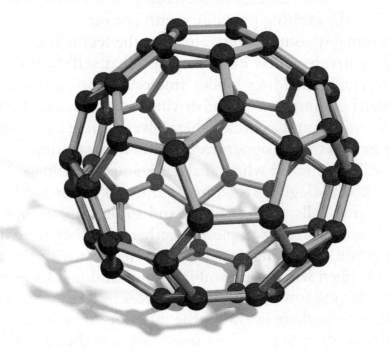

© ogwen/Shutterstock.com

Figure 2.   C60 carbon fullerene aka Buckyball.

as C70, nanotubes, and graphene were made in roughly that chronological order.[ii] In 2010, the physicists Andre Geim and Konstantin Novoselov received the Nobel Prize for discovery of graphene which is a single layer of carbon atoms with a

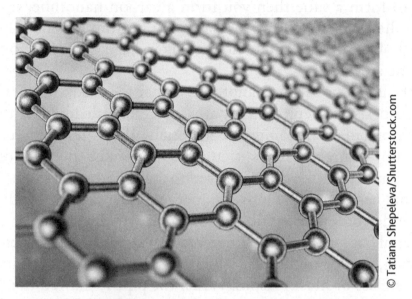

Figure 3.   Graphene molecular structure.

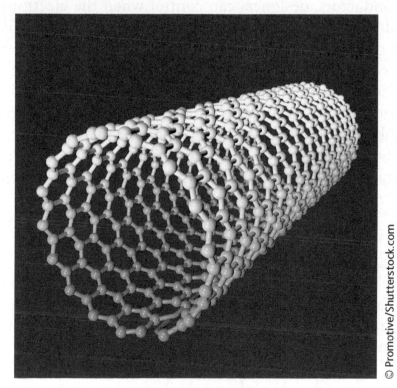

Figure 4.   Carbon nanotubes on a black background.

---

ii   https://www.nobelprize.org/nobel_prizes/chemistry/laureates/1996/.

hexagonal structure that looks a lot like chicken wire.[iii] This flat form of carbon has very high conductivity and may have a place in design of newer high speed computers that consume less power and produce less heat. If you take a piece of graphene and roll it up to form a tube then you form a carbon nanotube. Carbon nanotubes are not really rolled but they grow that way in a bottom–up manufacturing process. Nanotubes are hollow in the center and look like a cage with hexagon shapes running up and down the sides.

The neat thing about carbon nanotubes is that they look like little wires that you could use for nano-sized circuits or they also look like little fibers or threads that could be used for reinforcement in composite materials. The other really neat property of a nanotube is that by changing their structure you can make them conductive like house wiring or semiconducting like the parts used to make computer chips. Semiconductors are materials that can be coaxed into conducting under certain conditions or that can be insulating under others. Electrical conductivity or the ability of electrons to flow through a material has been crudely compared to water running through a pipe.[iv] When the pipe is frozen, water cannot flow which is similar to the condition of electrons in an insulator like glass or most plastics. When the pipe is thawed then water flows like it does in a conductor such as the metals copper and gold. In some carbon nanotubes like other semiconductors, designers can control when the electrons will or will not flow. This control allows us to build a material switch called a transistor. Believe it or not, such transistors are the basis of all computers and make these nanomaterials of high interest to computer chip manufacturers. Another intriguing property of carbon nanotubes and graphene is the very strong and stiff mechanical properties that it exhibits for its size. In fact, some engineers have thought the strength-to-weight ratio of carbon nanotubes would make it a great candidate material for constructing space elevators which would allow cargo to be easily lifted into space.[v] Product developers are also concentrating on using this strength in composites for more terrestrial purposes such as sporting goods and aircraft.

Nanowires

Not only can we utilize hollow carbon nanotubes, but solid carbon nanowires are building blocks for nanodevices. Solid carbon nanowires are grown or manufactured using a bottom–up manufacturing approach. One of the strategies for making nanowires is to use small dots of metals on a surface to serve as a starting point for wire growth using CVD. In the case of carbon nanowires, we can use nano-sized small iron metal droplets. Iron absorbs carbon at higher temperatures; in fact this is how we

---

make steel which is an iron alloy with primarily carbon in it. But iron has an absorption limit beyond which it cannot accommodate any more carbon. When I was a child, I learned about absorption limits while trying to mix punch from a dried drink mix. I loved sugar and thought the more the better, but I learned that I had added too much. No matter how much I stirred the mix, the sugar would not dissolve into the punch because I had exceeded the absorption limit of sugar in water. The undissolved sugar would just sink to the bottom of the plastic pitcher. In a similar manner when there is too much carbon in the small iron droplet it settles out of the iron droplet and forms a wire. As the iron absorbs more carbon then the excess is deposited on the end of the growing wire and it continues to get longer and longer. These carbon wires are structurally much different than their nanotube counterparts and have different conductivity and mechanical properties.

Similar strategies can be made to produce nanowires made out of silicon. All we need to do is to find a metal that absorbs silicon atoms. It turns out that gold does that nicely. At high temperatures, gold droplets absorb silicon in a CVD furnace until it reaches saturation. Similar to the iron/carbon situation, the gold deposits excess silicon and grows a silicon nanowire. Silicon nanowires are of significant interest because silicon is the main material used for transistors which serve as the foundation for all modern computer chips. In fact, computer chip manufacturers have been experimenting with silicon nanowires for three-dimensional transistors which might help in building higher capacity and faster computing devices in the future.[vi] Noble metals such as silver can be made into nanowires, as well. These wires are made in a bottom—up manner by reacting silver ions with copper droplets on a surface covered by a silver salt and water solution (Sanders et al., 2014). The reaction of silver under the proper driving forces causes the wires to grow. By applying external electric fields, the wire growth direction can be influenced. Noble metals such as silver, copper, and gold are some of the best electron conductors so they can be added to our arsenal of building blocks for electronics and sensor applications. Because of their small size metallic nanowires could be used as very sensitive electrochemical sensors.

Metal Nanoparticles and Quantum Dots

Gold and silver nanoparticles can be produced rather easily by reactions in solution which produce small metal nuclei. Under the right conditions of temperature and the right concentration of chemicals, small particles begin to form. These small particles under the growth conditions will continue to grow unless the reaction is stopped or terminated. Termination can occur because there are not enough chemicals for the

---

vi   http://spectrum.ieee.org/semiconductors/devices/nanowire-transistors-could-keep-moores-law-alive.

reaction to continue or because another chemical was added to stop the reaction. If properly controlled, nanoparticles with diameters even smaller than 10 nm can be achieved.[vii] For the noble metals an interesting thing occurs, the silver and gold metal particles are no longer silver and gold in color but red, green, yellow, or blue depending on two factors: size and shape. This is an unusual property that comes from the way the gold and silver particles interact with light which is in the area of the study of optics. Turns out that mankind has been using this interesting fact for centuries when decorating cathedrals and palaces with stained glass. The artisans of preceding millennia used the naturally occurring metals in their glass preparations, and as a result of their processes, small nanoparticles of the metals were suspended in the glass and adsorbed parts of the white light from the sun causing the transmission of colored light. Excellent examples of this property can be found in early Roman artifacts such as the goblet known as Lycurgis' cup. This cup is an example of dichroic glass. The small silver and gold particles suspended in the glass cause it to look green when viewed under reflected light shined from without the cup and to look red under transmitted light shined from within the cup (Freestone, Meeks, Sax, & Higgitt, 2007).

It turns out, also, that the energy absorbed by the metal nanoparticles can be controlled not only by the size of a metal nanoparticle, but also by the thickness a thin layer of the metal. This may seem like an insignificant difference, but it has allowed scientists to coat surfaces and small nano-sized glass beads with gold while they finely tune the color of light the beads will absorb. This type of structure is often referred to as a core-shell nanoparticle where the inside or core is one type of material and the coating or shell is another type of material. They remind me of peanut M&M's, one of my favorite candies. Naomi Halas, of Rice University, pioneered the work using core-shell nanoparticles coated with gold for treatment of cancer tumors. The idea is that you can program the gold surface of the particle to only seek and attach themselves to cancer tumor cells. When a lot of the particles collect on the tumors then the right color of light can be shined on the tumor and the particles will absorb the light, heat up, and kill the tumor cells selectively doing minimal damage to the surrounding tissues (Hirsch et al., 2003).

In addition to noble metals, compound semiconductor materials can be used to make small nanoparticles sometimes called quantum dots. Earlier it was mentioned that carbon nanotubes and silicon nanowires can be semiconductors. Both these materials achieve semiconducting capabilities with one type of atom, carbon in the case of nanotubes and silicon in the case of the nanowires. Compound semiconductors achieve the electrical properties of a semiconductor by mixing two different atoms. Specifically, the combinations of cadmium and selenium, or zinc and sulfur are used to form

---

vii   *The MRSEC education group website at the University of Wisconsin Madison shows how to make your own gold nanoparticles.* http://education.mrsec.wisc.edu/277.htm.

quantum dots. Again, by changing the diameter and shrinking these small particles, manufacturers can cause them to emit different colors of light. This has intriguing applications for my favorite technology: the television. Many of you have noticed the rapid innovations in television flat panel displays over the last 20 years. When quantum dots absorb light at one color or are electrically stimulated, they emit light at a visible color related to their size (less than 100 nm) allowing for very tightly controlled, tightly positioned and bright sources of light across the entire rainbow of colors. This allows television manufacturers to produce small pixels of light with significantly less input power saving on energy bills and consequently CO_2 emissions. It is also claimed that quantum dots may help solar cell (photovoltaic) manufactures achieve up to 45% higher energy collection efficiency. Similar to metal nanoparticles, quantum dots can be programmed on their surfaces to attach to specific cells or parts in the body which will mark them with little quantum dot tags. When light is shined on these quantum dot tags, they emit colors that can be photographed in an optical microscope.[viii]

Self-Assembling Monolayers

In the preceding sections we mentioned how developers were programming particles to attach themselves to cancer cells or other parts of the body. In nanotechnology, *programming* means making a coating on the nanoparticle that contains specially designed molecules which recognize other molecules on a cells surface. This process of using nanoparticles specially designed molecules to recognize other molecules is called biomolecular recognition and is often compared to a lock and key combination, where the molecule on a tumor cell surface called the target or receptor is much like a lock waiting for a specific key, the very key which will fit in it and only it. The specific biomolecule or key which seeks only this specific target can bind to the receptor and hold on tightly. Once the lock and key are engaged the attached nanoparticle can be activated to emit light giving away the location of the tumor, to heat up burning and killing the tumor, or to deliver a payload of medicine to kill the tumor. Coating the surfaces of metals with biomolecular keys can be done by carefully designing the biomolecule (our building block) so that part of the molecule loves to adsorb to a specific surface, for example gold. The area of the molecule that likes to stick to the surface is often called the *head group*. The remainder of the molecule, often referred to as the *tail*, is not important as far as the surface is concerned but is very important because it forms the key that will fit in the receptor on the tumor. Because of the preference of the head groups for a particular surface, when a solution filled with these molecules is pored over the surface they adsorb and assemble themselves into ordered layers ready to seek out the specified target in the patient's body.

---

viii   Private communications with Jacqueline Siy-Ronquillo of Navillum Nanotechnologies.

Let us review how this example fits our outline of the important components of a self-assembly process. The building block in this case is the biomolecule and specifically the head group and the tail which are both designed to seek the surface and tumor, respectively. The binding force is the attraction between the head group and surface which makes it want to adsorb and tightly stick to the surface. We discuss more about these binding forces later in the chapter. The driving force that moves the molecules is the preference for the head group to be next to the surface, the composition of the surface, and the random collision of the molecules in the solution that cause them to rotate, translate, and collide with the surface.

Molecular recognition is an example of molecules behaving selectively enough to only interact with a specific target. In medical and pharmaceutical industries such selectivity can open the door to personalized medicines which can be customized for patients suffering from a particular type of cancer or infection. The design of the selective parts of biomolecules (the key) is the task of protein and antibody engineering where sophisticated methods are used to manufacture these designer molecules. These manufacturing processes turn out thousands of candidate biomolecules, but each one needs to be tested so that only most selective and appropriate molecules are used for the treatment. Analysis of these molecules can be done in a multitude of ways but some of the newer and more interesting ways involve moving fluids of very small volumes back and forth between tests like a micro-sized laboratory. In fact these tests use lab-on-a-chip technology and the fluids containing the biomolecules are moved in small channels using microfluidic technology. The aim of lab-on-a-chip technologies is to perform laboratory tests—what might ordinarily take a counter top or a lab full of equipment—*on a chip* smaller than a microscope slide using miniaturization techniques. The goal is to improve the tests by doing them with a smaller amount of sample (patient blood, e.g.), in a portable format (allowing them to be done away from hospitals and labs), with lower cost, and/or with improved speed and accuracy. Medical lab-on-a-chip devices are an example of using nanotechnology components such as nanometer thick self-assembling monolayers (SAMs), engineered biomolecules, or nanoparticles, together with nanowire sensors, or MEMS devices, microelectronic logic and memory to form a system. In systems, the individual components combine their advantages to do tasks that could not be accomplished by the components individually.

Smart Materials

As far as building blocks go, smart materials are the ultimate goal. There are numerous ways to define smart materials, but I will restrict it to materials that can respond to changes in their environment and signal the response to other materials. Again some of the best examples come from molecular biology. Proteins are very large molecules which change their shape (called their conformation) when the environment changes.

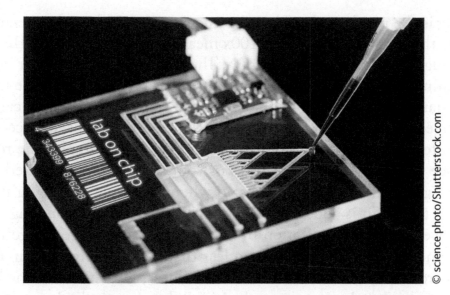

Figure 5. A lab on chip device integrates several laboratory processes in one device.

The body uses changes in proteins to accomplish housekeeping tasks in the cell. For example, DNA is a smart molecule and the process of protein production through gene expression involves an orchestrated interaction of DNA and protein complexes in smart ways. A large protein complex called RNA polymerase splits the DNA and copies the sequence of base pairs into a strand of messenger RNA. The messenger RNA carries this copy of the DNA "code" to a complex of proteins called the ribosome that "decodes" the message and produces a protein by combining the appropriate amino acids (the building blocks of proteins) (Nelson & Cox, 2005).

Tools of Nanoscience

Electron Microscopy

It turns out in a cruel twist of nature that the size limit of the objects we can see with light often called the resolution limit is related to the size of the light waves. Simply put we cannot see anything without a great deal of hassle smaller than 200 nanometers which is still two times bigger than our maximum size limit of nanotechnology. It is incredibly difficult to develop products if we cannot see what we have made and verify that they work. The way that has been employed for about the last century is to use a wave that is much smaller than 100 nm by substituting electrons for light in our microscopes. This discussion assumes that you have seen a light microscope. Light microscopes shine light through or reflect light off of your sample and use a series of lenses to magnify your sample so that you can see (resolve) the small details.

Magnifying is the process of making the image bigger and is often related by numbers followed by the letter x, for example 100x means that the features of the sample appear 100 times bigger than they are naturally. Therefore your typical hair would appear to be 5–10 millimeters in diameter which is easily visible. Changing the magnification is as easy as changing the combination of lenses in your microscope. Focusing the microscope changes the distance between the sample and the lens surface until the small details of interest are clearly visible and is also easily adjustable. The resolution is the ability to see the small parts or features of your sample and is limited by process you are using to get your image, and as I said when collecting visible light, is the process we are using this limit is about 200 nm no matter how much you magnify your image. Electrons overcome this limitation by the fact that we can converge them on a very small spot and get resolutions down to less than a nanometer. But in order to use an electron microscope, we must be able to have lenses, a way to focus the image, a way to magnify the image, and a way to see the image. In an optical microscope, I can use my eye to collect the light and form an image, but electron microscopes must use a detector and a computer. The lenses used to control electrons cannot be made out of glass like light microscopes but, instead, are tiny ring-shaped electromagnets that bend and direct the electrons as they pass through them. Focus is achieved by increasing or decreasing the strength of the electromagnetic lenses to get the electrons to all hit on the tiniest possible spot of the sample. The image is then formed by collecting the electron signal generated at that tiny spot; moving (rasterring) the spot and collecting the signal at thousands of different points allows the computer to produce an image. If I want to magnify my image, all I have to do is raster the small electron spot over a smaller section of the sample and the image will get much larger, in fact it can be as large as 1,000,000x or even 10,000,000x in some electron microscopes. The detectors are responsible for the brightness in each pixel of the image and this brightness is affected by the shape, height, and features of the surface. Therefore, the scanning of the electron spot gives us a magnified image of the sample surface showing its topographical features. The cool thing about electron microscopes is that by turning on a separate detector, we can collect an image where the features are due to the different types of atoms on the surface, and even the composition of each part of the surface can be analyzed (Flegler & Heckman, 1993; Goldstein et al., 2007).

The two most limiting things about electron microscopes are, first the fact that electrons repel each other so it is important that they do not pile up on the surface of our sample, and second that the samples must be viewed in a vacuum. Well, this first problem of electrons repelling each other is not a problem for metal samples because the extra electrons we are sending to the surface can conduct away, but ceramic and organic samples which do not conduct electrons are more difficult to deal with. Sample preparation and modern techniques are making it possible to overcome some of these challenges, and complimentary techniques have been developed that let us see things on the nanometer scale without electrons or vacuums.

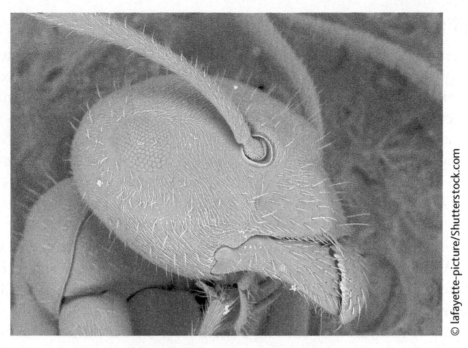

© lafayette-picture/Shutterstock.com

Figure 6. Ant as seen in a scanning electron microscope.

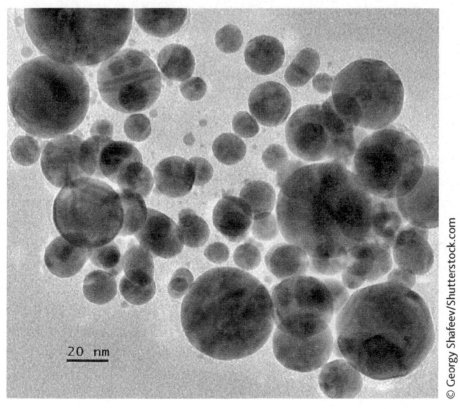

20 nm

© Georgy Shafeev/Shutterstock.com

Figure 7. Gold nanoparticles as seen with transmission electron microscope.

Scanning Probe Microscopy

The most important of these complimentary imaging techniques are done using scanning probe microscopes. In 1986, two scientists Binnig and Rohrer received the Nobel Prize in physics for developing the first scanning probe microscopy technique at IBM laboratories called scanning tunneling microscopy.[ix] Further work lead to the development of the atomic force microscope (AFM). As the name implies, AFM images are created by scanning the surface of a sample with a small physical probe and feeling the effects of the atomic forces between the tip and the surface. The probe is constructed of a micro-sized needle that is less than 10 nm in diameter at the very tip. This tip is connected to a long cantilever arm and resembles a small floppy needle and arm from a record player. The way it works is by bringing this probe tip into close contact with the surface. As the tip approaches the surface, it will feel the very small forces exerted by the surface atoms which will cause the tip to stop and the cantilever (record arm) holding the tip will bend and deflect slightly. Believe it or not, these small atomic-sized deflections can be measured by shining a small laser off the back of the cantilever and into a photodiode. Once the tip has made contact with the surface, it is scanned or dragged across the surface, and the small deflections are used to generate a map of the height changes due to features on the surface such as roughness, scratches, cracks, particles, adsorbed molecules, deposited layers, or even the tops of the atoms themselves. No light, electrons, or lenses are required, the computer just interprets the height and position changes to create an image or map of the surface. Again magnification of the image is accomplished by scanning smaller sections of the surface to generate the image (Eaton & West, 2010).

In addition to producing images, scanning probe microscopes can be used to actually manipulate and arrange atoms. By bringing the probe tip in contact with a loose atom on the surface, atomic forces can be turned on that fix the particle to the tip. The tip with its particle can then be moved to another spot on the surface and then when atomic forces between the tip and the atom are deactivated the atom will be left in the new position. IBM researchers used this to famously print the letters IBM out of 35 Xenon ions on a chilled nickel surface. More recently they used the techniques to make a stop motion animated movie entitled "A boy and his atom" by moving carbon monoxide molecules on a surface with a probe.[x] Moving atoms one at a time is the ultimate in bottom–up precision and control but the process can be very time consuming. Probe tips can also be used to scrape surfaces removing unwanted atoms. The tip can also be dipped into molecular solutions to coat it with molecules and then touched to the surface to deposit them in patterns. This process is called dip pin nanolithography.

---

ix   https://www.nobelprize.org/nobel_prizes/physics/laureates/1986/.

x   http://www.research.ibm.com/articles/madewithatoms.shtml#fbid=0DjABdJ_ZfL.

Uses of Nanotechnology

Nanotechnology is affecting many industries and sectors. They have the potential to enhance energy efficiency across all branches of industry, for example. This might come from things like nano-coated wear resistant drill probes to high duty nanomaterials used for lighter and more rugged rotor blades for wind or tidepower plants. Another highly likely energy area to be impacted is solar, whether from thin-layer solar cells, polymer solar cells, or dye solar cells (http://www.nanowerk.com/nanotechnology-in-energy.php). The National Nanotechnology Initiative (http://www.nano.gov/you/nanotechnology-benefits) notes that over 800 commercial every day products already rely on nanoscale materials and processes. This includes the following:

❖ Nanoscale additives in polymer composite materials for baseball bats, tennis rackets, motorcycle helmets, automobile bumpers, luggage, and power tool housings can make them simultaneously lightweight, stiff, durable, and resilient.

❖ Nanoscale additives to or surface treatments of fabrics help them resist wrinkling, staining, and bacterial growth, and provide lightweight ballistic energy deflection in personal body armor.

❖ Nanoscale thin films on eyeglasses, computer and camera displays, windows, and other surfaces can make them water-repellent, antireflective, self-cleaning, resistant to ultraviolet or infrared light, antifog, antimicrobial, scratch-resistant, or electrically conductive.

❖ Nanoscale materials in cosmetic products provide greater clarity or coverage; cleansing; absorption; personalization; and antioxidant, antimicrobial, and other health properties in sunscreens, cleansers, complexion treatments, creams and lotions, shampoos, and specialized makeup.

❖ Nanoengineered materials in the food industry include nanocomposites in food containers to minimize carbon dioxide leakage out of carbonated beverages, or reduce oxygen inflow, moisture outflow, or the growth of bacteria in order to keep food fresher and safer, longer. Nanosensors built into plastic packaging can warn against spoiled food. Nanosensors are being developed to detect salmonella, pesticides, and other contaminates on food before packaging and distribution.

❖ Nanoengineered materials in automotive products include high-power rechargeable battery systems, thermoelectric materials for temperature control, lower-rolling-resistance tires, high-efficiency/low-cost sensors and electronics, thin-film smart solar panels, and fuel additives and improved catalytic converters for cleaner exhaust and extended range.

❖ Nanoengineered materials make superior household products such as degreasers and stain removers, environmental sensors, alert systems, air purifiers and filters, antibacterial cleansers, and specialized paints and sealing products.

❖   Nanostructured ceramic coatings exhibit much greater toughness than conventional wear-resistant coatings for machine parts. In 2000, the US Navy qualified such a coating for use on gears of air-conditioning units for its ships, saving $20 million in maintenance costs over 10 years. Such coatings can extend the lifetimes of moving parts in everything from power tools to industrial machinery.

❖   Nanoparticles are used increasingly in catalysis to boost chemical reactions. This reduces the quantity of catalytic materials necessary to produce desired results, saving money and reducing pollutants. Two big applications are in petroleum refining and in automotive catalytic converters.

The Science behind the Technology

Atoms and Bonds

The materials and chemicals that we can use for making things can be separated into about 90 individual elements. The smallest possible pieces of the elements that we can use are atoms. It is important to remember that a gold atom is different than a carbon or iron atom. But atoms themselves are made from protons (positively charged), neutrons (no charge), and electrons (negatively charged). Atoms can actually be identified by the number of protons in them. If you add or take away a proton from an atom you have a different atom, for example if you remove a proton from an oxygen atom you change it to nitrogen. The same is not true when you take away an electron or a neutron that gives you an ion, or isotope, respectively. Atoms can stick or bond to other atoms in a number of different ways, the most important of which is by transferring or sharing electrons to form a bond. This new arrangement of electrons between atoms creates a lower total energy than each of the individuals possessed. The bigger the difference in the energy, the stronger the bond. Multiple atoms can bond together to form crystals or molecules. Molecules can consist of as few as two atoms and as many as thousands of atoms. Crystals can be as small as a few nanometers or as large as boulders.

The arrangement of electrons in atoms fluctuates randomly and under changing conditions. If an atom loses or gains an electron, it becomes an ion. The charge on this ion can attract other atoms or molecules with an electrostatic interaction. Electron distribution can be permanently unbalanced when two atoms bond with each other leaving a permanent dipole. The area of the molecule with excess electrons can be attracted to areas that are poor in electrons and is described as dipole–dipole interactions. Random unbalances of charges in atoms and molecules can induce a fluctuating dipole that attracts neighboring molecules and atoms temporarily. All of these forces are generated without transferring or sharing electrons in a bond so they are often called nonbonded or secondary bonding interactions. Forces between molecules due to these nonbonded interaction account for the self-assembly of large molecules like DNA where attraction between complimentary base pairs on one chain stick to the base pairs of the other chain to bind it together in the double helix. One of the strongest forms of nonbonded interaction occurs between water molecules and is so important to the biomolecular chemistry of life that they are called hydrogen-bonding interactions. Molecules that can bond with water easily with hydrogen bonds are called hydrophilic (water loving) molecules and those that repel water because they are incapable of hydrogen-bonding interactions are called hydrophobic (water hating) molecules. Hydrophobic and hydrophilic interactions are very important to the folding and shape changes in proteins in the body. Hydrophobic molecules are many times used to coat fabrics and materials to make them water repellent. Nonbonded forces between the AFM probe tip and the surface atoms are responsible for the images measured using that technique. Additional technologically useful forces can come from the interactions of atoms with magnetic fields, light (photons), temperature gradients, and the flow of fluids (Tro, 2012).

Scaling of Forces

For each and every one of us, we often do not think about nonbonded forces or bonds. We take for granted that molecules that make up our bodies will continue to function without too much conscious intervention on our part. But there are forces that we experience everyday like the force of gravity. Gravity affects us because of our mass or weight. The heavier we are the more force gravity exerts on us. But our mass is related to our size; in fact it is related to our height, width, and thickness or in other words all three of our dimensions. So what happens if we get much smaller? If we were able to shrink, the force of gravity would decrease rapidly with our size. If we could become 1000 times smaller, the force of gravity would actually be 1,000,000,000 times smaller. At this size, the effect of the nonbonded interactions with atoms and molecules around us would be much more significant than gravity. For example, if you sit in a darkened room with a window bringing in natural light, many times you see tiny dust particles floating in the air defying gravity. The force of gravity on these tiny particles is so small that the small thermal fluctuations and air currents keep them suspended for long periods of time. This makes it possible to move and manipulate nanoparticles using weaker forces such as capillary or electrostatic forces.

Other properties depend not on our size but on our surface area, which is a two-dimensional effect. One of these is the rate at which we lose or gain heat from our surroundings. Although, the rate at which we generate heat is related to the metabolism of our cells and our three-dimensional size. When I was younger, there was a movie where a scientist accidentally shrunk his children to a size smaller than an ant. It turns out that doing this would more rapidly affect their ability to generate heat (a

3-D effect) than their ability to lose it to their surroundings (a 2-D effect), so actually such an accident would have caused the children to freeze to death. Fortunately, for the plot of the movie, *Honey, I Shrunk the Kids*, the writers were not physicists, so shrinking the children just lead to a fun plot line. This same scaling effect can help small materials to heat and cool rapidly, which can be really useful when combined with another property of materials called thermal expansion. When a material is heated, the space between the atoms has to increase and the material expands. When the heat is removed, the material should return to its original size unless it melted. This motion can be used to actuate linear motion by expanding a micro-sized rod to push something, and it can pull it back by cooling it down. Typically, on a metal rod, the size of a pin, the heating and cooling can take minutes, but for micro and nano-sized materials, it happens in a fraction of a second.

Another important property to each one of us is the friction that keeps us from slipping and sliding every time we move. Classically, friction is calculated as a fraction of the force due to gravity working on us. For smooth surfaces, that fraction is small, and, for rough surfaces, that fraction is bigger. When objects get small, the force of gravity gets so small that our classical description of friction no longer explains the situation. Instead, objects experience a force often referred to as stiction where the attraction between surfaces is due to the nonbonded interactions between the atoms of the object and the surface and not the object's weight. In fact, geckos take advantage of stiction by using the small nano-sized hair like structures on their feet to maximize the contact of their feet with even smooth surfaces such as glass and granite. This allows geckos to climb directly up the vertical surfaces using only nonbonded interactions to support the weight of their bodies. Finally, the stiff and brittle nature of silicon is lessened with decreasing dimensions which means that it is much easier

Figure 8.   Fingers of gecko on glass.

to design, manufacture, and actuate small articulating parts on the micro and nanoscale. Silicon is still one of the preferred manufacturing materials because of the well understood top–down manufacturing processes developed by the computer industry (Rogers, Adams, & Pennathur, 2014).

Quantum Mechanics

Quantum mechanics deals with the energy distribution of electrons inside of atoms and molecules. It turns out that in very small objects like atoms there are some fundamental rules that limit the electrons in the atom. The combination of these rules describes the state of the electrons. The state of the electron dictates where it is most probable to find the electron in the atom and the energy of the electron. The probable location of the electron is called the electron orbital and, in general, the farther the electron is from the nucleus (where the protons and neutrons are located), the higher the energy. The energy of the electron can only assume certain values and is not continuous; in other words, energy is quantized. This is probably one of the weirdest but most wonderful results of electron behavior. So, rather than have any amount of energy between zero and infinity, electrons can only have the energies that correspond to the defined electron states. It turns out that electrons follow one of the most important thermodynamic principles, that the energy of a system is generally minimized. So, electrons preferentially fill the lowest energy states first. To remove the highest energy, electrons from an atom, require a certain amount of energy, called the first ionization energy. If this amount of energy is transferred to the electron in the atom, it will leave the atom; the atom then becomes an ion and takes on an overall positive charge. When multiple atoms bond together to form a molecule or crystal, all the electron states in all the atoms must be combined to form bonding states that allow the electrons to be shared or transferred and bonds to form between the atoms. These changes mean that the energy to remove an electron changes and is described by a value called the work function of the material. If energy equal to the work function is supplied to an electron near the surface of the material, then it will escape from the crystal and can flow through a circuit. When the energy transfer happens because a photon of light impacts the surface, this is called the photoelectric effect. In 1921, Albert Einstein won the Nobel Prize for this description of the photoelectric effect.[xi] If the absorption of a photon by an atom or molecule ejects an electron from a low energy state, the atom is in an excited state because thermodynamics says that the electrons should occupy all the lowest energy states. In order for one of the higher energy electrons to move down to one of the lower energy states, it must lose some of its energy. Electrons can jettison this excess energy by emitting a photon with energy equivalent to energy difference

---

xi   http://www.nobelprize.org/nobel_prizes/physics/laureates/1921/.

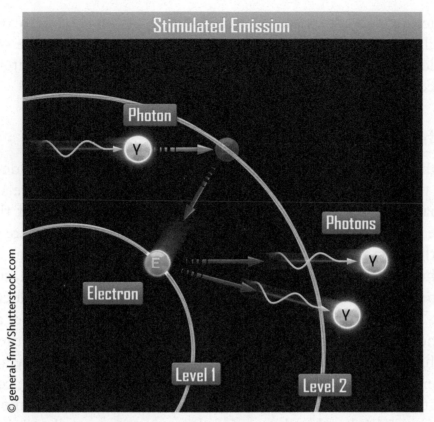

Figure 9.   Stimulated emission of a photon from an atom.

of the states. This emission of light due to the absorption of a photon followed by a transition of electrons between states is known as fluorescence. After this fluorescent event the atom has returned to its ground state and is more stable. Fluorescence is important in many biological tests, cellular microscopy, and in spectroscopic analysis of molecules and materials. Spectroscopy uses the light emitted from the material to analyze properties such as elemental composition, the elemental spatial distribution, bonding, surface condition, and so on (Engel, 2009).

Career Connections

Careers in nanotechnology are often difficult to articulate because they involve employment in the traditional science and engineering jobs where after being hired the employees are frequently asked to perform work designing and manufacturing at the nanometer scale. There are not many 4-year bachelor's degree programs currently concentrating on only nanotechnology and related theory and skills but many universities and colleges have elective tracks that allow some specialization in this area of study and nearly all tier one research institutions have graduate work which relate in one way or another to the topics discussed in this chapter. After completion of a bachelor's or master's degree most students find employment at the engineering level in a variety of companies including: Alcoa, Allied Electronics, Amgen Inc., BD (Becton, Dickinson), BP Solar, Bridge Semiconductor, Cabot, Cambridge, Celgene, Dana Corporation, Dow Chemical, Fairchild Semiconductor, General Dynamics Robotic System, General Electric, GlaxoSmithKline, IBM, Illuminex, IM Flash Technologies, Intel Corporation, Johnson & Johnson, Johnson Matthey, Kurt J. Lesker, Lockheed Martin, Micron, Merck, Northrup Grumman, Inc, On Semiconductor, Philips Medical Systems, Plextronics, PPG, Rohm and Haas, Seagate Technologies, Siemens Co., Thermo Electric PA, Tyco Electronics, Uniroyal Optoelectronics, and Western Digital. In addition to this partial list, there are numerous local and regional smaller businesses, startups, government agencies, and universities that offer jobs and careers related to nanotechnology.[xii]

At the two-year community and junior college level there are nanotechnology and microscopy specific degrees and certificates that can prepare you for technician level positions. Review of the job titles that many nanotechnology trained associate degree and certificate level students now work in include: biological laboratory technician, biofuels technician, chemical laboratory technician, cleanroom technician, deposition technician, device technician, equipment maintenance technician, engineering technician, etch technician, failure analysis technician, laboratory technician, lithography technician, materials science lab technician, medical devices technician, microfabrication technician, nanobiotech researcher, nanoelectronics expert,

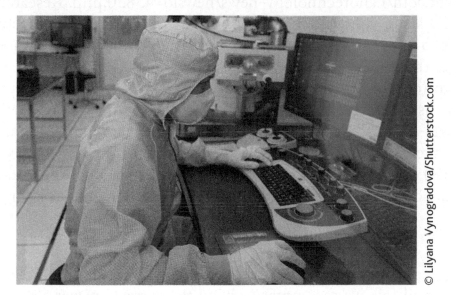

© Lilyana Vynogradova/Shutterstock.com

Figure 10.   Man working in clean room.

---

xii   Private communication with The Penn State Center for Nanotechnology Education and Utilization which has been a national leader in nanotechnology education and workforce development since 1998. This is a partial list of companies who have employed alumni of their program over the last 18 years.

nanofabrication technician, nanomaterials research associate, nanoscale fabrication technician, nanoscience technician, nanotechnologist, process technician, production scientist, quality control technician, real-time defect analysis technician, research assistant, scanning electron microscope operator, scanning probe operator, scientist specialist, solid state technician, test technician, thin films technician, and vacuum technician.[xiii]

Modular Activities

Discussions

❖ Post an original discussion in the online discussion board on the following topic: Which area(s) do you see nanotechnology playing a significant role in within the next 10 years? This might include areas such as medicine, environment, information technology, energy development and storage, medical, semiconductors, nanodevices, defense and security, or other areas. What are your thoughts and why? (200 words minimum). Next, comment on the post(s) of a minimum of one other student in a thoughtful and academic way that enhances the conversation. See rubric for grading and assessment measures.

❖ Post an original discussion in the online discussion board on the following topic: There is some concern that not enough attention is being paid to safety and environmental considerations for nanotechnology, such as in the article http://www.nanowerk.com/nanotechnology-news/newsid=43830.php. Research a bit on the topic and share your thoughts. (200 words minimum). Next, comment on the post(s) of a minimum of one other student in a thoughtful and academic way that enhances the conversation. See rubric for grading and assessment measures.

❖ Listen to (or read) the 5 minute audio segment at http://www.npr.org/sections/alltechconsidered/2015/11/23/457129179/the-future-of-nanotechnology-and-computers-so-small-you-can-swallow-them. What are your thoughts on the article and on ingestible technology generally? (200 words minimum). Next, comment on the post(s) of a minimum of one other student in a thoughtful and academic way that enhances the conversation. See rubric for grading and assessment measures.

Tests

❖ Online graded quiz on overall chapter content, written in multiple choice format. When submitted, we recommend giving the correct answer along with the page number in which it is found for questions students did not answer correctly.

---

xiii This list was composed by the Nanotechnology Applications and Career Knowledge (NACK) Network and the "industry advisory committees at colleges offering Nanoscience Technology associate degrees" and can be found at the link: http://nano4me.org/alumni/work.

❖ Take the nanotechnology awareness quiz at http://www.nanowerk.com/nanotechnology/quiz.php and submit your results to your instructor. You will be assessed on completion of the quiz, not on your overall results, since some of the topics are beyond what is covered in this class. It does, however, give you a feel for relevant technologies and social issues as they relate to nanotechnology.

Research

❖ Nanotechnology: Research and write a two-page informative report on a specific use of nanotechnology of your choice. This report must use proper citation and use a minimum of two sources.

❖ As you are likely aware, energy consumption versus energy creation is an issue here in the United States as well as worldwide. Research ways in which nanotechnology could have an impact and write a two-page informative report. This report must use proper citation and use a minimum of two sources.

❖ Research nanotechnology products that exist today or are in current development in a specific field, such as nanoconstruction, nanotransportation, nanotextiles, or nanoconsumer goods. Create an informative poster that communicates the details of nanoproducts in your chosen field. You will be assessed on the depth, accuracy, and quality of your work. You must also cite any sources you use.

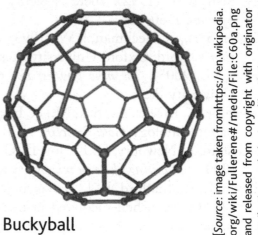

Buckyball

[Source: image taken fromhttps://en.wikipedia.org/wiki/Fullerene#/media/File:C60a.png and released from copyright with originator attribution and sharing alike if the image is modified]

Design and/or Build Projects

❖ Using materials available to you, build a three-dimensional buckyball model. You will be assessed on the accuracy and professionalism of your model.

Assessment tasks:

❖ Consider nanotechnology use in national defense. Choose one of the following nanotechnology areas in this field and assess the benefits and drawbacks of this particular technology as it relates to national defense: fabrics and materials such as armor or clothing intended to withstand extreme conditions, nano-robotics, nanotechnology in weapons, or nanotechnology in vehicles such as in passenger protection, fuel economy or operations. Write a two-page assessment of your findings using a minimum of two sources.

Terms

Fullerenes: A fullerene is a molecule of carbon in the form of a hollow sphere, ellipsoid, tube, and many other shapes.

BuckyBall: A spherical fullerenes are also called Buckminsterfullerene (buckyballs), as they resemble the balls used in football (soccer).

Nanotechnology: The branch of technology that deals with dimensions and tolerances of less than 100 nm, especially the manipulation of individual atoms and molecules.

MEMS: Micro-electro-mechanical systems (MEMS), is a technology that in its most general form can be defined as miniaturized mechanical and electromechanical elements (i.e., devices and structures) that are made using the techniques of microfabrication.

Bibliographys

Eaton, P., & West, P. (2010). *Atomic force microscopy.* Oxford: Oxford University Press, 2010.

Engel, T. (2009). *Quantum chemistry and spectroscopy* (2nd ed.). New York, NY: Pearson.

Feynman, R. P. (1992). There is plenty of room at the bottom. *Journal of Microelectromechanical Systems, 1,* 60–66.

Flegler, S. L., & Heckman, J. W. (1993). *Scanning and transmission electron microscopy: An introduction.* Oxford: Oxford University Press.

Fraser, G. (2005). *The new physics: For the twenty-first century* (p. 495). Cambridge: Cambridge University Press.

Freestone, I., Meeks, N., Sax, M., & Higgitt, C. (2007, December). The lycurgus cup—A Roman nanotechnology. *Gold Bulletin, 40*(4), 270–277).

Goldstein, J., Newbury, D. E., Joy, D. C., Lyman, C. E., Echlin, P., Lifshin, E., ... Michael, J. R. (2007). *Scanning electron microscopy and x-ray microanalysis* (3rd ed.). Springer.

Hirsch, L. R., Stafford, R. J., Bankson, J. A., Sershen, S. R., Rivera, B., Price, R. E., ... West, J. L. (2003). Nanoshell-mediated near-infrared thermal therapy of tumors under magnetic resonance guidance. *Proceedings of the National Academy of Sciences, 100*(23), 13549–13554.

Nelson, D. L., & Cox, M. M. (2005). *Leninger principles of biochemistry* (4th ed., pp. 1034–1080). W. H. Freeman.

Pelesko, J. A. (2007). A very thorough and interesting treatment can be found. In *Self assembly the science of things that put themselves together.* Chapman and Hall.

Rogers, B., Adams, J., & Pennathur, S. (2014). *Nanotechnology: Understanding small systems* (3rd ed.). CRC Press.

Sanders, W. C., Ainsworth, P. D., Archer, D. M., Jr., Armajo, M. L., Emerson, C. E., Calara, J. V., ... Swenson, J. D. (2014). Characterization of micro- and nanoscale silver wires synthesized using a single-replacement reaction between sputtered copper metal and dilute silver nitrate solutions. *Journal of Chemical Education, 91*(5), 705–710.

Tro, N. J. (2012). *Principles of chemistry: A molecular approach* (2nd ed.). Upper Saddle River, New Jersey: Pearson.

Chapter 13

Advanced Manufacturing and Production

(Continued)

(*Continued*)

- ❖ Terms
- ❖ References
- ❖ Further reading

Chapter Learning Objectives

- ❖ Define advanced manufacturing.
- ❖ Explain how Six Sigma and the DMAIC process can be used to improve quality in a business.
- ❖ Identify how computer-aided design and computer-aided manufacturing (CAD and CAM are used in advanced manufacturing and three-dimensional (3D) printing.
- ❖ Examine the role of the ISO 9000 Family of Standards in producing quality manufactured products.
- ❖ Describe how Lean Manufacturing and Supply Chain Management can reduce waste in a company and increase efficiency.
- ❖ Distinguish the difference between additive and subtractive manufacturing as well as the benefits and disadvantages of each.
- ❖ Define the characteristics of nanomanufacturing.
- ❖ Summarize how the Theory of Scientific Management contributed to the modern age of production technologies.
- ❖ Understand how RFID tags can be used in advanced manufacturing.
- ❖ Differentiate the role of robotics in advanced manufacturing.

Overview

Humans have been manufacturing things since the first time a caveman—or cavewomen—whittled a stick into a spear. It is in our nature to manipulate our environment. Innovations have allowed us to live in cold climates, feed our families, and fly across the ocean. In fact, almost every item we interact with daily has been manufactured. This includes the alarm clock that wakes us up in the morning and the light switch we turn off before we go to sleep.

Manufacturing often gets a bad rap, though, given its prevalence in society. Students find it a boring subject matter. They are less inclined to consider it as a profession. They imagine black and white photos of dirty shop floors and dreary assembly lines, similar to those seen in Figure 1. Something their great-grandfather did. Or something somebody else's great-grandfather did. People do not picture today's clean advanced manufacturing labs with state-of-the-art technology. In a survey conducted by Deloitte and the Manufacturing Institute, only 53% of the respondents thought US manufacturing jobs were both interesting and rewarding. In the same survey, though, 90% thought the manufacturing industry was important to America's economic prosperity. Almost as many (89%) thought it was significant to America's standard of living.

© Everett Collection/Shutterstock.com

Figure 1. Many women worked on assembly lines, especially during World War II when men went off to fight the war.

What, exactly, is manufacturing if it is recognized as important but has a sometimes negative reputation? Businessdictionary.com describes manufacturing as "the process of converting raw materials, components, or parts into finished goods that meet a customer's expectations or specifications. Manufacturing commonly employs a man-machine setup with division of labor in a large scale production." According to the Manufacturing Institute, the manufacturing sector is having a resurgence in the United States, is a strong contributor to the economy, and supports many high-wage jobs. Automation, three-dimensional (3D) printers, and advances in robotics have led to advanced manufacturing. This has taken manufacturing beyond the assembly line. Flexible, high-tech methods allow products to be produced that meet specific customer requirements rather than the one-size-fits-all model. Mass production, the standard of twentieth century production has become mass customization. Products, such as computers, can be produced in large quantities but contain software and hardware components individualized to specific users.

Interest in manufacturing is found both in the workplace and in the home. The number of people completing do-it-yourself (DIY) projects and attending makerspaces has increased. Makerspaces are community centers where people have access to materials, 3D printers, laser engravers, electronics, and so on, as well as experts who can assist them in designing and executing DIY projects. 3D printers became commercially available in 2009. The number of them bought for use in the home has been increasing since that time. Consumers can purchase or download blueprints off the internet to create items in their own home. Television shows that highlight American craftsmanship are common on cable television. Itsy.com is a popular website for purchasing handcrafted items. It appears that manufacturing by any other name has become popular with consumers. Understanding that it can be a fulfilling profession as well is important for students to recognize and appreciate.

Manufacturing for the masses first became widespread in the late 1800s and early 1900s. At this time there were many technological developments in machinery and the assembly line was introduced. As the industry matured, synthetic materials, such as plastics and lighter metals, started being used. To truly appreciate these innovations, imagine your life without plastic and aluminum. It would be a world in which convenience stores, if they even existed without the ability to package food easily, would not have bottles of pop and water available for purchase. New and more stable energy sources also contributed to modernizing production and expanding the capacity of assembly lines.

Most people are familiar with large manufacturing companies such as those who make cars (Ford or Honda), farm equipment (John Deere), or planes (Boeing). However, the number of small manufacturing firms far exceeds the number of large companies. The most recent data from 2014 show that there were 254,941 manufacturing firms in the United States. Over 107,000 of them employed a maximum of 4 people. In contrast, only 3524 had more than 500 employees (Manufacturing Institute). Manufacturing and production technologies make it possible for you to buy high quality, inexpensive items. They also offer you a way to a fun and exciting career.

History

Goods were once produced by skilled craftsmen, one at a time, in their own homes. But back in the day they were called cottages; hence the term, cottage industry. It was only through technological advancements that mass production was made possible. The industrial revolution changed everything. Mass production is the reason you can buy a new car, smart phone or whatever you want, whenever you want, and at a reasonable price. Imagine the cost and time if a single person was producing each item. Or, imagine having to build your own devices. Chances are your phones would not be very smart.

Many preconceived notions of what constitutes manufacturing have their roots in the industrial revolution, which began in the 1700s. This was a time when producing goods moved from small homes and villages to large cities and factories. Specialization of work developed with workers being skilled in a small segment of production tasks.

Workdays were long and grim. Labor unions and child labor laws were nonexistent or in their infancy. At the same time, being able to manufacture items quickly increased the volume available to customers while decreasing the cost of goods. One result was an improved standard of living for many segments of society. This lifestyle change has been pervasive and enduring. It has worked its way across generations and right down to you, the reader of this book.

Most students roll their eyes or yawn loudly when learning about the Industrial Revolution. It seems remote, old, and having nothing to do with them. Many can recall from memory that Eli Whitney patented the Cotton Gin in 1794, that there was mass migration from farms to city factories, and work conditions were often deplorable. One of the most important events was that Henry Ford invented the car and assembly line. Except that he did not. Henry Ford did what so many innovative entrepreneurs do both then and now. He took a good idea, researched it, revised it, and made it work more efficiently. The result was the mother-of-all-manufacturing-inventions that transformed everything: the assembly-line produced automobile. Not only did Ford alter how cars were produced, he paid his employees $5 a day, which at the time was unheard of. It did, however, enable them to afford the cars they were building, such as the Ford Motel T shown in Figure 2. Now everybody could buy one.

The automotive assembly line is one of the most complex production systems. Two technologies contributed to the success of the automobile and have had a lasting impact on mass production: interchangeable parts and the assembly line. The interchangeable parts system was first commercialized by Eli Whitney of the Cotton Gin fame. Whitney used the process to assemble muskets in the early 1800s, when the

Figure 2.   1920 Ford Model T.

United States was preparing for war with France. Even though he is given credit for interchangeable parts, he had built upon the previous work of French and English innovators. (As is so often the case, history likes to sanitize, categorize, and present neatly historical events. The truth is usually much more messy and interesting.) The key factor in interchangeable parts is precision. Interlocking pieces must always be the same dimension or they will not fit together. This is why quality management has emerged over the years as an integral component of production. The assembly line is generally the process in which interchangeable parts are added to a product, one at a time, until the product is complete. Ransom Eli Olds, who produced the Oldsmobile, designed the first automotive assembly line. Ford took his idea, improved upon it over the years, and created a superior process. His process was so efficient that one Ford Model T rolled off the assembly line every 3 minutes.

Modern assembly lines could not develop until there was a way to power them with a stable energy source. Before the age of mass production, craftsman used their own physical power to make their products or had assistance from their farm animals. One of the earliest sources of power for equipment was water and the waterwheel. Many early cities grew up around rivers for this very reason. Steam power was the first consistent energy source to fuel manufacturing machines. The first industrial revolution would not have been possible without this reliable power source. A steam engine converts heat energy of steam into mechanical energy, which propels pistons back and forth and powers an engine. Mechanical energy is the sum of potential energy and kinetic energy. A good way to remember potential energy is to picture yourself on a ledge with a large rock on the ledge above you. It may not be in motion yet (kinetic), but you keep your eye on it because of its potential energy. To view a 2-minute video on how mechanical energy is created from potential and kinetic, visit: https://www.britannica.com/science/mechanical-energy.

Steam engines can also power saws, huge hammers for shaping forced metals, and other industrial machines that are part of the production process (Steam Engine, Encyclopedia.com, 2002). To view a 30-second demonstration of a steam engine, visit https://www.youtube.com/watch?v=ESfSG2O1QYQ.

Once car ownership became universal, it changed where you could live, work, and play. Suburbs became the fashionable place to live, contributing to the decline of many downtown areas. School districts were expanded and one-room schoolhouses eliminated. The Sunday drive became a permanent fixture in many families. The car was also credited to the "invention" of the teenager. During the early years of the industrial revolution, this life stage was an unknown convention. People moved directly from childhood to adulthood at an early age. This quick jump contributed to the large numbers of child laborers in industry. However, starting in the 1920s, young people were spending more time in school and marrying at an older age. Child labor laws were also being passed in many states. The convergence of adolescent independence and mobile transportation created a new life stage. Young adults now had the ability to spend

Figure 3.   Teenagers in front of an old car.

time away from their homes and away from the watchful eye of parents. Who you could date expanded from the young women or man next door to whoever was within driving distance. By the mid-1940s youthful freedom from adult oversight changed the culture of the teen years and family unit forever. Figure 3 shows the newly labeled "teenagers" in front of an old car.

That is just one example of the broad impact of manufacturing on society and why it remains relevant to us now. The need for mass-produced items has not diminished. If anything, the need has increased in our consumer-based, global society. What has changed is the emphasis on cost, quality, and safety. While many processes are automated using computer and robotic technologies, human involvement is still necessary. After all, robots do not run themselves—at least not yet.

The modern assembly or production line has been in use since the early twentieth century. The foundation of any assembly line is the division of labor. Typically, a human stands in front of a conveyor belt and acts upon a product as it passes by. They perform specialized tasks such as adding a part, tightening bolts, or applying a coat of paint. The worker is dependent on the speed of the conveyor belt and must time their actions accordingly. They are not concerned with the steps that happen before or after the ones that they complete. One of the most popular examples of this process is the "I Love Lucy" episode in which Lucy and Ethel get jobs in a candy factory wrapping chocolates. To view a 2-minute video clip of the segment, visit the following website: https://www.youtube.com/watch?v=8NPzLBSBzPI.

So how did we end up with a manufacturing system where speed and efficiency became dominant over human capacities, where people became subordinate to machines? How did we become, in the words of Henry Thoreau, "tools of our tools"? One primary contributing factor is the Theory of Scientific Management.

Theory of Scientific Management

Frederick Winslow Taylor developed one of the most influential principles of the modern era: the Theory of Scientific Management. Although you have probably never heard of Taylor, his theory has had an unmistakable impact on your life. That is a pretty bold statement but true. The basic idea is that there is one best way to do any task; one best way to achieve maximum efficiency. That task can be any physical activity, such as shoveling coal, making a bed, playing golf, or turning a handle. It does not matter. According to Taylor, out of the possibly hundreds of ways to perform a job, only one was the most effective. Taylor made it his job to know exactly which one it was by observing, timing, and calculating.

Born in 1856, Taylor was an American mechanical engineer who dedicated his life to industrial organization. He famously conducted time studies of workers doing their jobs to document their every step. For example, he determined that the average number of bricks that could be laid in 10 hours was 480. He redesigned shovels so workers could lift the optimum load. Taylor meticulously measured each step in a procedure and dissected them down to seconds and inches. When he found barriers to an efficient workflow, he redesigned the steps. Workers could be repositioning closer to machines to reduce walking time or tools could be redesigned to tolerate more weight. Taylor scientifically standardized work. Man was made to work like a well-oiled machine.

The Taylorized system spread worldwide and had an enormous influence on industrialization. The confluence of scientific management and assembly-line production revolutionized manufacturing. Each worker could do one specific step, without extraneous movement or error, over and over again. The increased speed and efficiency in which manufactured goods could be produced reduced costs and made them accessible to many people. No longer was it just the well-to-do who could afford material items, they were now available to lower- and middle-class people as well.

Not everyone was impressed with scientific management. Workers felt overworked, enslaved, and devalued. They were just another cog in the machine. Many found their workplace to be rigid and confining. Over the years there has been an effort to make the shop floor more ergonomic and flexible. However, modern assembly lines still resemble the early ones in many ways with their individual workstations and emphasis on speed and productivity. In fact, the assembly-line mentality has permeated our everyday life with our rigid adherence to time (think about this next time you check your watch!).

> Today it is only modest overstatement to say that we are all Taylorized, that from assembly-line tasks times to a fraction of a second, to lawyers recording their time by fractions of an hour, to standardized McDonald's hamburgers, to information operators constrained to grant only so many seconds per call, modern life itself has become Taylorized (Kanigel, 1997, p. 14).

Quality Systems

Quality management has been emphasized in manufacturing as competition among companies has increased and business practices have matured. Customer satisfaction is often a main goal of any business. And what most customers want are high-quality items for reasonable prices. Satisfying customers should not be achieved, though, at the expense of a safe working environment or healthy profit. Quality management done right can reduce errors and increase productivity.

While the need for speed and quality has only increased since the early days of Taylorism, it has now taken on a more holistic approach. Measuring and standardization are still central tenants of manufacturing, but worker involvement is considered as well. Several established methodologies used to improve quality include Six Sigma, Lean Manufacturing, and International Organization for Standardization (ISO) 9001 among others.

Six Sigma

Six Sigma is a proven methodology that helps companies improve business processes by building quality practices into any system. These systems can include anything from manufacturing brake pads to routing customer service calls. The goal is to produce no more than 3.4 defects per million efforts. To reach this level of quality a rigorous process that requires training and experience is used. People can earn several levels of Six Sigma certificates. The highest level is the Master Black Belt. People who reach this level are outstanding leaders who have successfully completed numerous quality and continuous improvement projects.

The term, Six Sigma, has a specific meaning. It is derived from Sigma (σ), the Greek letter that represents standard deviation. Remember the bell curve you learned about in high school that represents a distribution of numbers (e.g., IQs, batting averages, test scores, or the number of gadgets produced each hour). See Figure 4 for a refresher. If there is normal distribution, the mean (average) is at the very center of the bell, and indicated by 0.5 in Figure 4. Each side is a mirror image of the other side. Numbers to the right are higher than average scores (hopefully that is where your grade will fall when you complete your next test). Numbers to the left are lower than average scores. When you gather and plot data as part of a quality procedure, your goal is to get very little variation, which would translate into a very skinny bell curve. This means every instance of what you are measuring (e.g., inches or time intervals) is exact or similar to every other one. A flat bell curve indicates lots of variation, and generally lower quality. If your machine is programmed to cut a board at a 90° angle to fit flush on a window, you want it to cut 90 degrees every time. Every 89° or 92° board will have to be scrapped. If this happens 3.4 out of a million times, it is probably not a concern. If it happens every

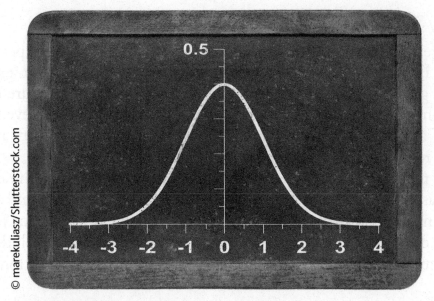
© marekuliasz/Shutterstock.com

Figure 4.   A bell curve figure written on a chalkboard.

15 or 20 cuts, this quality issue will cost your company money in terms of material, time, and possibly warehouse space.

One standard deviation on either side of the mean accounts for approximately 68% of the population you are studying. Two standard deviations out account for approximately 95% of your population. Three standard deviations account for over 99% of your population. Notice in Figure 4 how it also becomes flat at this point. There is not a lot remaining when you are over 99%. So if *Sigma* refers to the standard deviation, the *Six* indicates how many deviations from the median you should go. If three standard deviations comprise 99.7% of a population, six standard deviations encompass 99.9999998%. To achieve Six Sigma in a manufacturing process, you need to produce a gadget free from defects 99.9% of the time. At least that is the goal.

> To put this in perspective, if you were a publisher and a misspelled word was considered a defect, 99 percent quality would mean that for every 300,000 words that were read by the customers who purchased your books, 3,000 would still be misspelled. Six Sigma strives for near perfection; therefore, reaching Six Sigma quality would mean that for the same 300,000-word opportunity, only 1 word would still be misspelled (Islam, 2006, p. 16).

If we consider this book, with approximately 450 pages, and presume about 500 words a page, then of about 225,000 words none of the words should be misspelled. While this may be a hefty task, it is not one that is impossible.

So how does someone go about achieving near perfection in a procedure? They follow the Define-Measure-Analyze-Improve-Control (DMAIC) model, which is integral to Six Sigma. DMAIC is a process that contains the following five steps in a continuous loop: Define-Measure-Analyze-Improve-Control. See Figure 5. It is similar to the scientific method in that an activity is systematically measured, tested, and modified. While it is foundational to Six Sigma, it can also be applied independently. It is a logical, established method to achieve excellence. It is also important for any quality process. You must be able to gauge where you are to determine where you need to go.

In Figure 5, the image on each circle is a visual representation of that step. *Define* contains a magnifying glass, because the first step is to take a close look at what the problem is, to define it carefully along with the goals of the improvement activity. In many ways, this is the most important step. If you define the situation incorrectly, everything that follows will be for naught. Or worse, you may implement an "improvement strategy" for a situation that is not even faulty.

Measure contains a map location icon. Its significance is that you need to properly measure what is going on in a system to accurately know your position. Similar to

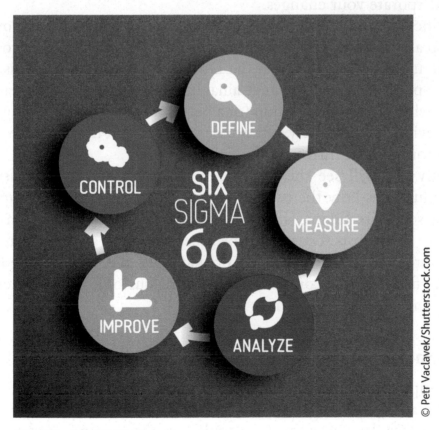

Figure 5. The Six Sigma DMAIC process.

a map, if you follow the wrong directions, you will end up in a place you may think is correct, but you will be mistaken. After you measure a process in your system over time (e.g., time to complete a task, ratings on a survey, length of a metal rod, etc.), you need to *analyze* it. The primary purpose of analyzing a process is to determine if and why there is any variation between occurrences of what you are measuring, and if there are variations are they at a level that is designated as defective. Ultimately, data analysis should get you to the root cause of why one product may have a defect while another one may not. (Remember: to achieve Six Sigma you can only have 3.4 defects per million.) For example, if a machine drills holes in a metal plate that will contain a rod, you might find that the size of the hole is too large for the rod to fit snugly after 50,000 plates have been drilled. You would take the results of this data analysis and apply them to the next step, *improve*.

In the DMAIC figure, *improve* shows an upward trending graph. At this point, you need to design and implement a solution to the identified variance or problem to increase your business success. Using the previous example, your company may decide to have maintenance work done on the machine after 50,000 uses, find another supplier for the drill bits, or maybe even buy a new machine. There are rarely single solutions to industrial problems. Once you decide on a course of action, you revise the system to incorporate your changes.

Control includes a picture of tightly bound gears. This shows how in the final DMAIC step an updated system is tightly regulated to meet quality goals. It also sets up the system up to be able to repeat the DMAIC process. The process is continuous, because while perfection really is unattainable, continuous improvement is not. Once you come up with a better mousetrap (or a better way of manufacturing a mousetrap), you need to review it continually. Otherwise your product may become stagnant while your competitors start to thrive.

Frequently you will see the term Lean Six Sigma instead of just Lean or Six Sigma. This illustrates how the two systems are both similar and complementary. As you have just seen, Six Sigma is focused on the cause of variation in a process and how to remove it. Lean is focused on the speed at which a process can be effectively completed by removing wasteful steps or materials. Both are focused on positively impacting the finances of a company and meeting customer requirements. They both also require continuous improvement strategies.

Lean Manufacturing

Toyota first implemented lean production in the 1980s. Its impact on manufacturing was groundbreaking. The goal for Toyota was to become lean by using less. This included less space, less investment in tools, less design time, and less inventory. It also required fewer, but more highly skilled, people to do the work. Machines became more flexible to be able to produce a wide variety of products at high volumes. This was a significant change from how assembly lines operated at that time. They were still based on Henry Ford's original model that required semiskilled workers to operate at fixed workstations. Product changes were costly and difficult to implement.

Although lean production philosophy emphasizes less, it still has a high requirement for quality and continuous improvement. One common tool of lean manufacturing is 5S. It is the process to organize a workplace. To do this, workers follow the five steps that each start with the letter "S". They appear to be commonsense behaviors but can speed up a process substantially.

1. *Sort*: Ensure needed items are in working order and remove any unnecessary ones.
2. *Set in order* (or *straighten*): Put items in logical order where they are easy to find.
3. *Shine*: Clean space for safety and to easily see when maintenance on tools is needed.
4. *Standardize*: Formalize the system so that it is properly completed the same each time.
5. *Sustain*: Train, organize, and communicate with team members to keep the process going.

Many of the lean manufacturing tools rely on visual cues. For example, posters and color coding can be used to quickly indicate when a machine is inoperative or when inventory is low. Part of this visual control is the Kanban inventory system (which means "sign" or "card" in Japanese). Cards are used to visually follow inventory components and processes until all are depleted indicating that inventory needs to be reordered. The term just-in-time (JIT) is used to describe this function of lean production. Inventory is kept to a minimum and only ordered when needed. This reduces inventory costs as items arrive just in time to be used. Sticky notes are also commonly used on a board to give a quick update on the status of a project. These low-tech approaches have been used successfully over many years and in numerous industries.

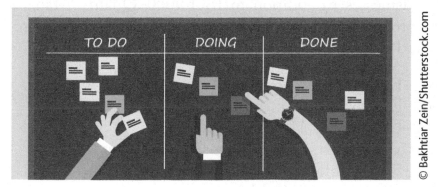

Figure 6. Kanban to-do list on a board.

These are a few of the strategies used in lean manufacturing to keep workflow streamlined. Removing waste of all types is what creates a lean workplace. Waste can include physical items, such as inventory or unused tools, or time used for unnecessary tasks. Similar to Lean Six Sigma, this is not a one-time activity but should be part of a continuous improvement cycle. Once you have created a new system, you repeat the quality steps to ensure that new waste has not creeped into the system. Lean manufacturing can also contribute to a green, eco-friendly manufacturing environment. By using less waste overall, energy consumption should be reduced and less material should be sent to landfills.

ISO 9000 Family of Standards

While Six Sigma offers certification for individuals, ISO 9000 is used to certify that a company meets quality management standards. ISO stands for the International Organization for Standardization and is an independent, nongovernmental organization with members in 161 countries. It is an internationally recognized association that has strict guidelines on how policies and procedures are created and implemented in organizations. Many businesses and government agencies will only do business with companies that are ISO certified. It has become one of the most well-known and respected standards. Over a million companies worldwide, both large and small, have adopted it.

To view a timeline of the ISO story, visit the following site: http://www.iso.org/iso/home/about/the_iso_story.htm#2.

ISO 9001:2015 is the standard that contains requirements for a quality management system. It requires a strong customer focus, input from top management, the process approach, and continuous improvement. Customers who buy from ISO 9001:2015 certified companies know they are getting consistent, high-quality goods and services. While ISO develops the standards, they do not directly certify companies. That is done by external bodies with ISO expertise.

The ISO family has expanded to include the following management system standards for businesses in addition to ISO 9000–Quality Management. They are tried-and-true processes that offer a roadmap to creating work environments that are efficient, safe, and superior.

ISO 50001–Energy Management
ISO 14000–Environmental management
ISO 22000–Food Safety Management
ISO/IEC 27001–Information Security Management
ISO 20121–Sustainable Events Management
ISO 45001–Occupational Health and Safety
ISO 37001–Antibribery Management Systems

High-Tech Manufacturing

Computer-Aided Design and Computer-Aided Manufacturing

Manufacturing has gone high-tech. Computer-integrated manufacturing (CIM) systems allow entire factories to be controlled automatically. Workstation operators, managers, and the machines they use can communicate and share data. This automation allows companies to centralize control of their operations. Computer numerical control (CNC) machines can receive their instructions directly from computer-aided design (CAD) and computer-aided manufacturing (CAM) systems. CAD/CAM systems are computer software programs that are used to design and manufacture products. A screenshot of one CAD/CAM system is shown in Figure 6. Early programs were two dimensional with limited capabilities. New software programs show the products in real 3D proportions. They can be rotated, tested (e.g., wind flow or tensile strength), integrated with other parts, reshaped, and manipulated similar to what you would do to an actual product. This streamlined process increases quality, reduces prototyping costs, and allows companies to respond quickly to customer requests. Once the CAD/CAM design is complete, it can be sent directly to a computer-controlled machine, such as a CNC machine or 3D printer. One main advantage of working digitally is that all steps can be stored, shared, and easily retrieved later.

CAD software packages are varied and can have custom tools and commands that are applicable to a certain job or discipline. Most CAD software programs have similar basic components. They all have a user interface, drawing commands, modifying commands, and options for exporting the digital object. The CAD software user interface allows the designer or drafter to interact with the various commands which generates a digital representation of any object they would like to create. The drawing commands allow the user to create shapes and features that define the object. The modify commands allow the user to manipulate the drawn shapes to further define the object. After the object is created the software provides a variety of options for sharing the created object with other people or exporting to another file format. Quality CAD software provides all the tools to detail out the object for others to understand and from which the object could be fabricated.

Advancement in society and technology has created the need for not only a paper copy of the object, but also a way to share the digital model. Today an architect can send the house file to a realtor who can use augmented reality to show their client what a house would look like on a plot of land, using a tablet, or what the views would look like through their windows even before the foundation is poured. A construction worker can access the digital model to get more information on a skyscraper level they are building. A machinist can access all of the quality control documents for a part they are machining with the touch of a few buttons or swipes.

CNC machines can complete many tasks that are used to being done by human machinists. They can run unattended and produce parts with tremendous accuracy. Tools that can be used include grinders, lathes, mills, and routers. CNC machines are flexible enough to produce different parts by receiving new directions from a CAD/CAM software program. While the technology may be new, the process is not. It starts with a piece of material, such as wood or metal, being manipulated into another shape by grinding or cutting. This is called subtractive manufacturing. As the name suggests, subtractive manufacturing uses machining processes to remove material from a piece of material stock. Milling, drilling, reaming, and tapping are all types of subtractive manufacturing processes. The advantage of using CNC machines is that they can be programmed to cut microscopically. Very few people can be this precise. One disadvantages of subtractive manufacturing is the amount of waste that is produced.

To view a 5-minute video of an American company using a CNC machine to create the head of an eagle, visit: https://www.youtube.com/watch?v=8CSwOebmb0A.

3D Printing and Prototyping

3D printing, on the other hand, is part of additive manufacturing. This is where materials, such as plastic, are deposited—or added—one layer at a time until a desired object is made. A 3D printer works similar to an ink jet printer. If working with plastics, as an example, you would use a type of plastic called thermoplastic, because it melts when heated and turns solid again when cooled. This is the most common type of 3D printing and the process is called fused deposition modeling (FDM). FDM takes a thin filament of thermoplastic, applies heat until it reaches its melting point, and then extrudes it one layer at a time until the object is complete. As the technology advances, the additive manufacturing process is able to not only create a rough working prototype, but also a quality finished product for the customer as well. 3D printing is not a new process but was developed in the 1980s. At that time it was used primarily in industry settings. Around 2010 desktop 3D printers became available, although they were still quite expensive for individuals. Today, you can purchase a quality desktop 3D printer for under $1,000. You can also purchase a 3D printing pen for less than $100. Similar to the regular printer, it uses thin strands of plastic that quickly warm up. Just draw, print, and manufacture your own creation.

The designs for objects to be 3D printed are created using CAD/CAM software. Most CAD software can save a digital file in a Stereolithographic, or STL, format. The STL file was created by the company, 3D Systems, for their software output. It has since been used by different software and devices becoming a universal file type to create 3D objects from a digital file. This is the universal file that can be read by 3D printers to create a physical prototype of the object you created in the design software. Your object will only be as smooth as the STL object is after it is saved.

The 3D printer's software takes the STL file and slices the object into many thin layers depending on the ability of the printer and the thickness of the material. The 3D printer's software will also identify where support material is needed to support the object's build material as it is extruded in thin layers. The support material will either dissolve when placed in a special bath or just break away from the object once it is completed. 3D printing is a type of additive manufacturing which is changing the way we think about manufacturing. See Figures 7 through 10 to see the progression of a

Figure 7. A simple mechanical part created using 3D CAD software.

Source: Courtesy of Jonathan Allred. © Kendall Hunt Publishing.

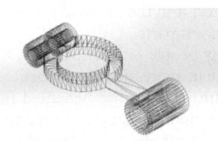

Figure 8. The same part as in Figure 7 but after being processed as an STL file. There are 1,224 triangles that now define this object. This is as smooth as the part will get.

Source: Courtesy of Jonathan Allred. © Kendall Hunt Publishing.

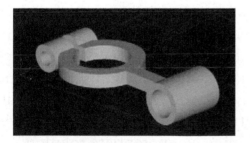

Figure 9. The same STL file as shown in Figure 8 above after being processed by the 3D printer software. You can still see the edges of the triangles.

Source: Courtesy of Jonathan Allred. © Kendall Hunt Publishing.

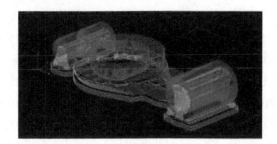

Figure 10. The same object as in Figure 9 after the 3D printer software has identified where it will create the geometry of the object using build and support material. It has also defined the multiple thin layers which it will use to build or print the object. The build material is RED, and the support material is GRAY.

Source: Courtesy of Jonathan Allred. © Kendall Hunt Publishing.

part using CAD technology. This technology is bringing the manufacturing plant into every home. The ability to replace broken items has become as quick as going to your desktop and printing out a new part.

Many available CAD and computer-aided engineering software have a simulation option where the computer will simulate different variables as the product is analyzed or evaluated. This could show the stress on the object from a given load, the pathway of plastic in an injection molding process, the fill time and cooling rates of different materials, and so on. Some types of software can create a simulation of the environment in which the object will be used furthering the opportunities of identifying constraints or weaknesses in the product or system. If a prototype has been created you can test it in the physical system to test the design, material used, and interaction with adjacent parts.

Many people only associate 3D printing with plastic parts, but many machines can print in more than just plastic. Some machines can print rubber components. The health industry has benefited from the 3D printing of human cells to create a body part. Some bakers have used the technology to print objects made of chocolate. Technology has also made it possible to print metal objects using an additive manufacturing process called sintering. This process uses metal powder and a heat source like a laser to weld the powder in many fine layers to create a finished part.

The advantage of using additive manufacturing is that prototypes are quickly produced in a process known as rapid prototyping. Scale models can be quickly made, reviewed, and revised before being manufactured in large quantities. Companies can save enormous amounts of time and money using this approach.

Prototyping is part of the problem-solving process when manufacturing new designs. Problem solving is a necessary skill for all employees in any industry. No matter what position you hold in a company you will find problems which need to be solved at every level. The basic steps of solving problems include identifying the problem, developing a solution, evaluating the solution, and communicating the solution to others. The problem solving process is not usually completed by a single individual. A team of professionals will share the responsibility of solving the problem allowing the team to share strengths and knowledge in obtaining an appropriate solution.

Finding a solution to a problem in the design industry depends on the nature of the original problem. You will not solve a problem in the construction industry the same way you might solve a problem with mechanical components. It is common in industry to have a design meeting with all of the members of the team to help come up with solutions to a defined problem. Sometimes the best way to identify or solve a problem is to see a physical or digital representation of the solution to that problem. An image of the solution can show additional constraints and issues which need to be addressed before creating a final model for analysis. This is also a powerful way to express ideas or communicate design intent. Numerous worldwide industries have already taken advantage of 3D printing capabilities in their businesses, including medical/dental, automotive, and aerospace.

* Local Motors, an innovative car company, built (or printed) the first 3D prototype car, Strati, with carbon-fiber-reinforced plastic.
* Veterinarians in Brazil used 3D printing to make a new shell for a tortoise burned in a fire.
* Scientists on the International Space Station printed a ratchet wrench from design files transmitted from Earth, while RUAG Space, a European supplier to the space industry, completed a pilot project to 3D print an antenna support tool for an Earth observation satellite.
* Researchers in General Electric's Additive Manufacturing Lab have used 3D printing to manufacture airplane engine parts that are lighter in weight and more durable.
* Dr. Matthew Bramlet, a pediatric cardiologist at the University of Illinois College of Medicine, and his team printed hearts of sick babies from MRI images to help surgeons understand the anatomy details in ways that were not possible before.
* Dutch researchers at the University of Groningen have developed an antimicrobial plastic that can be used to print teeth that resist bacteria.
* Technicians in the Netherlands used a 3D printer to create a titanium lower jaw for a woman's face by using a laser beam to melt successive layers of titanium powder.

While many companies are using 3D printing for prototyping purposes or single projects, and many medical products still need to undergo safety testing, future plans include producing actual merchandise. Current drawbacks of using 3D printing for mass production are its varying strength due to layer-by-layer construction, expensive equipment costs, and slow printing speed. However, the field has been maturing quickly, and the number of jobs related to the 3D printing industry has been rising steadily.

According to the BBC article, *Why 3D is more than a passing fad for jobs*, "The print anything revolution is real" (Borzykowski, 2016). Although it has not yet lived up to the hype of printing whatever you want easily and cheaply, it is on the verge of taking off. Professionals and new graduates have learned a lot more about additive manufacturing and materials, prices continue to fall, and improved technology has allowed companies to do shorter production runs. As the 3D printing business matures, just about every industry should benefit. Currently aerospace is one of the largest consumers of this technology with rapid growth also seen in healthcare. European and North American countries have been the biggest players, but China is positioning itself to lead as well.

The ability to easily and quickly manufacture items at will has drawn criticism. You may be familiar with the controversy surrounding the 3D printed gun. In May 2013, Defense Distributed (DD), a digital publishing company, put the blueprint design for this gun on its website. As stated on their history page, the company was formed to "create a political and legal vehicle for demonstrating and promoting the subversive potential of publicly-available 3D Printing technologies" (https://defdist.org/dd-history/). The gun designs were downloaded about 100,000 times in 2 days. However, the 3D plastic guns were found not to be very reliable when shooting. That same year, Solid Concept, a 3D-printing services company, used a laser sintering process to print a 3D metal gun that handled 50 rounds successfully. Not all concerns about 3D printing have to do with safety and national security. 3D printers have been accused of being energy hogs and a way to easily violate copyright-protected merchandise. There is also the worldwide environmental movement to reduce the use of plastics for health and ecological reasons. Instead of being reduced, the popularity of 3D printing may actually increase our consumption of plastic very rapidly.

Robotics

Robots have replaced many workers on assembly lines doing tasks that are monotonous and dangerous to humans. The first industrial robot was developed in the early 1960s. It was basically an arm that could pick up items and set them down in another location. Today, there are many types of robots completing a variety of tasks. Some operate automatically while others are under human control. According to researchers at World Robotics, robots sales increased an average of 17% per year between 2010 and 2014. The largest growth areas in 2014 were the automotive parts suppliers and the electrical/electronics industry. Five countries comprised 70% of global robot sales, including China, Japan, the United States, the Republic of Korea, and Germany.

Robots are as varied as the people who use them. Industrial robots or robotic arms can weld, paint, and handle parts. Autonomous robots can work in warehouses or on shop floors using a grid system to guide their movements. Consumers can even buy self-directed robots to vacuum their floors and mow their lawns. In the near future, you may even be able to buy a robot to serve as grandma's caretaker. Japanese researchers are working on humanoid robots to take care of their aging population. Regardless of the type, all robots have the following three basic elements:

1. A manipulator, such as an arm as shown in Figure 11, that does the work
2. A power supply, including electric motor, pneumatic (air- or gas-driven) or hydraulic (fluid-driven)
3. A controller, generally a computer, is the brain of the operation and controls the manipulator and sometimes the power supply

© Rainer Plendl/Shutterstock.com

Figure 11. Yellow robots welding cars in a production line.

A hydraulic system uses a pressurized liquid, usually some type of oil, to run equipment that moves and lifts. Force applied at one location is multiplied and transmitted to another location. The pipe connecting them can be wound around objects and split into separate paths. Hydraulics has more potential force than pneumatic and can lift heavier loads. A pneumatic system usually uses compressed air, although other inert gases can be used. It is good for providing rapid movement on a small scale but not lifting heavy loads as air can become compressed and less stable than hydraulic fluid. You can compress aid but not a liquid.

When robots were first introduced in industry, the jobs they did were separate from the workers. Now in some situations robots are working side-by-side with humans, acting more as smart assistants. There is a term for this arrangement, cobot. A cobot, or collaborative robot, helps people with repetitive tasks and those that require strength. Companies found that total automation did not necessarily mean increased efficiency or profitability. Robotic systems were expensive to purchase and maintain. They were not flexible. While they were good at singular tasks, it could be difficult and time consuming to reprogram them for different jobs. Managers discovered it could be more cost-effective to combine the advantages of robotic tools with human intelligence and abilities.

Robots and robotic manufacturing systems have become an established part of the industrial landscape. They have progressed from being used primarily for dirty and dangerous jobs to working cooperatively with humans. Researchers are working to make robotic sensors more refined to give them the dexterity and touch control of human hands. Advanced software application, using complex mathematical calculations and data processing, are making computer controllers more intelligent and, perhaps, even emotional. These skillsets may allow robots to work jobs that were not before possible.

Nanomanufacturing

Nanotechnology is the study of matter at the extremely small level, the nanoscale. A nanometer is one-billionth of a meter. To understand how small, there are 25,400,000 nanometers in an inch. A sheet of paper is approximately 100,000 nanometers thick. It takes advanced technology to view and manipulate material at this scale. With state-of-the-art equipment, such as the scanning tunneling microscope, scientists can see and manipulate individual atoms and molecules. They can put them right where they want them. This allows researchers to create nanomaterial that is very lightweight and incredibly strong. To learn more about the fascinating science of nanotechnology, read Chapter 12.

Nanomanufacturing is assembling products at the nanoscale. According to the United States Government Accountability Office report on nanomanufacturing, it is a future megatrend that may equal or surpass that of the digital revolution. While we are still in the early stages of any nano-revolution, research has been focused on fabricating material at the smallest level. There are basically two approaches to manufacturing at this level, top-down and bottom-up. Top-down construction is used to reduce large pieces of material to the nanoscale, similar to carving a statue out of a rock. Bottom-up is used to create objects by adding atomic-sized components one level at a time, similar to 3D printing. This can be prohibitively time consuming given how many atoms are needed to produce a single product. Scientists are exploring the prospect of creating self-assembling nanoscale components that will build themselves. But we are not there yet.

Manufacturers are currently using nanomaterials to create composite materials. Composite materials are synthetic and created by combining two or more other materials together. This merging allows the new material to have desired characteristics, such as superior strength or added waterproofing. Fiberglass and Kevlar® are two composite materials in which you may be familiar. Nanomaterials are already being added to sports equipment, such as tennis rackets and bicycles, to offer athletes strong and lightweight equipment. Researchers are even using nanomaterials to create superior bulletproof material that absorbs the energy of a bullet.

One company, Modumetal, has developed nanolaminated metals that are being used in oil fields to make equipment more resistant to corrosion. The company won the 2015 World Oil Awards in the New Horizons Idea category for their innovative technology. According to their President and CEO, Christina Lomasney, they are growing metals similar to the way that Mother Nature grows material such as rings in a tree. The process uses an advanced form of electroplating, which involves emerging a metal item in a bath containing different types of metal ions. By varying the amount of electrical current that passes through, separate layers are deposited on the metal item. Each layer is only several nanometers thick and of a different composition than the ones next to it. This helps strengthen the material and prevent cracks from being able to move through it (Bullis, 2015).

Radio Frequency Identification

The key to a successful, quality manufacturing system is the ability to track the products being made. See Chapter 10 for a full discussion of this technology. A radio frequency identification device (RFID) tag is one way to trace the pathway of individual products and becomes part of the CIM architecture. An RFID tag is a small, flexible label that contains a microchip and transmitter. An RFID receiver/scanner is used to obtain the information on the tag. An RFID tag can store and transmit data unique to that product, including identification number, creation and expiration dates, serial numbers of component parts, and shipping locations. Nearly any data you would want to track can be contained on the microchip. Information is transmitted to the receiver by radio waves and stored in a database management system, such as one used for inventory control. Stored data can be analyzed by salespeople for buying trends, notify purchasing when inventory is low, or quickly identify items that may be on a recall list, such as malfunctioning car brakes or fruit that was sold containing E-Coli. It is a good policy to notify consumers about defective products in a timely manner.

Supply Chain Management

Most manufacturing companies purchase components for their own products from outside suppliers and outsourcing. Supply chain management (SCM) is a computer software system that documents and tracks all the parts needed for production. When a company is dependent on external sources for their inventory, it is crucial that they maintain a system to specify where all the needed parts are at any time. Figure 12 shows different areas involved in a SCM. Inventory control is a key aspect of supply chains. When companies use JIT methods to keep their work processes lean and agile, supply chain systems are integral to them. SCM allows managers to follow the entire

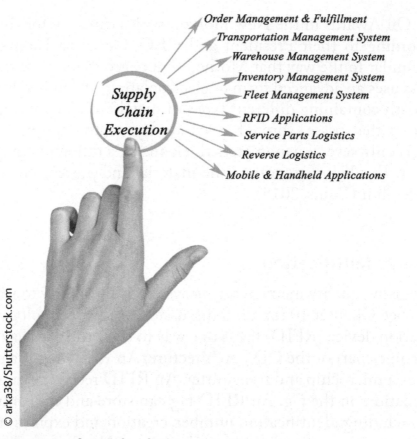

Figure 12.   Diagram of supply chain execution.

process from raw materials to final production to the grocery cart of the consumer who just bought the product.

Walmart has been a leading user of supply chain technology for years. It is one of the reasons they are able to streamline their workflow and keep prices low. Given their size and market share, they have a lot of influence with the companies who manufacture the goods they sell. In fact, many of these suppliers have changed the way they do business to reduce their own costs and meet shelf-space specifications of Walmart. To save time and increase accuracy, Walmart requires manufacturers to apply RFID tags to pallets and cartons. Paper purchase orders have been replaced with web-based forms, and software in stores tracks point-of-sale data. Walmart executives and their suppliers know exactly what products are selling and when. They are also notified when inventories get low and need to be resupplied. Merchants can use data trends to determine what items consumers are buying to plan future production runs. Walmart buyers can ensure they are acquiring trendy and popular items that their customers will want to purchase. Ultimately, Walmart's sophisticated SCM system guarantees that their customers can find the items that they want, when they want them, at a price they will pay.

Walmart's supply chain management and innovative use of technology paid off for the company in time savings, faster inventory turnover and warehouse efficiency. From the early 1990s through 2013, sales climbed from $1 billion per week to over $1.2 billion per day. (University of San Francisco, 2015, p. 6)

The Science behind the Technology

What is meant by the "science of manufacturing"? In a nutshell, it refers to the fundamental physics of a specific manufacturing process; not only the details, but how the parts of a production line work together as a system. (Nordstrom, Gawad, & Nowarski, 2006, p. 7)

Advanced manufacturing is an innovative process that has reached a stage of maturation. While science is embedded in many of its processes, manufacturers are still using human ingenuity to push the limit on how fast and efficient their factories can produce goods. New technologies, materials, and procedures will continue its advancement.

Electricity

As Nordstrom et al. (2006) articulately noted in the above quote, manufacturing is grounded in physics. The interaction between energy and matter is what powers the machines used to create the goods. You previously learned about pneumatic and hydraulic power in robotics. While steam power was the first energy source that allowed manufacturing on a large scale, electricity is the primary source now. In fact, electricity powers most of our modern society. At its most basic, electricity is the buildup or flow of an electric charge. Remember that an atom comprises electrons (negatively charged), protons (positively charged), and neutrons (no charge). An atom normally has electrons floating around a nucleus containing an equal number of protons and neutrons. The interaction between the electrons and protons produces an electrostatic force called Coulomb's Law, which states,

The electrical force between two charged objects is directly proportionally to the product of the quantity of charge on the objects and inversely proportional to the square of the separation distance between two objects.

Or

Equal charges repel; opposite charges attract

Only a proton and electron attract each other. Two protons together or two electrons together repel against each other. A good example is when you try to put two magnets together. The negatively charged electrons in each prevent you from being able to do this. Their closeness to each other also matters. If you hold a magnet in each hand about 6 in. apart, nothing happens. But the closer you move them to each other, the more they repel.

To make energy usable to us it must flow and not remain static. You are probably familiar with static electricity. It can give you fly-away hair but cannot be used to charge your phone. To make a charge flow, one of the electrons floating at the edge of an atom needs to be ejected by electrostatic force. The freed electron then attaches to new atom, but the negative charge of it then forces the ejection of an electron from that atom. This happens over and over to create an electric current, which is just electrons looking for a new atom. Now this electric current must be housed in a mechanism with a closed loop to keep it going. That mechanism is a circuit.

Figure 13 shows an example of a simple circuit. The battery is used to store energy in chemical form. While turned off, it is a form of potential energy (remember that from earlier in the chapter?). The switch starts or stops the electric current. When turned on, electric current powered by the battery (now kinetic energy) flows

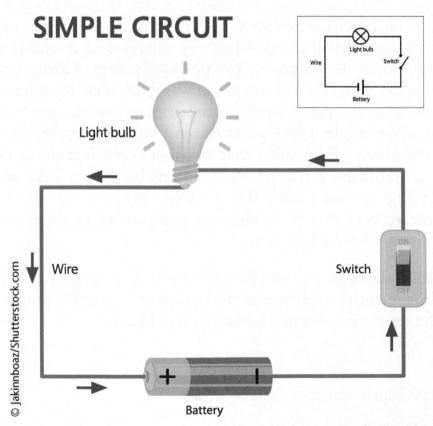

Figure 13.   A simple circuit that consists of a battery, switch, and lightbulb.

along the wire to provide power for the lightbulb. Other items can be added to this circuit (e.g., power plant in place of battery or factory in place of lightbulb), but the basic elements are the same.

Statistical Process Control

Manufacturing is also about measuring and math. Without calculations, it would be difficult to improve speed and efficiency in a process without error. Earlier you learned about standard deviation and its role in Six Sigma activities. Statistical process control (SPC) is another major scientific contribution to the manufacturing discipline. In the 1920s, engineers at Western Electric's Bell Laboratories (part of the original "Ma" Bell Telephone Company) were trying in vain to improve the quality of their transmission systems. One of their employees, statistician Walter Shewhart, discovered that there is variation in every manufacturing process, but some of it is natural (or controlled) and some uncontrolled. If adjustments are made for every fluctuation, including those that are within expected normal ranges, variation will increase and performance will degrade.

To document how a process is performing and whether or not it is in control, Shewhart invented the control chart. In a control chart, data are plotted on a time series graph that includes an upper-control level (UCL) and lower-control level (LCL). If variation occurs within these levels, it is controlled and stable. If data trend outside of the UCL and LCL, the process is uncontrolled and requires intervention. One primary goal of the control chart is to prevent wasting time and money revising processes that do not need adjusting. Control charts use the bell-curve model (although imagine it on its side) whereas the mean value is the center line and the UCL and LCL are generally three standard deviations. Notice that the data points in Figure 14 are within the UCL and LCL, indicating a normal range.

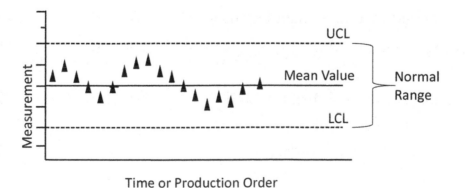

Time or Production Order

Figure 14. Example of a control chart.

Source: Courtesy of Cheryl Hanewicz. © Kendall Hunt Publishing.

Career Connections

There are many career options associated with advanced manufacturing. Skills can be developed at technical and community colleges or on-the-job training for entry-level positions. Management positions and engineering-based jobs generally require a bachelor's degree. According to industry experts at the US Bureau of Labor Statistics (BLS), employers are having difficulty filling jobs for machinists and maintenance technicians. One contributor to the shortage is the number of employees who are retiring. Machinists set up and operate computer-controlled machines, and their median annual wage is approximately $42,000. According to Payscale.com, CAD/CAM machinists average about $46,000 per year. Maintenance technicians earn over $36,000 per year and are responsible for fixing and maintaining machines, mechanical equipment, and buildings. Employment for both machinists and maintenance technicians is projected to grow 6% through 2024. Welders and other production workers are also needed and can earn an average of $38,000 (this amount can be much higher depending on position). Other jobs related to advanced manufacturing found at the BLS Occupational Outlook Handbook include quality control inspectors ($36,000/year), supply chain logisticians ($74,000/year), and materials engineers ($91,000/year).

Table 1 data were taken from the BLS Career Outlook page on manufacturing. Each industry shown is expected to need new workers by 2022.

Table 1. Employment in manufacturing industries that are projected to add jobs.

Industry	2012	Projected 2022
Motor vehicle parts manufacturing	479,600	507,500
Machine shops; turned product; and screw, nut, and bolt manufacturing	362,300	372,800
Architectural and structural metals manufacturing	341,400	410,400
Pharmaceutical and medicine manufacturing	270,800	283,800
Other fabricated metal product manufacturing	269,900	288,400
Household and institutional furniture and kitchen cabinet manufacturing	216,500	222,400
Other wood product manufacturing	189,800	211,500
Other food manufacturing	174,300	177,400

Industry	2012	Projected 2022
Motor vehicle manufacturing	168,000	176,200
Cement and concrete product manufacturing	161,600	218,900
Coating, engraving, heat treating, and allied activities	136,100	141,700
Motor vehicle body and trailer manufacturing	125,400	135,400
Office furniture (including fixtures) manufacturing	98,900	99,200
Boiler, tank, and shipping container manufacturing	96,500	98,900
Sawmills and wood preservation	84,300	93,500
Veneer, plywood, and engineered wood product manufacturing	63,800	83,500
Clay product and refractory manufacturing	40,600	44,400
Other transportation equipment manufacturing	32,500	33,800

There are also a range of professions using CAD/CAM skills. As of June 4, 2016, Payscale.com indicated that CNC operators and programmers were averaging almost $53,000 per year, while the salaries of manufacturing engineers and mechanical design engineers were in the mid-60,000 range. The US BLS includes CAD jobs under the title of "drafters." Similar to Payscale.com, they also list the median pay just a little under $53,000. The BLS indicates that the employment for drafters is expected to decline 3% from 2014 to 2024 due to improved software capabilities being more efficient.

According to a BBC article on careers in technology (Why 3D is more than a passing fad for jobs?), the most desired employees with 3D experience include industrial engineers, mechanical engineers, software developers, and marketing managers. Computer programmers are needed as well to develop software needed to run the machines. Artists will also be needed to help design and build the products. The BLS Occupational Outlook Handbook shows that multimedia artists and animators can earn almost $64,000 per year. The field is wide open as more and more businesses use 3D technology. "Those with some 3D printing knowhow are likely to be heavily recruited as the industry grows" (Borzykowski, 2016, para. 9).

Modular Activities

Discussions

❖ Post an original discussion in the online discussion board on the following topic: What students initially thought about manufacturing and manufacturing jobs and how that has changed as advanced manufacturing jobs are now clean, high paying, automated, and interesting (200 words minimum). Next, comment on the post(s) of a minimum of one other student in a thoughtful and academic way that enhances the conversation. See rubric for grading and assessment measures.

❖ Post a discussion on the similarities and differences between Lean Manufacturing and Six Sigma. Have students do research on work situations in which one methodology is more appropriate than the other. They can also debate the pros and cons of each method.

❖ Post a discussion for students on the ISO 9000 family of standards. Ask students whether the company they work for is ISO compliant. If so, how does that affect their work and what is their experience with it. If not, would they recommend to their manager that they should become ISO compliant. Why or why not?

❖ Find a recent article on 3D Printing Industry site at http://www.3dprintingindustry.com/. They have continuous updates on what is happening worldwide with 3D printing, so you can create a discussion on a really hot 3D topic. Have students find similar uses, debate the pros and cons of the technology, or consider how the technology may open new positions or replace workers.

Tests

❖ Online-graded quiz on overall chapter content, written in multiple choice format. When submitted, we recommend giving the correct answer along with the page number in which it is found for questions students did not answer correctly.

Research

❖ History of Lean/Six Sigma—who developed, why they were developed, how they have evolved over the years, and how companies are using them today to remain competitive.

❖ Research 3D printed prosthetics. What limbs have been developed, how are they being distributed worldwide, how does 3D printing benefit so many developing countries?

❖ Watch Charlie Chaplin's "Modern Times" and compare/contrast to today's manufacturing floors and businesses. Include the societal impact of assembly lines on working conditions; the morality of seeing humans as just another cog in the machine; the effect of industrialization on society; the need at the time for workers unions. How has technology made workers safer, cleaner, and working in better overall environments? While the movie is over an hour long, it is subdivided into several vignettes. You will only need to watch the first 20 minutes.

❖ Read the case study *3D Printing a Space Vehicle* at http://www.stratasys.com/resources/case-studies/aerospace/nasa-mars-rover. Students can read and/or download the information and view a short video that explains how 3D printing, additive manufacturing, and rapid prototyping helped created a lightweight vehicle. Conduct additional research on the Desert RATS (Research and Technology Studies) and the rover. Are there any updates to the technology? How did rapid prototyping help the team accomplish its goals? What related areas are using this technology? What other industries do you think would benefit from this technology?

❖ The company 3D Printing Industry has almost hourly updates on global news in the 3D world. Visit the site at http://www.3dprintingindustry.com/. Research an area of interest and write a paper on the topic using three other sources for an in-depth look.

❖ Research microfactories. What companies or industries are using them and why? Have they been successful and profitable or do you think they are a passing trend.

❖ Research nanomanufacturing to find the latest developments in how nanotechnology is revolutionizing manufacturing.

Design and/or Build Projects

❖ Design a pathway for earning a Six Sigma Green Belt, including classes to take, projects to complete, and desired jobs.

❖ Complete a 5S Lean process at your place of work: Sort, Straighten, Shine, Standardize, Sustain (since sustain is long term, write up a plan on how you plan to sustain your process, including getting buy-in from your supervisor if necessary).

❖ Apply ISO 9000 standards to your workplace. Many small companies can follow the guidelines without applying for certification. For example, they could select a few guidelines to following using the PDF, Quality Management Principles, available for download from ISO.org: http://www.iso.org/iso/home/standards/management-standards/iso_9000.htm.

❖ On the Internet you will find various free 3D printer templates and models such as at sites below.
  ➢ https://all3dp.com/best-sites-free-stl-files-3d-printing/
  ➢ https://grabcad.com/library/software/
  ➢ http://www.hongkiat.com/blog/download-free-stl-3d-models/
  ➢ http://3dprintingforbeginners.com/3d-model-repositories/

❖ If you have access to a 3D printer, obtain a template and print it. Submit the 3D print model along with the location of the template file to your instructor.

- ❖ Find a local makerspace. Learn about the facility and build a project there. A directory is located at http://spaces.makerspace.com/makerspace-directory. In the Salt Lake Area,
  - ➤ Make Salt Lake: https://makesaltlake.org/
  - ➤ Make Salt Lake, https://www.meetup.com/makesaltlake
- ❖ DIY hydraulic arm to learn how robots can be powered by air. This is detailed, so possible a class project or a project worth a lot of points/credit: http://ideas-inspire.com/syringe-hydraulic-arm/.
- ❖ Google Sketchup, http://www.sketchup.com/ is a free 3D modeling software the students can download and create a simple 3D object. Obtain this software and make a simple 3D model of a table. Video tutorials can be found at http://www.sketchup.com/learn/videos/ and other training resources can be found at http://www.sketchup.com/learn/.
- ❖ Build your own simple circuit and identify its parts. Submit a photo of the circuit (orthe circuit itself) along with a description. You may find resources such as below useful although we also recommend using a basic Internet search to find basic instructions and descriptions:
  - ➤ http://www.sciencefairadventure.com/Make_Electric_Circuits.aspx
  - ➤ http://www.dummies.com/how-to/content/how-to-build-a-simple-electronic-circuit.html
  - ➤ http://www.instructables.com/id/Build-a-Simple-Circuit-from-a-Pizza-Box-No-Solder/
  - ➤ http://www.energizer.com/science-center/make-a-simple-circuit

Assessment Tasks

- ❖ Review the sample American Society for Quality Six Sigma Black Belt certification exam found at http://asq.org/cert/resource/pdf/sample-exam/ssgb-sample-exam.pdf. Do you feel an exam such as this can adequately assess a person's knowledge and skills in this area? Why or why not? What other methods of assessment might be used? If you were a supervisor, how credible would you see a certification such as this? Write a two- to three-page assessment. This report must use proper citation and use a minimum of two sources.

Terms

CAD/CAM—Computer-aided design and computer-aided manufacturing are software applications used, often in combination, to design a product and then program a machine to manufacture it.

ISO 9000—This is an international quality management standard that companies can follow to develop and document a quality system and receive certification.

Lean Manufacturing—Lean is a systematic method used to eliminate or reduce waste from a process.

Nanomanufacturing—This manufacturing process is done at the nanoscale, which is one-billionth of a meter. Can be top–down to reduce large pieces of material to the nanoscale or bottoms–up to create objects by adding atomic-sized components one level at a time.

Taylor's Theory of Scientific Management—Taylorism is a methodology that analyzed workflow to find the most efficient way to complete a task or process.

Rapid Prototyping—This is a technique, often used with 3D printing, to create and evaluate a scale model of a product easily and inexpensively before starting large-scale manufacturing.

RFID—A radio-frequency identification tag is a small electronic devise that contains information, such as serial numbers, that can be stored and tracked.

Six Sigma—This data-driven method is used to improve quality in a process by eliminating defects and striving toward six standard deviations from the mean.

Statistical Process Control—SPC is used to measure quality in a process, often with a control chart, to determine if variation is within normal range or uncontrolled.

Supply Chain Management—SCM is managing and controlling the movement of merchandise from raw materials to point of consumption.

References

Borzykowski, B. (2016, January 28). Why 3D is more than a passing fad for jobs. *BBC*. Retrieved from http://www.bbc.com/capital/story/20160127-model-your-career-in-3-d

Bullis, K. (2015, February 16). Nano-manufacturing makes steel 10 times stronger. *MIT Technology*. Retrieved from https://www.technologyreview.com/s/534796/nano-manufacturing-makes-steel-10-times-stronger/

Islam, K. A. (2006). *Developing and measuring training the 6 sigma way*. San Francisco, CA: Pfeiffer.

ISO.org. *ISO 9000–Quality management*. Retrieved from http://www.iso.org/iso/home/standards/management-standards/iso_9000.htm

Kanigel, R. (1997). *The one best way: Frederick Winslow Taylor and the enigma of efficiency*. New York, NY: The Penguin Group Penguin Books.

Nordstrom, F., Gawad, P., & Nowarski. A. (2006). The science of manufacturing: Efficient manufacturing processes are based on fundamental factory physics laws. Contemporary Manufacturing Challenges. *ABB Review*. Retrieved from https://library.e.abb.com/public/a337191dea05a2cec1257126003432dd/Review_1_06_ENG72dpi.pdf

Steam Engine. UXL Encyclopedia of Science. (2002). *Encyclopedia*. Retrieved from http://www.encyclopedia.com/doc/1G2-3438100602.html

University of San Francisco. (2015). *Walmart supply chain low pricing strategy*. Made available by Bisk Education. Retrieved from http://www.usanfranonline.com/media/12786454/walmart-supply-chain-management-low-price-strategy-whitepaper.pdf

Further Reading

Berardinelli, C. (n.d.). A guide to control charts. *isixSigma*. Retrieved from https://www.isixsigma.com/tools-templates/control-charts/a-guide-to-control-charts/

Biography.com Editors (n.d.). *Eli Whitney biography*. Biography.com. A&E Television Networks. Retrieved from: http://www.biography.com/people/eli-whitney-9530201

Burt, V. (2016 Winter/Spring). 3D-printed satellite component presents a lesson in rethinking design. Concept to Reality.

Child Labor Public Education Project. (2011). *Project staff list of names and Labor Center*, University of Iowa. Retrieved from http://www.continuetolearn.uiowa.edu/laborctr/child_labor/

CNBC. (2015, September 24). *This start-up is growing metal. Interview with Modumetal President and CEO, Christina Lomasney.*

Columbus, L. (2015, March 31). 2015 roundup of 3D printing marketing forecasts and estimates. *Forbes*. Retried from http://www.forbes.com/sites/louiscolumbus/2015/03/31/2015-roundup-of-3d-printing-market-forecasts-and-estimates/#4075f24f1dc6

Deaton, J. P. (n.d.). How automotive production lines work. *How Stuff Works*. Retrieved from http://auto.howstuffworks.com/under-the-hood/auto-manufacturing/automotive-production-line.htm

Deloitte Development. (2014). Overwhelming support: U.S. public opinions on the manufacturing industry. *Manufacturing Institute*. Retrieved from http://www.themanufacturinginstitute.org/Research/Public-Perception-of-Manufacturing/Public-Perception-of-Manufacturing.aspx

Dyer, E. (2015, August 7). The world's first 3D printed car is a blast to drive. *Popular Mechanics*. Retrieved from http://www.popularmechanics.com/cars/a16726/local-motors-strati-roadster-test-drive/

Evans, J. R., & Lindsay, W. M. (2014). *Managing for quality and performance excellence* (9th ed.). Mason, OH: South-Western, Cengage Learning.

Feld, W. M. (2001). *Lean manufacturing: Tools, techniques, and how to use them*. Washington, DC: CRC Press.

GE Global Research. (2016). *3D printing creates new parts for aircraft engines*. Retrieved from http://www.geglobalresearch.com/innovation/3d-printing-creates-new-parts-aircraft-engines

Ghose, T. (2014, November 19). 3D-printed hearts help surgeons save babies' lives. *Live Science*. Retrieved from http://www.livescience.com/48816-3d-printed-hearts-improve-surgery.html

Giffi, C. A., Dollar, B., McNelly, J., & Carrick, G. (2014). *Overwhelming support: U.S. public opinions on the manufacturing industry*. Retrieved from http://www.themanufacturinginstitute.org/~/media/DD8C9A2E99B34E89B2438453755E60E8.ashx

Gilpin, L. (2014, March 5). The dark side of 3D printing: 10 things to watch. *Tech Republic*. Retrieved from http://www.techrepublic.com/article/the-dark-side-of-3d-printing-10-things-to-watch/

Greenberg, A. (2013, May 8). 3D-printed gun's blueprints downloaded 100,000 in two days (with some help from Kim dotcom). *Forbes*. Retrieved from http://www.forbes.com/sites/andygreenberg/2013/05/08/3d-printed-guns-blueprints-downloaded-100000-times-in-two-days-with-some-help-from-kim-dotcom/#2f4a018d88c6

Hall, N. (2016, May 23). Tortoise gets new shell, thanks for 3D printing. *3D Printing Industry*. Retrieve from http://3dprintingindustry.com/2016/05/23/tortoise-gets-new-shell-thanks-3d-printing/

History.com Staff. (2010). Interchangeable parts. *History*. Retrieved from http://www.history.com/topics/inventions/interchangeable-parts

Hollinger, P. (2016, May 5). Meet the cobots: Humans and robots together on the factory floor. Retrieved from http://www.ft.com/cms/s/2/6d5d609e-02e2-11e6-af1d-c47326021344.html#axzz49GOapqXg

Encyclopedia Britannica. (2016). Industrial Revolution. Retrieved from http://www.britannica.com/event/Industrial-Revolution

ISO.org. (2010, October 25). ISO 9001 certifications top one million mark, food safety and information security continue meteoric increase (Press release). Retrieved from http://www.iso.org/iso/news.htm?refid=Ref1363

ISO.org. (n.d.). Retrieved from http://www.iso.org/iso/home/standards/management-standards/iso_9000.htm

Kerr, D. (2013, November 7). Un-oh, this 3D-printed metal handgun actually works. *CNET*. Retrieved from http://www.cnet.com/news/uh-oh-this-3d-printed-metal-handgun-actually-works/

Knowles, E. (2015, December 19). 3D printed teeth could kill bacteria and prevent cavities. *The Science Explorer*. Retrieved from http://thescienceexplorer.com/technology/3d-printed-teeth-could-kill-bacteria-and-prevent-cavities

Manufacturing Institute. (2012). Facts about manufacturing. Retrieved from http://www.themanufacturinginstitute.org/Research/Facts-About-Manufacturing/~/media/A9EEE900EAF04B2892177207D9FF23C9.ashx

Manufacturing Institute. (2014, April). Small companies dominate the industrial landscape. *The Manufacturing Institute*. Retrieved from http://www.themanufacturinginstitute.org/Research/Facts-About-Manufacturing/Economy-and-Jobs/Company-Size/Company-Size.aspx

Markert, L. R., & Backer, P. R. (2010). *Contemporary technology: Innovations, issues, and perspectives* (5th ed.) Tinley Park, IL: The Goodheart-Wilcox.

Marshall, B. (n.d.) How hydraulic machines work. *How Stuff Works Science*. Retrieved from http://science.howstuffworks.com/transport/engines-equipment/hydraulic.htm

Milne, S. (2014, December 8). Developments in bullet proof materials using nanotechnology. *Azonan*. Retrieved from http://www.azonano.com/article.aspx?ArticleID=3934

Modumetal. (2015, October 21). Modumetal honored at 2015 world oil awards. *Press release* retrieved from https://www.modumetal.com/blogs/press-releases/115896199-modumetal-honored-at-2015-world-oil-awards

Nano.gov. (n.d.). Manufacturing at the nanoscale. *The National Nanotechnology Initiative*. Retrieved from http://www.nano.gov/nanotech-101/what/manufacturing

Nano.gov. (n.d.). What it is and how it worked. *The National Nanotechnology Initiative*. Retrieved from http://www.nano.gov/nanotech-101/what

Information Technology Laboratory. (n.d.) NIST. What are control charts? Retrieved from http://www.itl.nist.gov/div898/handbook/pmc/section3/pmc31.htm

Palermo, E. (2013, September 19). Fused deposition modeling: Most common 3D printing method. *Live Science*. Retrieved from http://www.livescience.com/39810-fused-deposition-modeling.html

Pneumatics vs. hydraulics: The pros and cons of each system. (n.d.). *Engineer Student*. Retrieved from http://www.engineerstudent.co.uk/pneumatics_vs_hydraulics.html

Pyzdek, T., & Keller, P. (2010). *The six sigma handbook* (3rd ed.). New York, NY: McGraw Hill.

Ross, A. (2016). *The industries of the future*. New York, NY: Simon and Shuster.

Rufe, P. D. (2013). *Fundamentals of manufacturing* (3rd ed.). Dearborn, MI: Society of Manufacturing Engineers.

Space station 3-D printer builds ratchet wrench to complete first phase of operations. (2014, December 22). *NASA*. Retrieved from http://www.nasa.gov/mission_pages/station/research/news/3Dratchet_wrench

SPC for Excel. (2016). Control Limited—Where do they come from? *SPC Excel*. Retrieved from https://www.spcforexcel.com/knowledge/control-chart-basics/control-limits

Statistical Process Control. (n.d.). *ASQ*. Retrieved from http://asq.org/learn-about-quality/statistical-process-control/overview/overview.html

The Physics Classroom. (2016). Coulomb's Law. *The physics classroom*. Retrieved from http://www.physicsclassroom.com/class/estatics/Lesson-3/Coulomb-s-Law

Torpey, E. (2014, June). United States Department of Labor, Bureau of Labor Statistics. *Got skills? Think manufacturing*. Retrieved from http://www.bls.gov/careeroutlook/2014/article/manufacturing.htm

Transplant jaw made by 3D printer claimed as first. (2012, March 8). *The BBC News*. Retrieved from http://www.bbc.com/news/technology-16907104

United States Department of Labor, Bureau of Labor Statistics. (2015, December 17). *Occupational outlook handbook*. Retrieved from http://www.bls.gov/ooh/a-z-index.htm

United States Government Accountability Office. (2014, January). Nanomanufacturing: Emergence and implications for U.S. competitiveness, the environment and human health. Retrieved from http://www.gao.gov/assets/670/660591.pdf

Ushistory.org. (2016). The invention of the teenager. *U.S. History Online Textbook*. Retrieved from http://www.ushistory.org/us/46c.asp

Sparkfun. (n.d.). What is electricity? Retrieved from https://learn.sparkfun.com/tutorials/what-is-electricity

Woodford, C. (2016, April 14). Electricity. *Explain that stuff*. Retrieved from http://www.explainthatstuff.com/electricity.html

Zhekun, L., Gadh, R., & Prabhu, B. S. (2004, September 28–October 2). Applications of RFID technology and smart parts in manufacturing. In *Proceedings of DETC'04: ASME 2004 Design Engineering Technical Conferences and Computers and Information in Engineering Conference*. The National Science Digital Library, Salt Lake City, Utah. Retrieved from http://nsdl.oercommons.org/courses/applications-of-rfid-technology-and-smart-parts-in-manufacturing/view

Chapter 14

Business Intelligence and Analytics

Outline

- ❖ Chapter learning objectives
- ❖ Overview
- ❖ Business intelligence
- ❖ Business analytics
- ❖ Data mining, big data, and massive data
  - ➢ What exactly is data mining?
  - ➢ Characteristics of data mining
    - ■ Volume
    - ■ Velocity
    - ■ Variety
- ❖ Mashups
- ❖ Data mining applications
  - ➢ Retail/marketing
  - ➢ Banking/finance
  - ➢ Manufacturing
  - ➢ Healthcare, pharmaceuticals, and scientific research
  - ➢ Business forecasting
- ❖ The science behind the technology
  - ➢ Relational databases
  - ➢ Predictive analytics
  - ➢ Artificial neural networks
- ❖ Career connections
- ❖ Modular activities
  - ➢ Discussions
  - ➢ Tests
  - ➢ Research

(Continued)

(*Continued*)

> ➤ Design and/or build projects
> ➤ Excel basics
> ➤ Common tasks in excel
> ➤ Doing more with excel
> ➤ Assessment tasks
> ❖ Terms
> ❖ References
> ❖ Further reading

Chapter Learning Objectives

❖ Define business intelligence and business analytics
❖ Describe big data and massive data
❖ Identify some of the application areas where data mining is used
❖ Examine how data mining is used to extract information from big and massive data
❖ Distinguish the differences between business intelligence and business analytics
❖ Summarize the three Vs of big data (volume, velocity, and variety)
❖ Compare structured and unstructured data
❖ Differentiate among the various capabilities and tools of data mining
❖ Explain how mathematical models can be used for predictive modeling

Overview

Think back to the last time you purchased a big ticket item such as a smart phone, computer, television, or car. Did you drive to a local store to learn about the item's pricing and capabilities? Probably not. Chances are you reviewed the item online, checked out customer comments, and compared costs among companies. It is possible that when you finally made your purchase, you bought the item from a business in another town, state, or even country. Technological advances have changed the way we do business. As consumers, it affects our buying habits. According to Internet Retailer, online sales increased 15%–16% at the onset of the 2015 holiday shopping season while brick-and-mortar stores saw a decline of between 4.7% and 10%. In the United Kingdom, 2015 was the first time internet-only stores ("etailers") posted higher online sales than websites of traditional stores (Davidson, July 2015).

The sharp increase in the number of consumers shopping online also influences how entrepreneurs set up new businesses. Business owners no longer have to rely strictly on local customers walking through the front door. Customers today can live just about any place in the world. Many businesses do not even have a physical location but conduct transactions online only (e.g., Amazon). However, they may still have office locations for the corporate side of the enterprise. All this digital business activity gives us the ability to gather, track, store, and analyze data; lots and lots of data. So much data that we need sophisticated software programs just to organize and make sense of it.

In 2007, Ian Ayres published the book, *Super Crunchers*. In the book, he discusses how data is being used to make decisions about everything from the quality of wine to consumer behavior. This is probably not new information to someone who follows the "customers who like this, also like this" feature on Amazon. What is significant about the super crunchers (or business analysts) is how they are using data to change the business landscape. Important decisions are being made using numbers and not experts. Numbers do not lie, right? Well that depends. To use numbers for business intelligence (BI) you have to know what the numbers are telling you. If you can interpret them correctly, they can reveal a lot. According to Jeremy M. Kolb, author of *Business Intelligence in Plain Language*, "these are the tools that enabled Target to identify pregnant women with focused marketing and drug companies to identify previously unnoticed side effects of their drugs" (p. 11). Figure 1 shows various elements that comprise business intelligence. Several trends driving BI are as follows:

1. The ease at which businesses can gather data through online forms on their websites
2. How easily people give up their information online
3. The ability to buy information directly from companies who consolidate and categorize data into market segments
4. The declining price for data storage

Figure 1.   Business intelligence is a process.

Related terms are competitive intelligence, data science, and predictive analytics. It has taken several decades for the convergence of data, computer power, and software abilities to reach this stage of maturation. In the beginning there was just the database. It probably was not a database yet but a word-processing document where secretaries or data entry clerks kept lists of thing such as employees, customers, sales territories, products, financials, and so on. That was the system used if your company was technologically advanced. Many companies in the 1980s (and even beyond) kept records on rolodex cards or on hand-written papers in hanging files which were located in file cabinets. Although it was not as sophisticated as today, companies always had a need to collect and maintain data. It is just more efficient to do so now. It is also easier to retrieve and filter the data. Managers are not necessarily smarter now; they just have access to better technology.

Business Intelligence

BI is about accumulating and evaluating information on your internal operations, customers, business partners, and competitors (Haag & Cummings, 2010). BI is a hot topic in the industry, and there are numerous definitions for BI. You already read one from Kolb in the Overview section. Another good explanation is from the website, CIO.com. This website serves Chief Information Officers (CIOs), who are the top executives responsible for the information technology (IT) systems of their respective companies. According to this online magazine, "Business intelligence, or BI, is an umbrella term that refers to a variety of software applications used to analyze an organization's raw data" (CIO.com, What is Business Intelligence, para. 1). The term "raw data" is important, because that is where all BI starts.

The first step in BI is to determine what constitutes raw data. If you were reviewing your own company, raw data could be anything related to your business. Items may not appear to be important individually, but part of BI is to look at the big picture to determine current trends and predict future ones. Virtually any information that you collect is raw data. It can include, but is not limited to, the following categories:

Human Resources: Employee names, positions, hired dates, salaries, social security numbers (SSNs), birthdates, medical insurance numbers, dependents, marital status, dependents, department, products and amount sold (for salespeople this will determine commissions)

Inventory: Merchandise names; merchandise numbers; merchandise manufacturers; merchandise color, size, and style; location-warehouse name and addresses; dates merchandise arrives and leaves warehouse; batch numbers

Sales: Merchandise names; merchandise color, size, and style; store number, name and address; salesperson

Customers: Customer names, addresses, products purchased, amount of products purchased, purchase dates, purchase price

Competitors: Competitor names and addresses, years in business, merchandise sold, yearly sales, top salespeople, financials

As you can see, raw data by itself is actually, well, raw. It might be interesting to view in a spreadsheet. You might be able to glean some information from it at a glance—or as much as your memory can hold. Now imagine that this information had been collected over the last 10 years and you had specialized software to find patterns and trends in your dataset. You might find out that employees in the mail department stay an average of 6 years longer than those in purchasing. You could discover what salesperson sells more merchandise than any other one, the average cost of the purchase order and what month the most sales happen.

Business Analytics

Although there are a number of terms associated with BI, and it covers a broad spectrum of activities, there are primarily two parts that are important. One of the one side is BI, which has already been addressed. The other side is the analytics piece. It goes beyond finding simple patterns but starts to make predictions about the future. According to the website Business Analytics, "Business intelligence provides a way of amassing data to find information primarily through asking questions, reporting and online analytical processes. On the other hand, business analytics takes advantage of statistical and quantitative data for explanatory and predictive modeling" (Difference between Business Analytics and Business Intelligence, 2013, para. 4). The article continues to explain that one other important distinction is that *analytics converts information into knowledge*. And that is why there is a chapter on Business Intelligence and Analytics in your *Understanding Technology* book. If BI was strictly gathering information, combining it, and answering simple questions, managers have been doing that for decades (albeit in a more time-consuming manner). Technology managers can now take big data, even massive data, and mine it for intelligent knowledge in ways that were not possible without technological assistance.

Data Mining, Big Data, and Massive Data

If you are an avid movie lover, chances are you have seen and enjoyed all five seasons of the award-winning TV series "Breaking Bad." After all, what can be catchier than the indisputably twisted story portraying a post-cancer diagnosis transformation of a brilliant scientist and once a humble high school chemistry teacher into a monstrous drug manufacturer? Perhaps, at some stage you even found yourself rooting for the overly prideful and greed-driven main character to be caught and prosecuted by the law for

Figure 2.   Data mining application areas.

Source: Courtesy of Elena N. Laricheva. © Kendall Hunt Publishing

the many crimes he committed. And, more likely than not, the idea of how great it would be to predict and prevent such crimes has at least once crossed your mind.

While in the fictitious world of "Breaking Bad" the bad guys could often escape the hands of justice, in the real world they are not always so lucky. Nowadays, drug and law enforcement administrations, as well as other federal agencies all over the world, successfully utilize the power of *data mining* to predict substance abuse, identify crime hot spots, detect terrorist threats, or uncover financial fraud.

Broadly defined as the *application of database technology, statistical analysis and modeling techniques to unveil hidden patterns and relationships in data with an aim to predict future results and outcomes*, data mining finds its applications in many other areas too—from retail, financial, and telecommunication industries to manufacturing, healthcare, pharmaceuticals, and scientific research, as shown in Figure 2. Such a wide applicability and apparent success can be largely attributed to the exponential growth in the volumes and availability of information collected by the public and private sectors in the structured, textual, spatial, Web, or multimedia formats (herein referred to as Big Data), as well as by recent advances in high-performance computing (HPC) and data storage capabilities.

Below we will first elaborate on what data mining is, identify major characteristics of Big Data and, finally, discuss some interesting examples of data mining applications as well as their societal impact.

What Exactly is Data Mining?

Four centuries ago, two fellows of the Royal Society—Robert Hooke and Antoni van Leeuwenhoek—forever changed the field of cell biology. By focusing light through a set of lenses in the optical microscope that van Leeuwenhoek had invented, they discovered the existence of microscopic organisms. Being the first to obtain the magnified images of the tiny objects that human eye, with its resolution of ca. 100 μm, could not see was an achievement. It opened doors to many groundbreaking discoveries. It enabled exciting observations. It was a revolution in measurement.

Data mining techniques that allow us to make sense of the large amounts of data generated as a result of our activities and interaction with the world is the twenty-first century equivalent of the light microscope. Data mining enables knowledge discovery. It allows making predictions. It drives progress—across multiple fields and disciplines.

In techspeak terms, data mining is the *exploration and analysis of large quantities (volumes) of data to discover meaningful patterns using* either **description or prediction methods**, or both. The purpose of descriptive data mining is to summarize. Predictive analytics is another step forward. By utilizing various statistical, modeling, and machine learning techniques, it allows people to make predictions about the future based on the recent and historical Big Data. Classification, regression, and deviation detection are all examples of predictive data mining tools, while clustering, association rule, and sequential pattern discovery belong to description methods.

Characteristics of Data Mining

In 2001, analysts working at Gartner, a Connecticut-based consulting company offering IT-related insights, defined three major characteristics of Big Data: volume, velocity, and variety. They are now commonly known as "three Vs" or, simply, V3. Figure 3 shows the variations. Let us start by discussing the first one.

Volume

We have previously defined data mining as the process of discovering hidden knowledge and gaining insights from large volumes of data. Now it is time to talk about the *volume* itself. So, how big is Big Data, really? What does the word "big" mean when referring to the amount of information? Is that amount measured in kilobytes, megabytes, gigabytes, terabytes, petabytes, or even more?

Let us use the large hadron collider (LHC), the world's largest particle accelerator, as an example. See Figure 4. Lying in a 17-mile-long tunnel a hundred meters

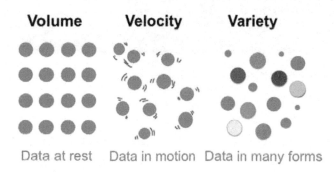

Figure 3. The three Vs of Big Data.

Source: Courtesy of Elena N. Laricheva. © Kendall Hunt Publishing

Table 1. Units of digital information.

Multiples of bytes	
Name (Symbol)	**Value**
kilobyte (kB)	10^3
megabyte (MB)	10^6
gigabyte (GB)	10^9
terabyte (TB)	10^{12}
petabyte (PB)	10^{15}
exabyte (EB)	10^{18}
zettabyte (ZB)	10^{21}
yottabyte (YB)	10^{24}

© Nicholas Greenaway/Shutterstock.com

Figure 4. Large hadron collider.

beneath the surface on the territory of the European Center for Nuclear Research (CERN) in Switzerland, this incredibly powerful atom smasher capable of generating temperatures 100,000 times greater than those of the Sun performs approximately 600 million particle collisions per second.

Not all of these collisions are effective but those that are generated from new particles that often decay in complex ways produce even more particles. As a result, the amount of digital information generated at the LHC site on a daily basis is ginormous. According to Sverre Jarp, a chief technology officer at CERN, this results in tens of petabytes annually that physicists have to sift through to extract valuable information. Back in 1970s, when he just started working at the facility, finding answers to the fundamentally important questions about the universe meant analyzing megabytes of data. Today's operations at the LHC generate data on a petascale.

But how big is a petabyte? Let us put things in perspective and try to actually visualize it:

❖ One petabyte is 10^{15} bytes, that is, 1,000,000,000,000,000 units of digital information. A number with 15 zeroes. It is such a large amount that to count it one byte per second would take 35.7 million years of your life. A pretty ambitious task unless you are immortal.

❖ A human body is estimated to contain 100 trillion cells. If you assume that one cell is a byte, then you might as well need to picture 10 people. A small crew.

❖ Count all the digital assets in the US Library of Congress, the largest library in the world, or clone the DNA of the entire US population twice and here you have it—a petabyte of data.

Now that you appreciate the scale, it is worth mentioning that the LHC is just one example of a Big Data source and, considering all other sources combined (e.g., social networks, genomics, astronomy, biomedical sciences), the petascale of data production is hardly the limit. In fact, the era of petabytes is already in the past and we have just recently stepped into a zettabyte domain quickly approaching yottascales, all of which speaks of a very high *velocity* of Big Data or the high and constantly increasing rate with which this data is being continuously generated.

Velocity

Based on the IBM reports, by 2012 humanity has already produced 2.7 zettabytes of digital information, 90% of which during the last 2 years. And while this number seems staggering, it is really nothing compared to the projected speed of data generation of 35 zettabytes per year starting 2020. Figure 5 shows this comparison.

According to Stephens et al., YouTube, Twitter, astronomy, and genomics are the four major producers of Big Data and the acquisition rates in these domains are expected to double in the next decade. For example, ever since the Australian Square Kilometer Array Pathfinder (ASKAP), the world's fastest radio telescope aiming at exploring the origins of the galaxies and mapping black holes, was first launched in 2012, it acquired loads of image data and keeps doing so at a rate of 7.5 terabytes per second. The latter number is expected to increase 100-fold by 2025.

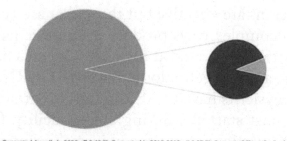

Figure 5.   Big Data volume and velocity.
Source: Courtesy of Armen Ilikchyan. © Kendall Hunt Publishing

Every minute, 300 hours worth of video is being uploaded on YouTube and the number keeps growing exponentially. By 2025, YouTube is expected to produce approximately 1–2 exabytes of data annually. Twitter is another big source as it generates about 500 million tweets per day, each one of which is about 3 kilobytes in size. Another 10 years—and we might as well expect 1.2 billion tweets per day. That is 1.36 petabytes worth of text, images, and videos in the format of regional and national news, hotel offering deals, corporate advertisements, real-life event updates, celebrity gossips, or political statements.

Genomics is another story in itself. In 2003, the humanity's quest for understanding the blueprint of life resulted in sequencing human genome—a 3-billion letter-long instruction for making a human body. Ever since that milestone was reached, genomic studies have flourished. The International HapMap project aiming to understand genetic variation in human population was launched. Scientists understood the passage of inherited traits, found cure for many genetic disorders, and learned to prevent them. They explained how proteins came forth, developed antibiotics and antivirals, and tracked down our geographic origins. They improved crops to better feed the world, advanced forensic and parenting analyses, and even put the British Royals in a very uncomfortable situation questioning their noble origins. But all of this came at a cost: the ginormous amount of digital information that had to be stored and processed. And now that the price for sequencing is continuously dropping, more and more genomes are being analyzed. Think 100 million to 2 billion sequenced genomes by 2025. That is 2–40 exabytes of Big Data—the type of data with both high volume and high velocity.

Variety

Big Data comes in various formats and can be broadly categorized into *structured* and *unstructured* data. *Structured* data is low maintenance. It is easy to enter, store, query, and analyze. *Unstructured* data requires special handling. It is the type of data that has to be cleaned, sorted, and extracted the value from. Examples of unstructured data include e-mails, social media posts, word-processing documents, audio, video and photo files, web pages, and more.

What are the various formats of structured and unstructured Big Data? Examples include the following:

1. Digital archives of scanned documents, statements, insurance forms, and medical records
2. Media files including videos, images, audio, flash, live streams, and podcasts
3. Data stored in the SQL, NoSQL, Hadoop, and other repositories
4. Social media such as YouTube, Twitter, Facebook, and LinkedIn
5. Machine log data
6. Sensory data (medical and car sensors, GPS tracking devices, road cameras, satellites)

Mashups

Various resources can be combined to create new information and products. Mashups are one popular way to take information from different areas of the Web and create something unique. They can be created as a web page or application and are part of the movement toward interactive Web 2.0 technologies. There are music mashups, video mashups, and even location mashups. Not everyone is using them for just fun and entertainment. Companies are taking advantage of this technology as well. For example, Starbucks is using them to help customers find a nearby store. Enterprise mashups allow businesses to easily combine disparate software and web applications. It started over a decade ago when Paul Rademacher created HousingMaps and overlaid Craigslist apartment and housing listings on Google maps of cities. Today, Google maps are commonly used in mashups as location data is a popular item on the Internet.

Data Mining Applications
Retail/Marketing

Area of application: Behavior patterns of potential byers, client segmentation for targeted marketing, client retention and risk management, and cross-selling/upselling.

Remember the last time you went shopping online for that new gadget that your favorite tech company has just released or a new pair of shoes and got stuck there looking over and over the features of the product while trying to decide whether or not you want to buy it? In the consumer world we live in, this situation should sound familiar to you. If you also own an account on Facebook or a similar network—which is more likely than not in our digital age—you may have noticed that, every time you login there after your online shopping adventures, the increasingly annoying ad for that same product you have been eyeballing for the past hour or so immediately pops up. This is the company trying to convince you to buy its product. This is data mining at work.

Nowadays, all major retailers—from grocery and department stores to investment banks and postal services—collect information about their customers to understand their shopping habits and create targeted ads. Data scientists working for such companies analyze the collected information computationally using data mining tools and, based on the insights gained during the process, come up with the most effective ways to sell the company's product and maximize its revenues.

One of the classical examples of how data mining works in retail and marketing evolves around Target, the famous US department store. Its data mining team has been faced with the task: to analyze a historical buying data of its female customers and identify women who are expecting a child. What for? As previous studies have shown, people undergoing life-altering events, such as having a baby, tend to alter their lifestyles as well. This includes changing their shopping habits, and, often, the very place they prefer to shop. If the company knew the approximate delivery dates of its pregnant customers, it would be able to build its advertising campaign in such a way as to prevent them from doing so and earn their loyalty. Target did just that.

Based on the history of purchases, the data specialists could determine that those ladies who had a baby registry in the Target system tended to buy larger supplies of scent-free soaps and lotions, as well as vitamin supplements, right around their third trimester of pregnancy. This helped Target to predict their delivery date with 83% accuracy and significantly increase its revenues. From the ethical perspective, however, there was a significant drawback. Not every pregnant female customer was happy to receive such ads or have their family notified about their pregnancy. But when it comes to companies maximizing their profits, your privacy is the least of their concerns.

Banking/Finance

Areas of application: Detection of financial fraud (e.g., fraudulent credit card usage), risk management, identifying stock trading rules based on historical market data, and personalizing financial services.

It may have happened to you in the past: a fraudulent payment on your credit card statement that you never authorized. In other words, you have been a victim of a cybercrime. This is not surprising at all considering how much data trails we all leave due to our daily activities online and frequent disregard of security measures. We make purchases in online stores, reserve hotels, book flights—all of which require us to input our credit card information.

Sometimes, this information gets stolen by the so-called "black hat hackers" who are skilled enough to compromise any computer system and install their key-logger and sniffer computer software that retrieve your personal information while keeping you totally unaware of it. Often times, these people also create "credit card sales" website where amateurs with no computer and programming skills can go and purchase your stolen credit card data.

To ensure the security of your personal information, many companies implement credit card fraud detection systems and to do so they rely on mining Big Data. The algorithms used vary but the logic behind is common. When people use credit cards online, they generate log files that can be mined for typical and atypical patterns. Things such as the geographical location of transactions or the e-mails and phone numbers associated with both the purchase and the user are examples of typical patterns. Whenever something suspicious happens, data mining algorithm will detect it.

The Royal Bank of Scotland used business analytics for a different, although important, reason. They used it to reconnect with their customers. In the 1980s, personal relationships with clients began to deteriorate as banks started to focus more on selling financial and insurance services to increase profits. Their analytics team wanted to turn that around and developed what they termed "personology." Using their vast database resources (think about how much information your bank knows about you), they notified customers about duplicate services they were purchasing without realizing it and specific offers that would save them money. "In sales and marketing terms, data is useless if it doesn't tell us something we don't already know about our customers" (Marr, 2016, p. 85).

Manufacturing

Areas of application: Quality assurance, warranty claim mitigation, and risk assessment.

Improving the quality and yield of a product is one of the ultimate goals in manufacturing. In the past years, many companies have achieved this goal by implementing lean and Six Sigma programs. However, some types of manufacturing are so complex that even after lean techniques are applied, there is still a room for variability. Examples include pharmaceutical, chemical, and mining industries, where a more intelligent approach to diagnosing and correcting process flaws is needed. This is when operation managers refer to Big Data analytics to analyze historical process data and identify factors that affect the quality and yield of the product the most.

In one particular example, an established European chemical manufacturer has successfully used neural networks—one of the data mining techniques that mimics the way the human brain processes information—to identify that the yield of their product was sensitive to the levels in carbon dioxide flow. This helped the company to correct the process accordingly, improve the overall yield, reduce the waste of raw materials, and minimize energy costs.

Rolls-Royce may be well-known for their high-end cars, but they also manufacture large engines used by airlines and militaries. They can generate 10 terabytes of data conducting simulation tests on one of their jet engines. Then they have to process it. Their manufacturing systems are becoming more connected and moving from a traditional manufacturing model toward the Internet of Things industrial environment. In fact, they are an industry leader. Rolls-Royce embeds sensors in their products through their engine condition monitoring program. This allows engineers and technician to track data in real-time operating conditions. In fact, if there is an issue with a plane in midair, the airline will be notified about its situation before it lands. While these capabilities allow engineers to monitor and respond to issues directly, "increasingly, Rolls-Royce expect that computers will carry out the intervention themselves" (Marr, 2016, p. 26). Adopting the Big Data approach has helped the company streamline production and reduce costs.

Healthcare, Pharmaceuticals, and Scientific Research

Areas of application: Development of personalized medicine and targeted therapies, prediction of patient's response to a newly developed drug and its dosage, computer-aided diagnosis, and patient segmentation for clinical trials.

Fighting the many types of cancer that humans suffer and die from has long been one of the main challenges in science. Many tumors are known to undergo genetic changes due to various external stress factors. As a result, they become particularly resistant to pharmacological interventions. To better understand the mechanism of such a resistance, scientists have launched several research efforts such as the Cancer Genome Atlas (TCGA) Project, Oncomine and the University of California Santa Cruz Cancer Genome Browser, the purpose of which is to study and document the changes that happen in various cancers genome-wide.

For example, one of the most successful medications with remarkable clinical activity against late stage metastatic melanoma is a Food and Drug Administration (FDA)–approved drug vemurafenib. Patients who get administered quickly show signs of improvement. However, in many cases, after an initial positive response to the drug, the signs of resistance occur. By mining genomic Big Data, scientists have been able to figure out the mechanism of this resistance and proposed a way to create more potent drugs.

Apixio, a data science upstart recently raised about $19 million to build upon its cognitive computing platform for health care. Most medical information is in the form of unstructured data, such as written physician notes, clinical charts, and PDF files. However, there is a vast quantity of data among all the different private and government health care providers and their various IT systems for storing it. Apixio has to extract these data and convert them into something a computer can read and analyze. The company has found a process to do this using machine-learning and natural language-process capabilities. Individual profiles can be created and grouped with similar individuals to determine what types of medicines and interventions work. This is the basis for personalized medicine. While still in its early stages, Apixio's system has already been helpful in identifying missing documentation and reviewing Medicare charts. This is an enormously important step in making the medical system more efficient for providers and helpful for patients.

Business Forecasting

Areas of application: Determining when customers will want to purchase particular items.

Every customer who walks into a store wants to have the item that they are looking for in stock at the exact time they want to buy it. Every company wants to make sure this happens. But figuring out how to do this has been a huge challenge.

Walmart, one of the world's largest businesses, has a Big Data and analytics department that stays on the cutting edge of technology. After Hurricane Sandy hit in 2004, they used data analytics to gain insights into what people purchased before the storm to be ready for the next one. They determined that people did not just buy what would be expected, such as flashlights and other emergency equipment. They also purchased strawberry Pop-Tarts. These breakfast treats have now been stocked in other locations and sell well ahead of large storms. According to their chief technology officer, they wanted to use predictive technology to get ahead of what people wanted to buy instead of waiting until after the fact. They currently process 2.5 petabytes of data every hour in an analytics hub to get sales information to business partners and suppliers. They have reduced problem-solving issues from about 2–3 weeks to 20 minutes.

The Science Behind the Technology

Relational Databases

To understand BI you have to understand the relational database. It is the reason that massive amounts of data can be stored, retrieved, and analyzed so efficiently. Earlier you learned about a simple database that stored records in rows and field names in columns. This is a single-table database. It is also limited in that you can only retrieve data within it. In a relational database, there are multiple tables that communicate with each other based on a unique field, called the primary key. For example, the primary key field could be a SSN. It is unique in that no two people share one. However, your SSN is probably stored in many databases (e.g., Internal Revenue Service, medical accounts, school archives, credit card companies, bank accounts, and employee records). It is possible to pull information from any database source that contains your SSN by creating a query that filters for it. A query is a logical way to ask a question of the data, which is often used for reporting purposes. When your bank sends you a statement each month, there is usually not a person who has typed it up and mailed it to you. The statement was created automatically by pulling data for your unique account number and all transactions related to it and filtered within a defined time period. When set up correctly, a relational database system will retrieve only queried data and never make a mistake.

Many big companies (and some smaller ones) maintain a data warehouse. This is a large relational database that generally contains historical data from different sources within a company. For example, there might be sales information from the West Coast, East Coast, and Midwest regions. In addition, different departments, such as inventory, advertising, and human resources can store information in a data warehouse. Do not forget that companies can also be storing information on their competitors, partners, and customers as well. Virtually any information that a company gathers or tracks can be put into a data warehouse. Data warehouses support online transaction processing (OLTP), which is the manipulation of data to support decision-making. Managers can data mine these sources to answer questions, determine current trends, and make predictions about aspects of their business using predictive analytics. It all sounds so easy. According to Thomas Davenport (2014) of Harvard Business Review, it is actually quite uncommon to have such good data available that you can start making predictions about what customers will do. "All in all, it's a fairly tough job to create a single customer data warehouse with unique customer IDs on everyone, and all past purchases customers have made through all channels. If you've already done that, you've got an incredible asset for predictive customer analytics" (para. 5).

Predictive Analytics

Predictive analytics is a computational process that uses statistics, probability, data mining, and other measurements to predict outcomes. It combines current and historical data to determine the probability of some event happening in the future. There are many software programs that can be used for this purpose, and no single predictive model works universally. Similar to the scientific method, predictive modeling is about creating, testing, and validating. According the website, Predictive Analytics Today, there are four steps to predictive modeling.

1. Create a model with software assistance and run various algorithms on your data set. An algorithm is simply a contained step-by-step process designed to solve a problem. Steps can include data mining and statistical analysis.
2. Test the model, including past data sets to see how its predicting capability works.
3. Validate the model for business understanding using visualization tools such as charts, reports, and graphs.
4. Evaluate the model to determine you selected the best fit for your data and business purposes.

Artificial Neural Networks

Artificial neural networks (ANNs) have also been used to evaluate Big Data. These synthetic systems mimic brain activity (or at least as close as we can get to simulating our most complex organ). ANNs do not follow a step-by-step process to identify a pattern or reach a conclusion. Instead, they classify items into categories. You do this every day. If you see something running at you on the street, you immediately know if it is a human or animal. If animal, you can probably figure out pretty quickly if it is cat or dog. If it is a dog, you might want to pet it or run away depending on your previous experience with them. And here is the interesting part. You can easily process all these steps in an instant. How you respond today may be different than how you responded yesterday or will respond tomorrow. You brain is quick, flexible, and makes various decisions based on your range of experiences. It is always learning and adapting, just like an ANN. See Figure 6 for an example of how ANNs can interact with each other. ANNs are commonly used for visual and speech recognition systems, because they learn to identify patterns. Part of this recognition is determining the probability that something is what it appears to be. Have you ever had a representative from a credit card company contact you about illegal activity on your account? You can thank an ANN.

> Neural networks are most useful for identification, classification, and prediction when a vast amount of information is available … For example, if you provide a neural network with the details of numerous credit card transactions and tell it which ones are fraudulent, eventually it will learn to identify suspicious transaction patterns. (Haag & Cummings, 2010, p. 109)

ANNs work in three layers with connections among them. The input units are artificial neurons, or nodes, that are designed to receive information from the outside. The output units are artificial neurons that signal how to respond to the input information. In between the input and output unit layers is the hidden unit layer. Connections between units are given a weight, or number, that is either positive (excited) or negative (not excited). The strength of the connection weights determine their influence and are modified based on what the network learns. Using a mathematical calculation, the nodes use the weights to determine whether a threshold has been reached and whether they should fire (similar to real neurons in the brain). There is a feedback loop called backpropagation, where outputs are compared

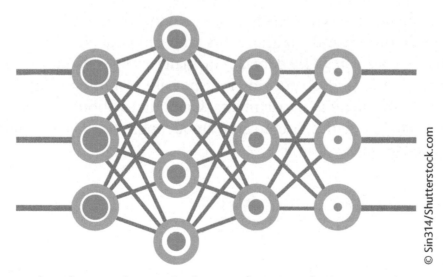

© Sin314/Shutterstock.com

Figure 6. Example of a mathematical neural network. Input units are location on the left, output units are located on the right, and the hidden layer units are in the middle.

and weights are modified through an iterative process. During backpropagation the ANN learns. Eventually new data can be entered and the ANN can work on its own.

Billions, if not trillions, of dollars have already been spent globally trying to unravel and recreate the workings of the brain. There are four major investments underway to develop a complete understanding of it: (1) Human Brain Project co-funded by the European Union; (2) Brain Research through Advancing Innovative Neurotechnologies [BRAIN] Initiative, a public–private partnership sponsored by the United States National Institute of Health; (3) Brain Mapping by Integrated Neurotechnologies for Disease Students [Brain/ MINDS], a 10-year Japanese project to map the brain of marmosets; and (4) China Brain Project, part of their 5-year plan that makes brain science a priority. The applications for this knowledge are enormous. It would open up vast new possibilities in neuroscience, artificial intelligence, gerontology, and a wide range of medical and technical fields.

Career Connections

There are a wide range of career opportunities available in the BI and analytics field. A CIO is the executive in charge of the information and, often related to that, the computer technology at a company. According to an article in CIO Update, a person in this position needs to know information about all company operations to make appropriate strategic decisions. They have to be business savvy as well as technically savvy. They have to interpret business needs, communicate with various stakeholders, and be familiar with regulatory compliance and security. It is one of the most difficult jobs but one that is well paid. CIO salaries average between $157,000 and $262,000 in the United States (If you work for a Fortune 500 company, it is closer to $2 million).

Big Data employment opportunities continue to increase. The largest growth sectors are ad tech, financial services, ecommerce, and social media. Companies worldwide are looking for engineers and programmers who can code and utilize data analytics. Data scientists are also in high demand in many industries and can earn between $90,000 and $180,000. This job position is responsible for integrating Big Data among a company's IT and business departments. Candidates should be educated in computer science, math, statistics, architecture/modeling, and/or data analytics. Having business knowledge is also desirable.

Modular Activities

Discussions

❖ Post an original discussion in the online discussion board on the following topic: Some say there are situations where BI goes too far based on how it is used and by whom. What are your thoughts and why? (200 words minimum). Next, they can comment on their thoughts about the practice and how it has affected them positively or negatively when using Facebook (200 words minimum). Next, comment on the post(s) of a minimum of one other student in a thoughtful and academic way that enhances the conversation (100 words minimum). It is suggested that they include additional information from their research and not simply agree/disagree comments. Personal opinions and observations are welcome but should be backed up with some supporting documentation.

❖ Conduct online research about how Facebook uses Big Data to make money and keep its services free to users. Students should first create an initial post on what they found and how it is designed to help users. Next, they can comment on their thoughts about the practice and how it has affected them positively or negatively when using Facebook (200 words minimum). Next, comment on the post(s) of a minimum of one other student in a thoughtful and academic way that enhances the conversation (100 words minimum). It is suggested that they include additional information from their research and not simply agree/disagree comments. Personal opinions and observations are welcome but should be backed up with some supporting documentation.

❖ Conduct online research on which companies, governmental agencies, or other entities are the biggest users of Big Data. What data is being collected and by whom, how is it being used, and what future implications does the student think are possible? (200 words minimum). Next, they can comment on their thoughts about the practice and how it has affected them positively or negatively when using Facebook (200 words minimum). Next, comment on the post(s) of a minimum of one other student in a thoughtful and academic way that enhances the conversation (100 words minimum). It is suggested that they include additional information from their research and not simply agree/disagree comments. Personal opinions and observations are welcome but should be backed up with some supporting documentation.

❖ Students can conduct a brainstorming session in class to determine all the places their personal information is kept in a database (e.g., doctor's office or bank). They should each identify five places and include what information is being held and what they believe is the unique primary key identified with it. When they make their post, though, they cannot repeat any place that has already been mentioned by another student. Or, depending on the size of the class, if they do need to repeat one, they should add different personal information that may be tracked.

Tests

❖ Online graded quiz on overall chapter content, written in multiple choice format. When submitted, we recommend giving the correct answer along with the page number in which it is found for questions students did not answer correctly.

❖ Use the website Socrative found at http://www.socrative.com/ (free of charge) to set up a live interactive quiz where students can instantly see the overall results for the class. Ask the following questions:
  ➢ I feel that increasing amounts of data and information tracking for consumers is overall a positive thing (true/false)
  ➢ I feel that healthcare facilities should be not only permitted to but also expected to share patient information with one another (true/false)
  ➢ If I were a business manager I would rely heavily on BI (true/false)

Discuss the results as a class

Research

❖ Research how the Wildlife Conservation Society and their partners are using Big Data to track animals and their migration patterns; human activity and urban expansion; and plants and their biodiversity for conservation purposes. The report should be a minimum of three pages and include at least three academic resources.
❖ Interview a CIO, IT Director, or similar position. Ask questions about their day-to-day job responsibilities; educational requirements for the job; most rewarding and most challenging aspects of the job; and what they had wished they had known before starting their current position. Also ask three original questions of your own. Write a two-page paper on the interview. Also include your own observations about the work environment and whether you are more or less interested in pursuing a similar profession.
❖ Predictive Analytics Today includes descriptions for 13 existing algorithms at the following website: http://www.predictiveanalyticstoday.com/predictive-modeling/. Select two of the algorithms listed and do additional research on what they are used for and the math behind them. Include specific examples of how they were applied to real-world problems.
❖ Compare and contrast the four current, large-scale projects that are underway to study the brain: (1) Human Brain Project, (2) BRAIN Initiative, (3) Brain/MINDS, and (4) China Brain Project.
❖ A data scientist is a new and evolving profession. Research companies who hire them and determine what types of degrees and experiences are needed to get a job in this field. Find a data scientist and interview him/her.

Design and/or Build Projects

❖ Create your own mashup. Find a tutorial on one you find interesting and follow the directions such as a weather mashup. Some will require more technical knowledge than others, so select one that matches your skills. Send web link to your instructor for him/her to view.

❖ Understanding Excel is a good skill for students to have, since many databases start (or even remain) in spreadsheets. Also, if students are obtaining data from online sources, many files are downloaded in spreadsheet formats. You will complete the Excel 2016 lessons using the following link. http://www.gcflearnfree.org/excel2016/

❖ I also suggest you review the related videos at the website for topics that are new to you. Most of the lessons have a challenge assignment located at the end, which you will complete using an existing file that they provide within the lesson.

Assessment Tasks

❖ Read the article http://www.npr.org/2016/06/10/481262383/can-web-search-predict-cancer-promise-and-worry-of-big-data-and-health. What are your thoughts on Big Data and health as described in this article? Do you feel the use of Big Data in health care is generally positive or negative? Write a two-page report expressing your thoughts and opinions about both the article itself and the topic it addresses.

❖ Consider this question: "Can Big Data resolve the human condition?" Ponder ways in which Big Data could affect the human condition now and in the future. Write a two-page commentary on your thoughts. If you use other sources, they must be properly cited. You will be assessed on the depth and thoughtfulness of your answer.

Terms

Artificial Neural Network—ANN is a computer model designed to mimic the way the brain works by identifying patterns instead of processing data step-by-step.

Big Data—Extremely large sets of stored data that is usually analyzed for some purpose.

Business Analytics—Exploration of Big Data using statistical measures for decision-making.

Business Intelligence—Broad term that covers all aspects of raw data, software applications, and data mining.

Data Mining—Extracting and analyzing data from large data sets for a specific purpose or to answer a business question.

Data Warehouse—A central repository for data collected throughout an organization.

References

Davenport, T. H. (2014, September 2). A predictive analytics primer. *Harvard Business Review*. Retrieved from https://hbr.org/2014/09/a-predictive-analytics-primer

Haag, S., & Cummings, M. (2010). Management information systems for the information age. New York, NY: McGraw-Hill Irwin.

Hays, C. (2004, November 14). What Wal-Mart knows about customers' habits. *New York Times*. Retrieved from http://www.nytimes.com/2004/11/14/business/yourmoney/what-walmart-knows-about-customers-habits.html

Marr, B. (2016). *Big data in practice*. West Sussex: Wiley.

Further Reading

Akhilomen, J. (2013). Data mining application for cyber credit-card fraud detection system. *Proceedings of the World Congress on Engineering*, London, Vol III.

Auschitsky, E., Hammer, M., & Rajagopaul, A. (2014). How big data can improve manufacturing. Retrieved from http://www.mckinsey.com/business-functions/operations/our-insights/how-big-data-can-improve-manufacturing

Ayers, I. (2007). *Super crunchers*. New York, NY: Bantam Books.

Barnes, R. (2013). Big data issues? Try coping with the Large Hadron Collider. Retrieved from http://www.campaignlive.co.uk/article/1185012/big-data-issues-try-coping-large-hadron-collider#

Bednarz, A. (2015, July 27). How much do CIOs really make? Pay packages of 25 Fortune 500 execs revealed. *Network World*. Retrieved from http://www.networkworld.com/article/2951799/careers/how-much-do-cios-really-make-pay-packages-of-25-fortune-500-execs-revealed.html

Bloomberg. (2005, July 24). Mix, match, and mutate. Retrieved from http://www.bloomberg.com/news/articles/2005-07-24/mix-match-and-mutate

Business Analytics. (2013, March 15). Difference between business analytics and business intelligence. Retrieved from http://www.businessanalytics.com/difference-between-business-analytics-and-business-intelligence/

Chen, H., Chiang, R., & Storey, V. (2012). Business intelligence and analytics: From big data to big impact. *MIS Quarterly, 36*(4), 1165–1188.

CIO. (n.d). Business intelligence definitions and solutions. Retrieved from CIO.com at http://www.cio.com/article/2439504/business-intelligence/business-intelligence-definition-and-solutions.html

Collins, H. (2013). Predicting crime using analytics and big data. *Government Technology Magazine*. Retrieved from http://www.govtech.com/public-safety/Predicting-Crime-Using-Analytics-and-Big-Data.html

Cyranoski, D. (2014, October 8). Marmosets are starts of Japan's ambitious brain project. *Nature*. Retrieved from http://www.nature.com/news/marmosets-are-stars-of-japan-s-ambitious-brain-project-1.16091

Cyranoski, D. (2016, March 18). What China's latest five-year plan means for science. *Nature*. Retrieved from http://www.nature.com/news/what-china-s-latest-five-year-plan-means-for-science-1.19590#/brain

Davison, L. (2015, July 31). Internet-only stores to overtake retail giants' online sales this year. *The Telegraph*. Retrieved from http://www.telegraph.co.uk/finance/newsbysector/

retailandconsumer/11773421/Internet-only-stores-to-overtake-retail-giants-online-sales-this-year.html

Duhigg, C. (2012). How companies learn your secrets. *The New York Times Magazine*. Retrieved from http://www.nytimes.com/2012/02/19/magazine/shopping-habits.html

Dupati, A., & Gill, L. (2014). Vemurafenib: Background, patterns of resistance, and strategies to combat resistance in melanoma. *Medical Student Research Journal, 3*, 36–43.

GAO Report. (2004). Data mining: Federal efforts cover a wide range of uses. Retrieved from http://www.gao.gov/assets/250/242240.html

Gest, H. (2004). The discovery of microorganisms by Robert Hooke and Anthonie van Leeuwenhoek, fellows of the Royal Society. *Notes Records of the Royal Society London, 58* (2), 187–201.

Goldman, M., Craft, B., Swatloski, T., Ellrott, K., Cline, M., Diekhans, M., … Zhu, J. (2013). The UCSC cancer genomics browser: Update 2013. *Nucleic Acids Research, 41* (D1), D949–D954. doi:10.1093/nar/gks1008

Guy, S. (2015, December 1). Online sales surge while in-store sales drop to start the holidays. *Internet Retailer*. Retrieved from https://www.internetretailer.com/2015/12/01/online-sales-surge-as-stores-sales-drop-start-holiday

Hooke, R. (1665). Micrographia, or, some physiological descriptions of minute bodies made by magnifying glasses: With observations and inquiries thereupon. London: Jo. Martyn and Ja. Allestry.

Human Brain Project. (n.d). Overview. Retrieved from https://www.humanbrainproject.eu/2016-overview;jsessionid=y1fv8bh2l9a1c5vdklk52fmo

IBM Research. (2013). Square kilometer array: Ultimate big data challenge. http://www.skatelescope.org/uploaded/8762_134_Memo_Newman.pdf

Ide, I. (2014, October 5). Big data: Career opportunities in tech's hottest field. *Mashable*. Retrieved from http://mashable.com/2014/10/05/big-data-careers/#mPFk6IV5s8qx

International HapMap Consortium. (2003). The International HapMap Project. *Nature, 426*(6968), 789–96.

King, R. (2006, November 13). When companies do the mash. *Bloomberg*. Retrieved from http://www.bloomberg.com/news/articles/2006-11-13/when-companies-do-the-mashbusinessweek-business-news-stock-market-and-financial-advice

Kolb, J. M. (2012). *Business intelligence in plain language*. Chicago, IL: Applied Data Labs.

Laney, D.(2001). 3D data management: Controlling data volume, velocity and variety. *Gartner*. Retrieved from http://blogs.gartner.com/doug-laney/files/2012/01/ad949-3D-Data-Management-Controlling-Data-Volume-Velocity-and-Variety.pdf

Laricheva, E. (2012). *Turning on fluorescence in silico: From radical cations to 11-cis locked rhodopsin analogues*. Electronic thesis or dissertation. Retrieved from https://etd.ohiolink.edu/

Leavline, E. J. (2015, September). Artificial neural network design flow for classification problem using MATLAB. *International Journal of Advanced Research in Biology, Ecology, Science and Technology, 1*(6). Retrieved from http://ijarbest.com/mm-admin/journal/issue6/document_2_gBOc_18112015.pdf

Lohr, S. (2012). The age of big data. *The New York Times*. Retrieved from http://www.nytimes.com/2012/02/12/sunday-review/big-datas-impact-in-the-world.html?_r=0

National Institute of Health. (2003). International Consortium Completes Human Genome Project: All goals achieved; new vision for genome research unveiled. Retrieved from https://www.genome.gov/11006929/2003-release-international-consortium-completes-hgp/

Nations, D. (2016, March 26). The 10 best mashups on the web. *About Tech*. Retrieved from http://webtrends.about.com/od/webmashups/tp/10-best-mashups-on-the-web.htm

Newman, R., & Tseng, J. (2011). *Cloud computing and the square kilometre array*. Retrieved from http://www.skatelescope.org/uploaded/8762_134_Memo_Newman.pdf

Ola. (2015, November 22). Rolls Royce: Internet of things in aviation. *Harvard Business School*. Retrieved from https://digit.hbs.org/submission/rolls-royce-internet-of-things-in-aviation/

Pimentel, F. (2014). *Big data + mainframe*. Retrieved from file:///Users/Elena/Downloads/BIG%20DATA%20+%20MAINFRAME%20(1).pdf

Predictive Analytics Today. (n.d.). What is Predictive Modeling? Retrieved from http://www.predictiveanalyticstoday.com/predictive-modeling/

Rajdeepa, B., & Nandhitha, D. (2015). Fraud detection in banking sector using data mining. *International Journal of Science and Research, 4*(7), 1822–1825.

Rhodes, D. R., Yu, J., Shanker, K., Deshpande, N., Varambally, R., Ghosh, D., … Chinnaiyan, A. M. (2004). ONCOMINE: A cancer microarray database and integrated data-mining platform. *Neoplasia, 6*(1), 1–6.

Sarker, A., O'Connor, K., Ginn, R., Scotch, M., Smith, K., Malone, D., & Gonzalez, G. (2016). Social media mining for toxicovigilance: Automatic monitoring of prescription medication abuse from Twitter. *Drug Safety, 39*, 231–240. doi:10.1007/s40264-015-0379-4

Scheuerman, M. (2011, February 8). Just what is the CIO's role? *CIO Update*. Retrieved from http://www.cioupdate.com/insights/article.php/3924041/Just-What-is-the-CIO146s-Role.htm

Stephens, Z. D., Lee, S. Y., Faghri, F., Campbell, R. H., Zhai, C., Efron, M. J., Ravishankar, I., Schatz, M.C., Sinha, S., Robinson, G. E. (2015). Big data: Astronomical or genomical? *PLoS Biology*, 13(7), e1002195. doi:10.1371/journal.pbio.1002195

Sullivan, T. (2016, May 25). Apixio raises $19 million venture capital to advance cognitive computing for healthcare. *Healthcare IT News*. Retrieved from http://www.healthcareitnews.com/news/apixio-raises-19-million-venture-capital-advance-cognitive-computing-healthcare

Tadjdeh, Y. (2015). Big data helping to pinpoint terrorist activities, attacks. *National Defense Magazine*. Retrieved from http://www.nationaldefensemagazine.org/Pages/default.aspx

The Cancer Genome Atlas. (n.d.). Retrieved from https://cancergenome.nih.gov/.

The National Institute of Health. (n.d.). What is the brain initiative? Retrieved from http://www.braininitiative.nih.gov/

Ungureanu, H. (2015). 'Predictive policing' police program banks on data mining to anticipate and prevent violent crime. *Tech Times*. Retrieved from http://www.techtimes.com/articles/88679/20150927/predictive-policing-police-program-banks-on-data-mining-to-anticipate-and-prevent-violent-crime.htm

van Leeuwenhoek, A. (1977). *The select works of Antony van Leeuwenhoek: Containing his microscopical discoveries in many of the works of nature; History of ecology*. New York, NY: Arno Press.

Woodford, C. (2016, March 18). Neural networks. *Explain that Stuff*. Retrieved form http://www.explainthatstuff.com/introduction-to-neural-networks.html

Zahedi, F., & Zare-Mirakabad, M. R. (2014). Employing data mining to explore association rules in drug addicts. *Journal of AI and Data Mining, 2*(2), 135–139. Retrieved from http://jad.shahroodut.ac.ir/article_308_2f33fd480ca2908921534fca50dda004.pdf

Chapter 15

Robotics and Artificial Intelligence

(*Continued*)

(Continued)

- ❖ Concerns about robotics development
  - ➢ Circumstance and consequence reasoning
- ❖ The science behind the technology
  - ➢ Points of articulation
    - ■ Degrees of freedom
  - ➢ The physics of robotics
    - ■ Work
    - ■ Friction
    - ■ Torque
    - ■ Electromagnetism
    - ■ Current
- ❖ Career connections
  - ➢ Mechanical engineers
  - ➢ Electromechanical technician
  - ➢ Electrical and electronics engineering technicians
  - ➢ Electrical and electronics engineers
  - ➢ Computer hardware engineers
  - ➢ Computer and information research scientists
- ❖ Modular activities
  - ➢ Discussions
  - ➢ Tests
  - ➢ Reports or papers
  - ➢ Design and/or Build projects
  - ➢ Assessment tasks
- ❖ Terms
- ❖ References
- ❖ Further reading

Chapter Learning Objectives

- ❖ Define artificial intelligence and robotics
- ❖ Identify learning systems
- ❖ Explain machine learning
- ❖ Examine the role of artificial intelligence as it relates to robotics
- ❖ Describe deep data mining as intelligence

Overview

Interest and concern for robotics and artificial intelligence has been around for centuries. Consider Leonardo da Vinci. Around 1495 he designed a mechanical knight that uses humanoid automation which is known as Leonardo's robot. While not the only robotic invention of his, it is perhaps the most well known.

From that time to today the potential role of robotics has been visible and growing. As an example of how relevant robotics is today, in 2011 President Obama launched the National Robotics Initiative via the National Science Foundation (NSF), the National Institutes of Health (NIH), National Aeronautics and Space Administration (NASA), and the US Department of Agriculture (USDA) to stimulate robotic development in industrial automation, elder assistance, and military application. The goal of the initiative is to accelerate the development and use of robots in the United States that work beside or cooperatively with people.

The field of robotics has transcended industrialized use and is being researched in commercial, educational, and recreational areas as well. There are now even self-directed robotics kits for children such as the ReCon 6.0 programmable rover. At a cost under $50 it teaches basic programming, early mathematics, and basic problem solving. Another is the Revolution Six Robot Kit, which while more expensive at around $400, is a hexapod robot that can walk on six legs, has detection and tracking abilities, and is highly customizable.

In some cases there are robots intended to either look or function as close to human beings as possible. These are commonly known as androids. One such model is known as Sophia and is made by Hanson Robotics which was related in early 2015. Sophia has lifelike silicon skin and can emulate over 62 facial expressions. She can also see (via eye-like designed cameras and computer algorithms) as well as speak and learn over time. Hanson, CEO and founder of Hanson Robotics notes, "The artificial intelligence will evolve to the point where they will truly be our friends, not in ways that dehumanize us, but in ways

Artificial Intelligence.

Source: © NicoElNino/Shutterstock.com

the rehumanize us, that decrease the trend of the distance between people and instead connect us with people as well as with robots" (Taylor, 2016). Another robot of similar concept is Gemini, a robot created by Hiroshi Ishiguro. The idea for this type of robots is to have humans and these super intelligent computers coexist in ways in which they are almost indistinguishable.

Artificial Intelligence

Artificial intelligence has vast meaning across technological fields. Most literally it means something made or produced by human beings that the ability to acquire and apply knowledge and skills. The knowledge of relevance is of human nature. We aim to make technology-based systems that have awareness and understanding of items, both tangible and intangible, such as facts, information, descriptions, and skills, which are acquired through experience or education by perceiving, discovering, or learning.

Generally, we see Artificial Intelligence (AI) as semi-intelligent machines that can do work and solve problems for us. However, that perception is shifting as these machines become more complex. A part of the debate is determining what intelligence might mean in a technological machine-based environment. To address this, one might look at the meaning of intelligence in as it relates to humans. However, this also does not have a clear definition. Legg and Hutter (2007a) attempted to create a universal definition which is as follows:

If we scan through the definitions pulling out commonly occurring features we find that intelligence is:

❖ A property that an individual agent has as it interacts with its environment or environments.

❖ Related to the agent's ability to succeed or profit with respect to some goal or objective.

❖ Depends on how able to agent is to adapt to different objectives and environments.

Putting these key attributes together produces the informal definition of intelligence that we have adopted, "Intelligence measures an agent's ability to achieve goals in a wide range of environments." Features such as the ability to learn and adapt, or to understand, are implicit in the above definition as these capacities enable an agent to succeed in a wide range of environments.

That said, when trying to determine what it is exactly that one would be automating, intelligence conversations move in the direction of statements such as these: intelligence is the ability to learn new things, intelligence is the ability to solve new problems, and intelligence is the ability to combine resources or information to come to a logical conclusion. Of particular significance is the question whether a strong ability to search, filter, and statistically place odds can constitute intelligence. If so, then if a computer system is able to search filter and statistically validate odds on an answer with a confidence of, say, 90% or more, is it intelligent?

Machine Learning

As Legg and Hutter (2007b) point out, "Machines can have physical forms, sensors, actuators, means of communication, information processing abilities and environments that are totally unlike those that we experience. This makes the concept of 'machine intelligence' particularly difficult to get a handle on. In some cases, a machine may display properties that we equate with human intelligence, in such cases it might be reasonable to describe the machine as also being intelligent. In other situations this view is far too limited and anthropocentric. Ideally we would like to be able to measure the intelligence of a wide range of systems; humans, dogs, flies, robots or even disembodied systems such as chat-bots, expert systems, classification systems and prediction algorithms."

Interactive learning is used by humans to accumulate the knowledge needed for making future decisions as well continuously getting feedback to improve the learning process. Knowledge accumulation and improvements in learning have not only helped human beings survive and build civilizations over time, they have also helped in evolving the brain and make humans smarter. By the twentieth century, the invention of computers opened the doors for new methods of communication as well as information saving, retrieving, and manipulation. Scientists and engineers have developed machines that are faster, lighter in weight, and more efficient. It would be challenging to find a field that has not benefited from the advancement in technology. Scientists and engineers program machines to perform specific tasks in applications that deal with health systems, transportation, communication, and so on.

One drawback is that machines are not intelligent and need human intervention to provide an acceptable level of performance. Machines are programmed by humans, and we still cannot trust machines in decision-making in critical areas such as dealing with human health. The challenge is to have machines learn and build knowledge in a similar fashion to humans. With machine learning, researchers are exploring many possibilities such as the following:

* Machines making decisions without explicitly being programmed
* Machines providing solutions for unsolved problems
* Machines making inferences from given data
* Machines analyzing very large scale data to extract information

Machine learning is a concept that researchers have been working hard to accomplish even prior to the beginning of computer era. After the invention of computers, more attention has been given to machine learning as the possibilities increase. Traditional programming languages require user intervention. We want intelligent machine to help in improving our health system, build cars, play games, and much more. Such intelligent machines also need to interact with the environment to deliver the needed service. The science that deals with machine learning without being programmed is known as the "intelligence system." That being said, training machines to learn is the first step in making machines intelligent. Machine learning is a field evolved from the study of pattern recognition and computational learning theory in artificial intelligence.

What Is Machine Learning?

Learning from current experiences to make better decisions for future input is the main concept of machine learning. Researchers and scientists are applying sophisticated mathematical models to mimic the human brain's learning process to achieve that concept. Training the human brain to perform tasks such as classification, recognition, and accurate predictions is done since the early phases of childhood. In most cases, the human brain performs well for future decisions based on previous similar scenarios in the past. For instance, in a given natural scene, it is not a challenging task for the brain to recognize (classify) each object such as trees, sky, and so on. This is true even if a given object is partially hidden by another object. Automated object classification is not that trivial and may be accompanied by errors. Training machines to perform the same task may require a *classifier* that would accurately classify each pixel to its corresponding object. Usually, a classifier is trained using a *training set*. In the training phase, prediction of the decision boundary that separates objects from background is done by modifying the model weights until a *cost function* is of minimized error. Based on the inferred knowledge from the training phase, future data could be classified to its corresponding classes (objects). The future data is defined as the *testing set*. Figure 1 shows a machine learning diagram for classification.

Any machine learning model should have an accurate decision-making function with minimized false negative and false positive rates. A false negative could be defined as a data point that should have been classified into a class but got rejected while a false positive is a data point that has been classified into a class but should have been rejected. For example, when performing a Tuberculin skin test, a false positive is when the patient is diagnosed with having active tuberculosis while he/she is actually healthy. A false negative is when the patient declared healthy while he/she has active tuberculosis disease. It has been found that wrongly classified data are more harmful

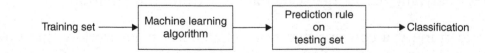

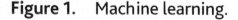

Figure 1.   Machine learning.

than unclassified data. Also, a false positive is found to be less harmful compared with a false negative. False positives may expose the patient to extra emotional and physical procedures but it is still less costly than declaring a patient healthy when he/she is not. Because of these factors, some learning algorithms are tuned to produce as few false negatives as possible even if they produce more false positives.

Scenarios where machine learning would be of help are as follows:

1. Repetitive tasks that do not need a high level of precision
2. Tasks with methods that need tuning
3. Tasks of inexact methods
4. Customized tasks based on the input

Types of Machine Learning Algorithms

Machine learning algorithms can be categorized into three main categories:

1. Supervised machine learning
2. Unsupervised machine learning
3. Semisupervised machine learning

Supervised learning is one of the most commonly used learning models. In supervised learning, the classifier is trained using a set of labeled training set. Data labeling usually is manually done by an expert. One important factor with this learning model is selecting the right training set to build a classifier that accurately classifies the testing dataset with minimal false positive and positive negative rates. Figure 2 displays the

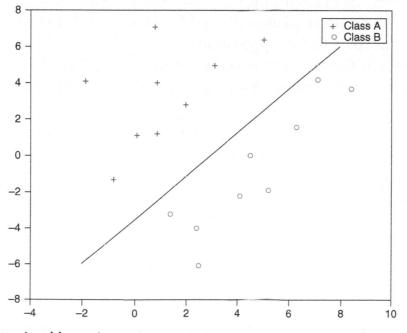

Figure 2. Supervised learning.

results of classification using supervised learning. The testing dataset is classified into two classes, Class A and Class B. This is done based on the labeled input dataset. Note that because the data points are highly separable, a linear decision-making provided two well separable classes.

Manual data labeling by an expert is a time-consuming process and may not even be possible in some high precision applications such as in the medical field. For instance, tissues such as skin or fat may not even have clear margins in projections such as magnetic resonance imaging (MRI).

Other limitations for supervised learning are algorithm generalization and dealing with variations. To overcome this limitation, the unsupervised learning model is proposed. The advancements in technology have produced a massive amount of unlabeled data that could be applied to a learning algorithm. With an unsupervised model, labeled data are not needed and the classifier learns from the unlabeled dataset by finding similarities or differences in the data attributes or features. Similar data points based on some previously defined attributes are clustered together into different classes. In this scenario, the attributes that best differentiate the data are needed to have an optimized outcome. When the data points are clustered in a hierarchical manner, it produces a hierarchical clustering. Nonhierarchical clustering group data points into a number of disjoint clusters. Figure 3a shows an example of set of unlabeled data points. Figure 3b displays the results of unsupervised learning. In Figure 3b, the data points are clustered into three clusters. A k-means algorithm is applied to cluster the data where k represents the number of centroids and is equal to three in Figure 3b. Each centroid is initiated as the center of a cluster and the distance between each point and the centroid points is computed to classify the point to its corresponding cluster. With unsupervised learning, user interaction is not required. It uses unlabeled data for the decision-making based on some attributes. This is an advantage over the supervised algorithms. However, it needs to be tuned to produce a satisfactory outcome for different applications.

The third model for learning is semisupervised learning. This model combines both the supervised and unsupervised learning models. Semisupervised learning is

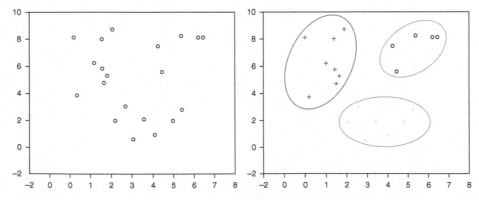

Figure 3.   (a) Unlabeled data points and (b) clustered points in their corresponding classes.

more practical for real-life applications: it can make use of the available labeled dataset even if it is of small size and at the same time learn in an unsupervised fashion. Weakly labeled data can also be used with the semisupervised model. Weakly labeled data is labeled data but with incomplete information. There has been an increased interest by researchers in semisupervised learning during the last years because of the availability of a very large amount of unlabeled data that can be combined with a few labeled data (supervised model) to increase the accuracy of detection. Semisupervised approaches can be categorized into two main types: clustering or manifold assumption.

Applications of Machine Learning

Having intelligent machines along with a fast decision-making environment that could interact with the user would help having more advanced technology. We want machines to drive cars, detect abnormalities and diagnose diseases, retrieve information, authenticate users, perform speech recognition, predict weather forecasting, prevent fraud, and much more. Researchers in well-known companies and high-ranking educational institutions are continuously working on developing accurate novel machine learning algorithms for different applications. Currently, ecommerce and technology companies are using machine learning algorithms to detect credit card fraud. Email providers such Yahoo use machine learning algorithms to detect spam emails. Google applies these algorithms to detect spam websites. In terms of authentication, Amazon launched a contest for developing an algorithm that automates employee access granting and revocation. In the medical field, machine learning would provide unlimited services. Some of the current researches include wait list time prediction for emergency rooms, strokes and seizures prediction, heart failure identification, and so on. Some mobile applications have been developed with machine learning algorithms. JustShakeIt is an application that sends emergency messages when strokes occur.

Researchers such as statisticians, data scientists, and psychologists have been working for more than six decades to make machines intelligent, accurate, fast, and responsive. Intelligent machines interact with the environment to learn and make inferences for future predictions. To accomplish that, psychologists study human learning and behavior as it is still not quite clear how the human brain functions. Having a clear understanding of the human brain would help to mimic its functions using machines. Data scientists work on analyzing the available data to find the attributes that best differentiate a given piece of data. Finally, statisticians are working on understanding patterns given within data. Understanding the patterns enables us to pick the one that best separates them. Machine learning algorithms have already been applied in many applications and the future is promising. Our ultimate goal is to have machines playing games, training players, retrieving information, and possibly even helping in complex medical diagnoses and DNA sequence identification. It is clear that machine learning would be immensely useful for the development of technology and in providing better services.

Object Recognition

Object recognition is the ability to perceive an object's physical properties and apply meaning to that object (e.g., its function and how it relates to other objects). In the area of artificial intelligence it has to do with an AI program's ability to pick out and identify objects from visual inputs that come in the form of video or still images. Some methods used might include three-dimensional modelling, component identification, edge detection, or analysis from multiple viewpoints (angles). Essentially, object recognition systems find objects in the real world based on an image or images of that world. While humans can do this without effort or thought, computers much use algorithms to find their answers.

Ranachar Kasturi, a professor of Computer Science and Engineering at the University of South Florida wrote a book called *Machine Vision* in 1995. In the book, he addresses basic aspects of object recognition including the architecture and main components. He noted that there are particular components an object recognition system must have. These include a model database, a feature detector, a hypothesizer, and a hypothesis verifier. He notes,

> All object recognition systems use models either explicitly or implicitly and employ feature detectors based on these object models. The hypothesis formation and verification components vary in their importance in different approaches to object recognition. Some systems use only hypothesis formation and then select the object with highest likelihood as the correct object. Pattern classification approaches are a good example of this approach. Many artificial intelligence systems, on the other hand, rely little on the hypothesis formation and do more work in the verification phases. In fact, one of the classical approaches, template matching, bypasses the hypothesis formation stage entirely. (p. 461)

He then addresses some factors that affect the complexity of recognizing images. These include scene consistency, image-models spaces, number of objects in the model database, and number of objects in an image, and the possibility of occlusion. Much of what Kasturi described over 20 years ago remains the case today.

Smart Systems

Smart systems have the potential to play many key roles in our every day lives. These systems are able to sense their environment, act upon the information received and control their circumstances in a predictive or adaptive manner. Consider, for example, smart homes. Smart home environments could be instrumental in our daily living. They could alter our perceived environment, help us with decision-making, and help us control our local environment (such as lighting, temperature, and sound). Google Home, released in later 2016, is one such system. Google Home is a voice-activated home product which is similar to Amazon's device known as the Echo. Amazon Echo, released to the public in 2015, is capable of omnidirectional audio, far field voice recognition, and can answer questions, read audio books to end users, give news, report on weather, provide spore scores, set alarms or timers, and control lights or thermostats among other things. Like Echo, Google

Amazon Echo.

Source: © Peppinuzzo/Shutterstock.com

Home can to do similar things. It can also control your television, manage your Google calendar, sent texts, make reservations, and handle more complex search questions.

Robotic Cars

One significantly growing area for smart systems is in automotive. Robotic cars, also known as autonomous cars (although automated may be a better term), driverless cars, self-driving cars or vehicular automation, involve the use of mechatronics, multi-intelligent systems, and artificial intelligence to assist and inform vehicle's operators. These are discussed in greater detail in the autonomous and semiautonomous chapter. It is important to note, though, that smart systems are the key to their functionality. A smart car has advanced electronic detection systems, advanced micro processing, and advanced control systems used to interpret sensory information and then act on that information in some fashion.

As of 2013, the national highway traffic safety administration has established the following classifications for vehicle automation. This includes cars that have automated features all the way to those which are fully automated.

No-Automation (Level 0): The driver is in complete and sole control of the primary vehicle controls—brake, steering, throttle, and motive power—at all times.

Function-Specific Automation (Level 1): Automation at this level involves one or more specific control functions. Examples include electronic stability control or pre-charged brakes, where the vehicle automatically assists with braking to enable the driver to regain control of the vehicle or stop faster than possible by acting alone.

Combined Function Automation (Level 2): This level involves automation of at least two primary control functions designed to work in unison to relieve the driver of control of those functions. An example of combined functions enabling a Level 2 system is adaptive cruise control in combination with lane centering.

Limited Self-Driving Automation (Level 3): Vehicles at this level of automation enable the driver to cede full control of all safety-critical functions under certain traffic or environmental conditions and in those conditions to rely heavily on the vehicle to monitor for changes in those conditions requiring transition back to driver control. The driver is expected to be available for occasional control, but with sufficiently comfortable transition time. The Google car is an example of limited self-driving automation.

Full Self-Driving Automation (Level 4): The vehicle is designed to perform all safety-critical driving functions and monitor roadway conditions for an entire trip. Such a design anticipates that the driver will provide destination or navigation input, but is not expected to be available for control at any time during the trip. This includes both occupied and unoccupied vehicles (National Highway Traffic Safety Administration, 2013).

As can be seen from the above classifications, the level of complexity, information acquisition and sharing, decision support capacity, and automation varies significantly based on the system in question. Regardless of the level, however, robotics and artificial intelligence play a role.

Robotics

Robotics deals with the design, operation, and application of robots and their related systems. As a field it involves mechanical engineering, electrical engineering, systems engineering, computer science, and mechatronics overall. Essentially, it is the study of robots.

Mechatronics

Mechatronics is an interdisciplinary field of engineering that involves mechanical, instrumentation, electronics, automation, control systems, telecommunications, and computer engineering. It is a relatively new field of engineering that developed as needs have evolved. The origin of the word is credited to Tetsuro Mori, an engineer of the Japanese company Yaskawa Electric Corporation and was trademarked in 1971. Later, when the company released the right of using the word in public in the early 1980s, it caught on and spread worldwide.

Robots Defined

To understand robotics, we must first understand robots. Definitions of robots vary, but at their core they are programmable machines that can be used to complete jobs/tasks. In some cases they can do work by themselves, and in other cases a person must instruct them about what to do. Often they are used for either repetitive or mundane jobs, or for jobs that might be dangerous for humans. The word robot originates from a Czech playwright and novelist named Karel Capek who introduced the term in his play Rossum's Universal Robots in 1920. It comes from the Slavonic word rabota which means servitude or forced labor.

Robots usually have not only a computer to control its actions, but also sensors that can detect things like heat or light as well as end effectors that are designed to interact with the environment via hands or feet or similar. The sensors help the robot collect information used in decision-making either by the robot itself or by its operator(s) and the effectors help the robot have some type of effect on the environment. In this way they can be seen as being sensitive or adaptive to the environment.

While we often envision robots as having human-like characteristics (commonly called androids), this is not necessarily the case. If anything, this is down for social acceptance as there is little practical or technical reason for following this type of structure. In some cases robots look nothing like humans. Consider, for example, Universal Robots products which are robotic

Universal Robots robotic arm.

Source: © Ociacia/Shutterstock.com

arms intended to be incorporated into production environments. They have multiple articulation points, wide flexibility, varied sizes, and easy collaboration into existing systems. While it functions in ways similar to a human arm, it has no such appearance. One other benefit of robots like the Universal Robots arm is that they are built to be able to safely work side by side with humans. This is possible due to their refined sensors and software along with standards for movement, speed, and force.

There are also robots that operate in environments that would be nearly impossible for humans. Space exploration is one such arena, but so are many areas of underseas exploration or scientific exploration such as in nuclear facilities. One such robot is the ultra-maneuverable Sea Wasp, built by Saab. This remote-controlled robot works underwater and is designed to dismantle explosives.

Robotics in Medicine

We could choose almost any industry and discuss the role of robotics. For now we will address some types of robotics used in medical fields. You can learn more about these and similar technologies in the medical technology chapter. We have had telemedicine, also known as telepresence robots, for some time, which is when there is remote diagnosis and treatment of patients via telecommunication technology. Now, we have medical robots that take it one step further. These medical robots can perform routine rounds, check in on patients, manage their charts, and check vitals. The first robot of this type to obtain Food and Drug Administration (FDA) clearance for hospital use is the RP-VITA Remote Presence Robot produced jointly by iRobot Corp. and InTouch Health (MacRae, 2013).

Vita comes equipped with a two-way video screen and is a mobile cart. These types of robots not only can perform routine patient monitoring, but can also allow physicians and other staff to work remotely to diagnose patients or offer medical advice via teleconferencing. The remote physician simply uses a computer, laptop, or iPad to log in to the RP-Vita on-site. This is particularly useful in specialized medicine (Jaslow, 2013).

Indeed, robots are vast and varied. When determining robot use there are any number of factors to consider, but some of the basics are: functionality, ease of programming, reliability, independence, mobility, technical durability, adaptability, and serviceability. If you have a job that needs doing, there is likely a robot that can help do it. If not, the technology likely exists to build it.

Vertical and Horizontal Artificial Intelligence

Nowadays, robots usually have advanced sensory systems that process information and appear to function as if they have brains, a form of computerized artificial intelligence of sorts based on how you define it. The robot can perceive conditions and can potentially determine its course of action based on these conditions.

In considering the artificial intelligence of robotics it pays to consider the markets being served and the functions for which they are used. Based on these factors we end up with two general types of systems—vertical and horizontal. Vertical systems generally are very specific markets and are commonly business-to-business. They are built for precise functions for equally precise situations. One might think of systems such as Wayblazer, a travel process and sales planner for businesses, as a vertical system. They tend to do one job and do it very well. Horizontal systems, on the other hand, serve broad audiences, are generally more utility-like, and are usually end-user/consumer based. One might think of systems such as Cortana or Siri as horizontal systems.

Robot Learning

Robot learning combines machine learning and robotics. Machine learning, as covered earlier in this chapter is combined with a variety of data and learning methods for training the robots. They might learn from observation of people, from research, from human feedback, or even from gaming environments. This is important because we sometimes want robots operating in very dynamic situations. As Trevor Darrell of the Berkeley Vision and Learning Center notes:

> Most robotic applications are in controlled environments where objects are in predictable positions. The challenge of putting robots into real-life settings, like homes or offices, is that those environments are constantly changing. The robot must be able to perceive and adapt to its surroundings. (Yang, 2015)

University of California Berkeley, where Darrell teaches, is working to refine and enhance how robots learn. Instead of attempting to pre-program for a vast array of situations, researchers are working to use deep learning, trial-and-error (experience), and reinforcement learning. Robots recognize patterns and categories among the data it is receiving to help its learning process. Willow Garage Personal Robot 2 (PR2), nicknamed BRETT for Berkeley Robot for the Elimination of Tedious Tasks, is one such robot (Yang, 2015). Researchers already have programs designed to allow robots to learn after observing a human complete a task wherein it is able to mimic the movement almost exactly. Scientists have now also successfully created robots that can build better versions of themselves.

Sample Systems

Artificial intelligence includes, for some, the ability of a computerized system such as IBM's Watson to use deep data mining and algorithms to find what seem to be the most likely correct answers to questions. Others say that Watson is not really actually thinking—as in creating original thoughts or formulating and testing hypotheses—and thus is not really artificially intelligent.

IBM Watson

Watson is a computer system capable of answering questions posed in natural language, developed in IBM's Deep QA project by a research team led by principal investigator David Ferrucci and named after IBM's founder, first CEO, and industrialist Thomas J. Watson. Development began in 2007, during which time the computer system was specifically developed to answer questions on the quiz show Jeopardy. In 2011, Watson competed on Jeopardy!, a show that centers around answering clues in the form of answers, against the former winners Brad Rutter and Ken Jennings and received the first prize of $1 million. By played against former winners Brad Rutter and Ken Jennings and received the first place prize of $1 million. By playing against two human who were known for competing and winning at Jeopardy, Watson was able to display its capacity for learning. It won nearly every round, showing that artificial intelligence was able to solve complex questions and problems in the blink of an eye.

Operating at a rate of 80 teraflops, which is a trillion floating-point operations per second, Watson had access to 200 million pages of structured and unstructured content consuming 4 terabytes of disk storage. It uses a common cognitive framework to make decisions and goes through a four-step process to solve problems. The first is "observe," where Watson receives data input. The next step is "interpret," where it categorizes what the data are as both segments of data and as a whole. As an example, it would attempt to identify nouns and verbs as well as a context for them. Next, it would determine if the current sentence structure makes sense as well as establish the type of sentence structure it is. The third step is "evaluate" where, with a combination of interpretation and evaluation, Watson analyzes the patterns of the data and arrives at the fourth step "decide" by giving an answer based on its extensive database of information. In this last step, it also weighs its odds that its answer is correct and acts accordingly in its responses. Since its performance on Jeopardy!, Watson has gone on to serve in medical, food, and technological industries, proving that artificial intelligence can be enormously helpful to the world at large in a great variety of areas.

Siri and Viv

Siri, Apple's iconic person digital assistant available on smartphones, was created in early 2010 and was one of the first applications to capably handle natural query languages. While Siri was originally a standalone iPhone app, it was acquired by Apple in late 2010. After the Apple acquisition Siri was enhanced to handle multiple languages, was scaled to serve large audiences, was integrated with other iPhone tools, was given a voice with which to communicate its answers, and was re-released in late 2011. Siri was able to connect to Web services such as Yelp or Rotten Tomatoes and return basic search results.

Siri is followed by Viv. Viv made its public debut in 2016 after 4 years of development based on the work of Dag Kittlaus and Adam Cheyer (artificial intelligence developers behind Siri). It is seen as the next generation of smartphone artificial intelligence while at the same time an enhanced version of Siri. One of Viv's strength is its ability to handle dozens of complex requests using not only its own resources but also those of third-party merchants. One other difference between Siri and Viv is that Viv is based on a much more open platform. Viv is open to all devices and all services.

Viv has competitors or can be seen as a competitor in areas such as Google Now, Amazon Echo, Microsoft's bot engine and Facebook's bots for messenger to name a few. SoundHound's Hound app is an example. Hound is a voice-power virtual assistant launched in early 2016. It lets users complete tasks and search for answers to questions by using only voice commands. Hound's goal is to be able to interpret what users are asking better and faster than its competitors.

Baxter

Baxter, made by Rethink Robotics of Boston, Massachusetts, is what we might call a smart robot. Baxter is an artificial intelligence system that can learn from its environment as well as its experience and can build on its understanding and capabilities based on this new knowledge. Baxter is able to work alongside humans and can learn from their behavior. Essentially, it is an industrial android (humanoid robot).

Part of the appeal of Baxter is that it allows nontechnical people to operate it with relative ease. Users can simply physically guide Baxter through the motions they want it to perform without the need for hired programmers or application developers. Baxter can also adapt easily and, being only about 3 feet tall, can easily be moved from place to place. This also means it can fit through standard doorframes. Added to that, Baxter has interchangeable end effectors at the end of each arm.

Baxter.

Baxter has a cost of about US$36,600 and in 2013 won both Massachusetts Institute of Technology Review's 10 Breakthrough Technologies of 2013 but also was a 2013 Edison Award's gold winner in Applied Technology Innovation. The year before, it won Time Magazine's Best 25 inventions of 2012 award as well (Active Robots, 2016).

Emotions, Perception, and Subjectivity

A robot is essentially just a complex system of code and sensors that can observe and execute what it senses. So, with that in mind, what are some considerations as they relate to emotions, perception, and subjectivity? First, let us address the issue of sensing the environment. A robot cannot "see" or "feel" as we do. It must process information in bits and bytes. This means that it used sensors for vision and acoustics as well as stereo and range. It also means it needs sensors for tactile, which is sense of touch, information acquisition. In the world of computer-based applications this science of applying touch sensation is known as haptics. Haptics, when it comes to computer-based systems gathering information, using haptic perception (recognizing things through touch), includes measures such as force, contour following, or the like. Haptics, when it comes to computer-based systems giving an end user a sense of touch such as with a controller in a gaming environment, known as haptic feedback, makes use of vibration, force, motion, or temperature.

In haptic perception as well as in vision and acoustics, robotic systems have to resolve issues of modeling, classification, and recognition. They need to computationally understand structure and range, materials, size, and weight of items around them. They then need to be able to navigate, manipulate, control, and/or learn from that environment. Some ways this occurs is with feature detection and matching, or with motion tracking and visual feedback. Another method, particular to vision, is with computational stereo. With this method, a single point is viewed from various viewpoints (two or more), which are then triangulated to allow for depth perception necessary for understanding of three-dimensional spatial relationships.

In regard to emotions, robots and artificial intelligence do not have these features as would be defined for humans. Emotion, after all, is an instinctive state of mind derived from one's circumstances, mood, or relationships with others. Emotions are based on what we feel, perceive, or experience subjectively, which is known as sentience. As humans, we have the ability to think (reason) and the ability to feel (sentience). What robots do have, however, is the potential capacity for artificial sentience and artificial awareness inclusive of self-awareness.

Belief Space

Belief space "refers to a mathematical framework that allows us to model a given environment statistically and develop probabilistic outcomes" (Ross, 2016). It gives robots a way to achieve greater situational awareness. While previously too complex, advances in data analytics combined with exponentially greater sets of experiential robot data enables programmers to develop robots who can intelligently interact with environment.

Cloud Robotics

Cloud robotics works to use cloud-based technologies (cloud computing, storage, etc.) to create shared infrastructure and services for robotics. This requires convergence of information and resources. Robots or other agents could share information and processes as well as physical technological resources via cloud-based services or other networks. One might think of Google Car as cloud robotics. After all, it uses the network to map locations, load data on traffic, determine special location in relation to objects around it, and so on. Data from each Google car is also used for machine learning via grid computing in the cloud (Goldberg, n.d.). As Goldberg (n.d.) of Berkeley describes it, "Cloud Robot and Automation systems can be broadly defined as any robot or automation system that relies on data or code from a network to support its operation, i.e., where not all sensing, computation, and memory is integrated into a single standalone system."

James Kuffner, then a professor at Carnegie Mellon, later the lead of Google's robotics program and as of 2016 an employee of Toyota Research Institute, gave this description in 2010 which still applies today, "Imagine a robot that finds an object that it's never seen or used before—say, a plastic cup. The robot could simply send an image of the cup to the cloud and receive back the object's name, a 3-D model, and instructions on how to use it" (Ackerman & Guizzo, 2016; Guizzo, 2011).

OpenAI

OpenAI is a nonprofit artificial intelligence research company, associated with business magnate Elon Musk, that aims to carefully promote and develop open-source friendly AI in such a way as to benefit, rather than harm, humanity as a whole. As one Wired article notes, "Musk and Altman worry that if people can build AI that can do great things, then they can build AI that can do awful things, too. They're not alone in their fear of robot overlords, but perhaps counterintuitively, Musk and Altman also think that the best way to battle malicious AI is not to restrict access to artificial intelligence but expand it. That's part of what has attracted a team of young, hyper-intelligent idealists to their new project." Whether this is the platform that becomes the standard for AI or not only the future will tell. The fact that companies, programmers, and larger communities are grappling with how to handle AI development is without question.

Concerns about Robotics Development

The continued refinement and capabilities of robots do not come without some concerns as well. Professor Stephen Hawking, Director of Research at the Center for Theoretical Cosmology (science of the origin and development of the universe) at the University of Cambridge, has said that efforts to create thinking machines pose a threat to our very existence. In 2014 he told BBC News, "The development of full artificial intelligence could spell the end of the human race." He notes that while the primitive forms of artificial intelligence developed so far have proved very useful, he fears the consequences of creating something that can match or surpass humans (Cellan-Jones, 2014).

Some others with similar concerns include Elon Musk, founder, CEO and CTO of SpaceX, co-founder and CEO of Tesla Motors, and co-founder of Paypal; Bill Gates, co-founder of Microsoft; and Steve Wozniak, co-founder of Apple, Inc. At an International Joint Conference on Artificial Intelligence in 2015, Hawking, Musk, and others wrote an open letter, "Autonomous weapons: an open letter from AI & robotics researchers" that can be found at http://futureoflife.org/open-letter-autonomous-weapons/. In this letter, they voiced concerns about military use of AI. They write, "autonomous weapons have been described as the third revolution in warfare, after gunpowder and nuclear arms" and later continue, "If any major military power pushes ahead with AI weapon development, a global arms race is virtually inevitable, and the endpoint of this technological trajectory is obvious: autonomous weapons will become the Kalashnikovs of tomorrow. Unlike nuclear weapons, they require no costly or hard-to-obtain raw materials, so they will become ubiquitous and cheap for all significant military powers to mass-produce. It will only be a matter of time until they appear on the black market and in the hands of terrorists, dictators wishing to better control their populace, warlords wishing to perpetrate ethnic cleansing, etc" (Future of Life Institute, n.d.). As a society and as individuals we must consider the role we see robotics and artificial intelligence playing in our futures across all areas, positive and negative.

Circumstance and Consequence Reasoning

Consequence reasoning is reasoning from consequences of prior actions or knowledge. We know not to touch a hot stove burner because: (a) we understand already that the heat of the burner can easily get hotter than is necessary to injure human skin in a very short time and (b) we may have been burned at some time in our past or seen someone else be burned. There are markers we use to help us identify stove burner that might be hot including its current color (if it is red we presume it is hot), radiating heat, prior use, current circumstances, and so on. Being able to consider circumstances and consequences and make decisions accordingly is playing an increasing role in robotics and artificial intelligence.

Let us consider the Atlas robot, created by Boston Dynamics (now a Google owned company). Atlas is a 6-foot tall, 330-pound robot capable of traversing across complex terrain and completing equally complex mobility tasks. It specialized in mobile manipulation, as can be seen in the Boston Dynamics video at https://www .youtube.com/watch?v=rVlhMGQgDkY. It understands when the circumstances around it change and adjusts accordingly. If it is travelling by foot and hits an ice patch or obstacle, it adjusts its path, weight balance, or similar. If it is picking up a box and someone knocks that box out of its hands, Atlas will identify where the box is and pick it up again. Now, robots are being developed that are capable of reflection, consideration, and decision-making. One example is in the use of DIARC, a distributed integrated affect, reflection, cognition architecture for robots (Scheutz et al., 2013). So now, we are at the point of making robots that can even appropriately reject commands it is given based on reasoning regarding the effects of those actions. The Human-Robot Interaction Laboratory at Tufts University works with these types of robots. You can learn about the current work and see examples (including videos) of some consequence reasoning robots at http://hrilab.tufts.edu/.

The Science behind the Technology

There is much that could be said of the science behind the technology of robotics and artificial intelligence, particularly in relation computer science (the science of information processes and their interactions with the world) and physics (the study of matter and its motion through space and time, along with the related concepts of energy and force). There is equally as much that could be said in relation to engineering—particularly mechanical and electrical. We will cover a few areas below that have been mentioned in the reading and will be expanded upon here.

Points of Articulation

Articulated robots have joints that allow for the rotation of united parts on a pivot point of some sort. Imagine an arm that moves at the elbow, wrist, and finger joints. Each of those points is articulated. The same would go for a robot. These joints allow for a range of motion. In some cases each joint might have varying degrees of freedom.

Degrees of freedom.

The number of degrees of freedom a system possesses is equal to the number of independent parameters (measurements) that are necessary to uniquely define its position in space at any instant of time. A higher number indicates increased flexibility in positioning. A knee joint, for example, has one degree of freedom in humans while an ankle joint has two degrees of freedom and the human hip (a ball and socket joint) has three degrees of freedom.

Robots, based on their number of points of articulation and degrees of freedom at each point, can have immense flexibility. Added to that, with robots we can exhibit a great amount of control and precision. As a matter of fact, some surgeons prefer to use robotic systems due to their accuracy, flexibility, and visual capabilities. As Dr. Kenneth Meredith, Gastrointestinal Oncology and Robotic Surgery of Sarasota Memorial Hospital notes, "I prefer to do the robotic surgery because of the degrees of freedom and articulation and the three-dimensional view that it allows us to have over conventional laparoscopy, which is absolutely phenomenal for either benign or malignant diseases" (Lederer, 2016).

The Physics of Robotics

We already mentioned degrees of freedom earlier. There are a variety of other components of physics involved in robotics and artificial intelligence. Examples include work, friction, torque, electromagnetism, and current, among others. We briefly describe each below.

Work

Work is the transfer of energy that results from applying a force over a distance. In order for work to have been done, the object that the force is being applied on needs to move. Here is the basic formula: work equals force times distance.

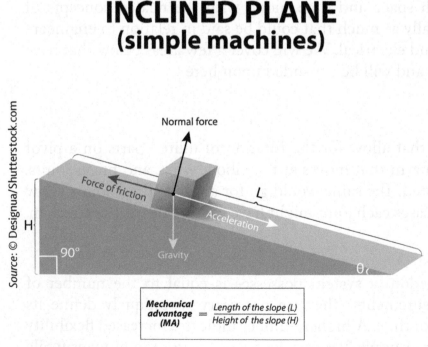

INCLINED PLANE
(simple machines)

Normal force

Force of friction

L

Acceleration

H

90°

Gravity

θ

$$\text{Mechanical advantage (MA)} = \frac{\text{Length of the slope (L)}}{\text{Height of the slope (H)}}$$

Source: © Designua/Shutterstock.com

Friction.

Friction

Friction is a force that resists the motion of objects or surfaces—it opposes the direction of a direct force. Imagine a robot with wheels that are in contact with the ground. The wheels spin and create friction which permits the robot to move.

Friction can be found any time objects are in contact with one another. Think of it as the forces that cause object to slow down when it is touching another object.

Torque

Torque is a rotational force—it is what causes objects to spin, rotate, turn, or twist. Torque has dimensions of distance × force; the same as energy. The equation for torque is r times F, where F is the force vector and r is the vector from the axis of rotation to the point where the force is acting. And what is a vector, you ask? A vector is a mathematical object that has a size, called the magnitude, and a direction.

Electromagnetism

Electromagnets are used in electric motors to attract and repel magnets on the rotor at exact times to make the motor spin. It is based on electromagnetic force, a type of physical interaction between electronically charged particles.

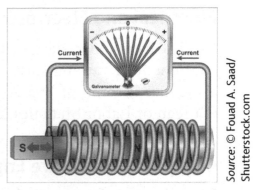

Source: © Fouad A. Saad/ Shutterstock.com

Electromagnetism.

Current

Current is what carries energy in a circuit and can only flow if there is a difference in voltage between two locations—it is the rate at which a charge passes by a point on that circuit.

Career Connections

Mechanical Engineers

Mechanical engineers design, develop, build, and test mechanical and thermal sensors and devices, including tools, engines, and machines. This is perhaps the broadest of all engineering disciplines although persons in this field work most often with engineering services, research and development, and manufacturing. A person in this field could expect, as of 2015, to make a median income of $83,590 per year ($40.19 per hour) with a steady job market (Bureau of Labor Statistics, 2015a).

Electromechanical Technician

If one decided to be an electromechanical technician, a person who combines their knowledge of mechanical technology with electrical knowledge and operates, tests, or maintains unmanned, automated, robotic, or electromechanical equipment, they could expect, as of 2015, to make a median income of $53,340 per year ($25.65 per hour) with a steady job market (Bureau of Labor Statistics, 2015b).

Electrical and Electronics Engineering Technicians

If one went in to the career field of electrical and electronics engineering technicians, they could expect to help engineers design and develop a variety of electrical and electronic equipment (such as navigational or communications equipment) and could expect, as of 2015, to make a median income of $61,130 per year (29.29 per hour) but this field is in a slight decline market (Bureau of Labor Statistics, 2015b).

Electrical and Electronics Engineers

Electrical engineers design, develop, test, and supervise the manufacturing of electrical equipment of varying sorts. A person in this field could expect, as of 2015, to make a median income of $95,230 per year ($45.78 per hour) with a steady job market (Bureau of Labor Statistics, 2015b).

Computer Hardware Engineers

If one became a computer hardware engineer who researches, develops, and tests computer systems and components such as processors, circuit boards, memory devices, and the like, they could expect, as of 2015, to make a median income of $111,730 per year ($53.72 per hour) with a steady job market (Bureau of Labor Statistics, 2015b).

Computer and Information Research Scientists

Computer and information research scientists invent and design new approaches to computing technology and find innovative uses for existing technology. This area often requires advanced education (doctoral or professional degree). A person in this field could expect, as of 2015, to make a median income of $110,620 per year ($53.18 per hour) with a growing job market (Bureau of Labor Statistics, 2015c).

Modular Activities

Discussions

❖ Post an original discussion in the online discussion board on the following topic: Some people say artificial intelligence will have a negative influence on society while others say it will have a positive influence. What are your thoughts and why? (200 words minimum). Next, comment on the post(s) of a minimum of one other student in a thoughtful and academic way that enhances the conversation. See rubric for grading and assessment measures.

❖ Post an original discussion in the online discussion board on the following topic: Watch the one minute video about a robot that uses consequence reasoning. As this type of technology continues to develop, do you see at as mostly positive or mostly negative? Why? (200 words minimum). Next, comment on the post(s) of a minimum of one other student in a thoughtful and academic way that enhances the conversation. See rubric for grading and assessment measures.

❖ Technology Fair: Find a current or upcoming technology (in the current news) that is of interest to you and research it. After researching a minimum of three sources (each less than a year old) pertaining to the specific technology of your choice (without duplicating other student topic choices) inspired by current events and/or the preceding lessons, share with the class the following: (a) Name and detailed overview of the technology, (b) diagram/images or other visuals of the technology, (c) demonstration of how the technology works, and (d) affect or potential affect this technology may have on society at large. Use your creativity to share this information with the class. Treat it as if you were the teacher as you are teaching your classmates about this particular technology. After you have completed your own post, read the posts of others and then make a substantial/meaningful response to your favorite three.

Tests

❖ Online graded quiz on overall chapter content, written in multiple choice format. When submitted, we recommend giving the correct answer along with the page number in which it is found for questions students did not answer correctly.

❖ Use the website Socrative found at http://www.socrative.com/ (free of charge) to set up a live interactive quiz where students can instantly see the overall results for the class. Ask the following questions:

➢ If I were given a robo-nanny to watch over my eight-year old son while I went to work, I would use it [true/false]

➢ I would trust a robotic doctor to perform a surgery on me as long as a human doctor was there as a guide and resource [true/false]

➢ I believe we should be permitted to have robotic cars that do not need a human navigator present who can take over the controls at any time [true/false]

➢ I am worried that artificial intelligence could potentially cause robots to revolt against man in some manner in the future [true/false]

Discuss the results as a class afterward.

Reports or Papers

❖ IBM's Watson: Research IBM's Watson and write a two-page report. This report must use proper citation and use a minimum of two sources.

❖ Professor Murray Shanahan describes embodied versus disembodied artificial intelligence systems in his two-minute video at http://www.bbc.com/news/technology-30296999. Watch this video and then write a two-page informative report that describes these two types of AI. This report must use proper citation and use a minimum of two sources.

❖ Robotics is increasingly used with children, such as Anki's Cozmo (https://anki.com/en-us/cozmo). Choose a child or youth-based robot or robotic platform and then write a two-page informative report that describes the technology. This report must use proper citation and use a minimum of two sources.

Design and/or Build Projects

❖ You have been given the opportunity to help create the physical design of a robot that is intended to be sold commercially to household consumers. This robot is known as "Mr. Walkyourdog." The idea is that Mr. Walkeryourdog can do the dog walking for the owner, without the owner present for the walk. Sketch your design and then explain in detail why you added the elements you did. Things to consider: number of dogs, type of terrain, how it handles obstacles, size (height/width/length), weight, how it is powered, how it moves, what a standard consumer might want it to look like, and so on.

Assessment Tasks

❖ Consider the artificial intelligence field of robotic nannies (also known as robotic childcare, robo-nanny, or robotic babysitter). Assess the benefits and drawbacks of this particular technology. Write a two-page assessment of your findings using a minimum of two sources.

❖ Watch the five-minute video at http://www.bbc.com/news/technology-30299992 and write up to 500 words minimum (approximately two-double spaced pages) on your assessment of the system that Stephen Hawking uses in the interview as well as where you think the future of speech synthesizers or similar systems may be going.

Terms

Actuator—Motor that is responsible for moving or controlling a mechanism or system.

Cloud robotics—Cloud robotics works to use cloud-based technologies (cloud computing, storage, etc.) to create shared infrastructure and services for robotics.

This requires convergence of information and resources. Robots or other agents could share information and processes as well as physical technological resources via cloud-based services or other networks.

Computational stereo vision—Using two or more vantage points in order to extract three-dimensional information for digital images. The 3D information can be extracted by examining the relative positions of the objects from the varying viewpoints.

DIARC—A distributed integrated affect, reflection, cognition architecture for robots.

Haptics—Refers to sense of touch (tactile).

Haptic feedback—Haptic or kinesthetic communication recreates the sense of touch by applying forces, vibrations, or motions to the user.

Horizontal artificial intelligence—Horizontal systems serve broad audiences, are generally more utility-like, and are usually end-user/consumer based. One might think of systems such as Cortana or Siri as horizontal systems. Compare to vertical artificial intelligence.

Machine learning—A subfield of computer science that evolved from the study of pattern recognition and computational learning theory in artificial intelligence.

OpenAI—A nonprofit artificial intelligence research company, associated with business magnate Elon Musk, that aims to carefully promote and develop open-source friendly AI in such a way as to benefit, rather than harm, humanity as a whole.

Sentience—The capacity to feel, perceive, or experience subjectively.

Vertical artificial intelligence—Vertical systems generally are very specific markets and are commonly business-to-business. They are built for precise functions for equally precise situations. One might think of systems such as Wayblazer, a travel process and sales planner for businesses, as a vertical system. They tend to do one job and do it very well. Compare to horizontal artificial intelligence.

References

Ackerman, E., & Guizzo, E. (2016). Toyota AI team hires James Kuffner from Google Robotics, will have Rodney Brooks as adviser. *IEEE Spectrum*. Retrieved from http://spectrum.ieee.org/automaton/robotics/industrial-robots/toyota-ai-team-technical-and-advisory-teams

Bureau of Labor Statistics. (2015a). Mechanical engineers. *Occupational Outlook Handbook*. Retrieved from http://www.bls.gov/ooh/architecture-and-engineering/mechanical-engineers.htm

Bureau of Labor Statistics. (2015b). Architecture and engineering occupations. *Occupational Outlook Handbook*. Retrieved from http://www.bls.gov/ooh/architecture-and-engineering/home.htm

Bureau of Labor Statistics. (2015c). Computer and information research scientists. *Occupational Outlook Handbook.* Retrieved from http://www.bls.gov/ooh/computer-and-information-technology/computer-and-information-research-scientists.htm

Cellan-Jones, R. (2014). Stephen Hawking warns artificial intelligence could end mankind. *BBC News.* Retrieved from http://www.bbc.com/news/technology-30290540

Future of Life Institute. (n.d.). Autonomous weapons: An open letter from AI & robotics researchers. Retrieved from http://futureoflife.org/open-letter-autonomous-weapons/

Goldberg, K. (n.d.). *Cloud robotics and automation.* University of California Berkeley. Retrieved from http://goldberg.berkeley.edu/

Guizzo, E. (2011). Cloud robotics: Connected to the cloud, robots get smarter. *IEEE Spectrum.* Retrieved from http://spectrum.ieee.org/automaton/robotics/robotics-software/cloud-robotics

Jaslow, R. (2013). RP-VITA robot on wheels lets docs treat patients remotely. *CBS News.* Retrieved from http://www.cbsnews.com/news/rp-vita-robot-on-wheels-lets-docs-treat-patients-remotely/

Lederer, P. (2016) Mechanical medic. *SRQ Magazine.* Retrieved from http://www.srqmagazine.com/articles/368/Mechanical-Medic

Legg, S., & Hutter, M. (2007a). A collection of definitions of intelligence. In B. Goertzel, P. Wang (Eds.), *Advances in artificial general intelligence: Concepts, architectures and algorithms* (Vol. 157, pp. 17–24). Amsterdam: IOS Press. Retrieved from http://www.vetta.org/documents/A-Collection-of-Definitions-of-Intelligence.pdf

Legg, S., & Hutter, M. (2007b). Universal intelligence: A definition of machine intelligence. *Minds Mach, 17,* 391–444. Retrieved from http://arxiv.org/pdf/0712.3329v1.pdf

MacRae, M. (2013). Top five medical technology innovations. *American Society of Medical Engineers.* Retrieved from https://www.asme.org/engineering-topics/articles/bioengineering/top-5-medical-technology-innovations

National Highway Traffic Safety Administration. (2013). U.S. Department of Transportation releases policy on automated vehicle development. Retrieved from http://www.nhtsa.gov/About+NHTSA/Press+Releases/U.S.+Department+of+Transportation+Releases+Policy+on+Automated+Vehicle+Development

Ross, A. (2016). *The industries of the future.* New, York, NY: Simon & Schuster.

Scheutz, M., Briggs, G., Cantrell, R., Krause, E., Williams, T., & Veale, R. (2013). Novel mechanisms for natural human-robot interactions in the DIARC architecture. Retrieved from http://hrilab.tufts.edu/publications/aaai13irsfinal.pdf

Taylor, H. (2016). Could you fall in love with this robot? *CNBC.* Retrieved from http://www.cnbc.com/2016/03/16/could-you-fall-in-love-with-this-robot.html

Further Reading

Bengio, Y., Lamblin, P., Popovici, D., & Larochelle, H. (2007). Greedy layerwise learning of deep networks. *Proceedings Neural Information Processing Systems, 19,* 153–160.

Campbell, M. (2015). Apple invention uses vibrations and temperature to simulate different materials on touchscreens, trackpads. *Apple Insider.* Retrieved from http://appleinsider.com/articles/15/04/23/apple-invention-uses-vibrations-temperature-control-to-simulate-different-materials-on-touchscreens

Chen, Y., Lin, Z., Zhao, X., Wang, G., & Gu, Y. (2014). Deep learning-based classification of hyperspectral data. *IEEE Journal of Selected Topics in Applied Earth Observations and Remote Sensing, 7*(6), 2094–2107.

Duda, R., Hart, P., & Storck, D. (2001). *Pattern classification* (2nd ed). Hoboken, NJ: Wiley Interscience.

Fergus, R., Weiss, W., & Torralba, A. (2009). Semi-supervised learning in gigantic image collections. *Proceedings of Advances in Neural Information Processing Systems*, 522–530.

Fukushima, K. (1980). Neocognitron: A self-organizing neural network model for a mechanism of pattern recognition unaffected by shift in position. *Biological Cybernetics, 36*, 193–202.

Hamilton, L. (2014). Six novel machine learning applications. *Forbes Woman*. Retrieved from http://www.forbes.com/sites/85broads/2014/01/06/six-novel-machine-learning-applications/#3fe3730267bf

Hinton, G., Osindero, S., & Teh, Y. (2006). A fast learning algorithm for deep belief nets. *Neural Computation, 18*, 1527–1554.

Kasturi, R. (2005). *Machine vision*. Retrieved from http://www.cse.usf.edu/~r1k/MachineVisionBook/MachineVision.files/MachineVision_Chapter15.pdf

Kulkarni, P. (2012). Reinforcement and systemic machine learning for decision making (1st ed). Hoboken, NJ: Wiley-IEEE Press.

Lee, H., Ekanadham, C., & Ng, A. (2008). Sparse deep belief net model for visual area v2. *Proceedings of Advances in Neural Information Processing Systems, 20*, 873–880.

Wu, F., Wang, Z., Zhang, Z., Yang, Y., Luo, J., Zhu, W., & Zhuang, Y. (2015). Weakly semi-supervised deep learning for multi-label image annotation. *IEEE Transactions on Big Data, 1*(3), 109–122.

Yu, D., Hinton, G., Morgan, N., & Chien, J. (2012). Introduction to the special section on deep learning for speech and language processing. *IEEE Transactions on Audio Speech Language Processing, 20*(1), 4–6.

Chapter 16

Energy and the Environment

Outline

- ❖ Chapter objectives
- ❖ Overview
- ❖ Nonrenewable energy sources
  - ➢ Oil
  - ➢ Coal
  - ➢ Natural gas
  - ➢ Nuclear energy
- ❖ Renewable energy sources
  - ➢ Hydroelectric power
  - ➢ Solar power
    - ▪ Solar thermal energy
    - ▪ Photovoltaic cells
  - ➢ Geothermal energy
  - ➢ Wind power
- ❖ Nanotechnology and energy
- ❖ Climate change
- ❖ Greenhouse effect
  - ➢ Intergovernmental panel on climate change
  - ➢ Paris agreement
  - ➢ U.S. National aeronautics and space administration
- ❖ The science behind the technology
- ❖ Career connections
- ❖ Modular activities
  - ➢ Discussions
  - ➢ Tests
  - ➢ Research

(Continued)

(*Continued*)

> ➤ Design and/or Build projects
> ➤ Assessment tasks
❖ Terms
❖ References
❖ Further reading

Chapter Objectives

❖ Examine nonrenewable energy sources
❖ Compare different sources of renewable energy
❖ Distinguish the differences between renewable and nonrenewable energy sources.
❖ Differentiate how various energy sources generate electricity
❖ Explain how photovoltaic cells work
❖ Define how nanotechnology is being used in energy research
❖ Describe climate change and the greenhouse effect
❖ Summarize why climate change is controversial

Overview

Energy makes the world go round. Or at least it seems that way. Think about how often you plug in each day and the consequences of when you do not. Nothing strikes more fear into a modern-day heart than when a battery icon turns red. The truth is that the energy network your lifestyle depends on is largely invisible to you. It runs behind your walls, is stored underground, and is packaged in convenient little batteries. You probably do not pay attention unless the electricity goes out or your car runs out of gas. Even then it is more of a temporary nuisance, because there is no doubt that it will be resolved quickly. A world without an endless energy supply is likely inconceivable to anyone reading this book. However, approximately 1.2 billion people—17% of the world's population—do not have access to energy. Many others have sporadic or low-quality in access. Energy distribution is also unequal throughout the world; with rural areas sub-Saharan Africa and developing Asian countries at the limited ends.

There are many predictions about what global future energy needs will look like. But almost unanimously, experts predict that energy demands will continue to grow. Although, interestingly, over the past 2 years there have been similar or decreased energy use in many countries. Some experts believe a record warm winter in Europe, economic recessions in Brazil and Russia, as well as increased hydropower output in China contributed to this stagnation. Or, perhaps, better efficiencies and energy-conscious consumers are reversing the trend. We will know more over the next few years when data are collected, analyzed, and published.

China has experienced economic growth and now contains the world's largest middle class. India, with a population of approximately 1.2 billion, is also developing a vibrant economy and resultant middle class. Although these are two of the largest countries in the world, many other moderate-sized ones have seen an improved quality of life as well. People with money to spend want material goods to buy. Middle-class staples include computers, cell phones, electronic gadgets, and cars. All of these items take energy to manufacture, energy to transport, and energy to use. Oil, coal, and gas continue to be the top three global energy sources. They are also all nonrenewable energy sources, meaning that when they are gone, they are gone. We keep developing better technology and processes to reach these sources in places that were not possible before. But these developments can also have a negative impact on the environment and quality of life for citizens. Who consumes the most energy? The top 10 and bottom 10 total energy users for 2015 are found in Table 1.

Since experts predict an increase in energy consumption over the long run, and nonrenewable sources will dry up at some point, renewable energy might be the answer to our energy needs. However, it is still too soon to determine whether this will turn

Table 1. Ten highest and ten lowest energy-consuming countries in 2015. Data retrieved from Enerdata total energy consumption site. Mtoe is an acronym for million tonnes of oil equivalent.

Ten Highest Energy Consumers		Ten Lowest Energy Consumers	
Country	**Mtoe**	**Country**	**Mtoe**
China	3,101	New Zealand	21
United States	2,196	Portugal	22
India	882	Norway	32
Russia	718	Romania	33
Japan	435	Colombia	34
Germany	305	Kuwait	38
Brazil	299	Chile	38
South Korea	280	Czech Rep.	40
Canada	251	Uzbekistan	45
France	246	Sweden	47
Iran	244	Algeria	53
Indonesia	227	Belgium	54

out to be the case. As shown in Figure 1, only 10% of the United States' energy came from renewable sources in 2014. Of that, only about half came from the sources you are probably most familiar with, including solar (4%), wind (18%), and hydroelectric or water based (26%). The other half came from biomass, which is plant-based material and includes wood. While wood may be renewable, many species of trees take a long time to renew (regrow) relative to other sources. Wood can also pollute the air. Renewable energy does not necessarily equate to clean energy.

Companies researching and working in the alternative-fuel industry are also subject to the twists and turns of the economy like any other business. It may seem like they have a loftier purpose, reducing our dependence on nonrenewal energy sources. They may even have a superior product. But there are other factors that influence the development, use, and distribution of new energy sources. Research and development (R&D) expenses can be very high for companies developing new technologies. It can take years to recover costs in an uncertain industry. It is also difficult to predict which

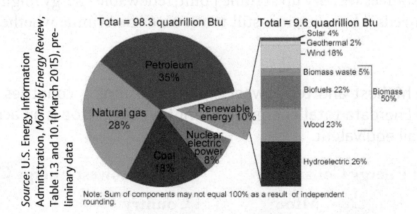

Source: U.S. Energy Information Adminstration, Monthly Energy Review, Table 1.3 and 10.1(March 2015), preliminary data

Figure 1. U.S. energy consumption by energy source in 2014.

Source: File downloads are approved for public use when displayed with the appropriate credits. https://www.eeremultimedia.energy.gov/solar//photographs/solar_dish_sets_world_record_efficiency.

Figure 2. This was the Stirling Energy Systems dish that was located in Arizona. The company went bankrupt in 2011 even though the dish set a world's record for solar-to-grid conversion efficiency. Part of the reason was the declining costs of solar voltaic cells. This is an example of the unpredictable nature of alternative-fuel technologies.

alternative-fuel technologies may resonate with consumers. Figure 2 is a picture of the Stirling Energy Systems disk. The company went bankrupt in 2011 even though the dish set a world's record for solar-to-grid conversion efficiency. Part of the reason was the declining costs of solar voltaic cells Generally, new technologies are more expensive when introduced to the general public and have a limited number of early adopters. It is only over time that new cultural standards emerge and become commonplace. What energy sources will become most prevalent in the future is difficult to predict (nanotechnology, anyone?), but there is a good body of research to help us review options critically for guiding our way.

Nonrenewable Energy Sources

Energy sources are considered nonrenewable if they cannot be replaced within a short time period. The four major nonrenewable energy sources are crude oil, natural gas, coal, and nuclear energy. Oil, coal, and gas are referred to as fossil fuels, because they are the by-product of plants and animals that lived millions of years ago. Nuclear energy relies on uranium, a common metal found in rocks.

Oil

Crude oil, or unprocessed oil, is found in underground reservoirs, within sedimentary rocks and in tar sands near the Earth's surface. It is a useful product because it contains hydrocarbons, which are molecules of various lengths. Hydrocarbons can take on different forms, including gas (e.g., methane), wax (e.g., paraffin), and liquid (e.g., kerosene) depending on their structure. Because crude oil contains so many different types of hydrocarbons, they have to be separated in a refinery. Different chain lengths have different boiling points. So one part of the refinery process is to heat the oil, let it vaporize, and then condense the vapor. This is called *fractional distillation*. *Conversion* is a newer chemical process to break longer chains into shorter ones. For example, this could turn diesel fuel into gasoline. Once items are separated into their various fractions, they must be treated to remove impurities. Then they are combined to produce desired products based on demand.

Gasoline is by far the main petroleum product in the United States. This is followed by distillate fuel oil (diesel fuel and heating oil), hydrocarbon gas liquids, and jet fuels. Propane is one of the most commonly used hydrocarbon gas liquid. In 2014 the top three gas-consuming states were California (11%), Texas (10%), and Florida (6%). New York and Ohio tied for fourth place with 4% each. While the United States consumed nearly 1.5 million barrels a day, the world consumed 91.2 million. The three largest petroleum-consuming countries are United States (21%), China (11%), and Japan (5%). According to the U.S. Energy Information Administration, the global supply of crude oil, liquid hydrocarbons, and biofuels should meet the world demand through 2040.

Coal

There are four classifications of coal: (1) anthracite, (2) bituminous, (3) subbituminous, and (4) lignite. The differences among them are the amount of carbon in the coal and the amount of heat energy it can produce. Anthracite is ranked highest and contains 86%–97% carbon with high heat value. It accounts for less than 1% of coal production and is used mainly in the metals industry. All anthracite mines are located in northeastern Pennsylvania. Bituminous contains 45%–86% carbon and is most abundantly found in the United States. It accounts for 48% of production and is used to generate electricity and in iron and steel production. Subbituminous contains 35%–45% carbon with 90% of production occurring in Wyoming. Lignite contains 25%–45% carbon and has the lowest energy content of all classifications. It accounts for 8% of production. It is primarily produced in Texas and North Dakota where it is used at power plants to generate electricity.

Coal is mined using one of two methods. It depends on the geology of the coal deposit which one is selected. Both are safe and efficient due to technological advances. *Underground mining* is generally used more often. Workers can use the longwall mining technique to extract over 75% of all the coal from a section of the coal seam. Self-advancing, hydraulically powered supports contain the roof until all coal is extracted. Then these machines are removed and the roof collapsed. The room-and-pillar technique extracts less coal, since workers cut a network of areas or rooms, leaving pillars of coal behind. The *surface* or opencast mining is most economical when a coal seam is near the Earth's surface. More than 90% of the coal can be recovered this way. First, explosives are used to break up the soil and rock, which is then removed. The exposed coal seam is broken up (fractured) and removed using large pieces of equipment such as power shovels and trucks.

In 2014, the largest majority of all coal consumption (92.8%) was used by the electric power sector to make steam. The steam created the energy to turn turbines, which activated generators and produced electricity. In fact, coal was used to produce 39% of all electricity in the United States. Coal can also be converted into gas or liquid to serve as a fuel sources or combined to make other products. Although overall consumption declined slightly between 2013 and 2014, along with a 6.8% decline in related jobs, coal production had a slight increase of 1.5%. While the United States produced 19.4 billion short tons in 2015, most recent world data (2011) show 979.8 billion were mined worldwide. Nearly 73% of the world's coal reserves are concentrated in the following five countries: United States with 26%, Russia with 18%, China with 13%, Australia with 9%, and India with 7%. It is difficult to determine just how much coal is left, because it is buried underground. Even so, experts estimate that there are enough coal servers to last about 256 years.

Natural Gas

Natural gas is located deep beneath the Earth and consists mainly of methane. It also contains a small amount of hydrocarbon liquids and nonhydrocarbon gases. It is not always easy to determine where gas reservoirs are located. Geologists are needed to locate suitable rocks and use seismic surveys to find the right locations for drilling. Once they find them, a production well is bored down and the gas flows up to the surface. Most natural gas that Americas use is drilled within the states. Texas, Pennsylvania, and Oklahoma are the top three states that contain natural gas proved reserves. Gas is considered to be a clean energy source, especially when compared to other fossil fuels such as oil and coal. It is generally low in price and works with the current power infrastructure.

Not all natural gas is located within reservoirs. Some gas is found within the pores of shale and other rock formations. Hydraulic fracturing is used in these cases and has been used for many years. A mixture of water, chemicals, and sand is forced at high pressure down a well to release this gas. This process is also referred to as fracking and has ignited a firestorm of protest in the United States. Environmentalists and concerned citizens believe that fracking is detrimental to the environment and quality of life for those who live near the fracking sites. They are worried about the large amount of water needed for the process and claim that the many chemicals that are used will contaminate nearby groundwater. Another concern is that the process could cause earthquakes, and some believe this has already happened. The United States Geological Survey recognizes that fracking does cause extremely small earthquakes, but claims they are not a safety concern due to their slight size. Their research is ongoing.

Nuclear Energy

Nuclear power has been used for many decades in the United States, and according to recent data, provides nearly 20% of its electricity. It was developed in the 1940s during World War II. It has been used to provide residential energy since the 1950s. There are approximately 100 nuclear facilities in 30 states powering more than 18 million homes. Worldwide there are about 440 commercial nuclear power plants providing over 11% of the world's electricity needs.

There are two different types of nuclear power plants in the United States, the *boiling water reactor* and the *pressurized water reactor*. Similar to other energy power plants, steam is produced by nuclear energy to activate a turbine/generator system to create electricity. In a boiling water reactor, water surrounding the nuclear fuel rods in the reactor vessel is heated until steam is created. Pipes then carry the steam to the turbine/generator system to produce electricity. Pressurized water reactors are more common. In these types of nuclear power plants, water surrounding the nuclear fuel rods in the reactor vessel is heated but prevented from boiling. This heated water is them pumped to a steam generator, where its heat is used to boil a separate water supply that makes the steam. This steam then drives the turbine/generator to produce electricity.

Nuclear power plants obtain the heat they need to create steam through a process called *fission*. Fission is when the nucleus of an atom, such as uranium, splits apart (fissions) and releases energy. Energy is produced because the mass of the uranium nucleus is heavier than the sum of masses of the fragments. There are two types of uranium in nuclear fuel, U-238 and U-235. Most of the fuel comprises U-238. However,

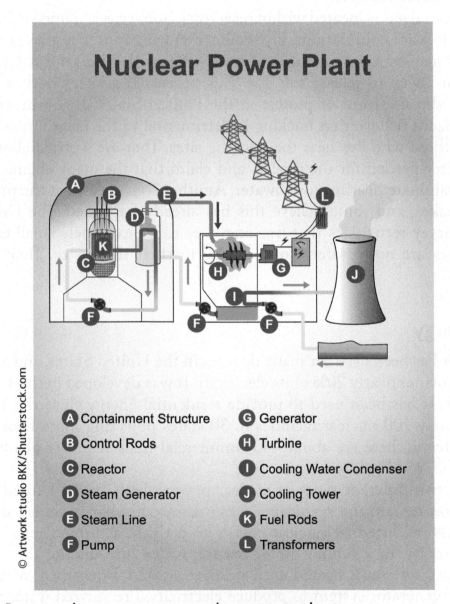

Nuclear Power Plant

A Containment Structure	**G** Generator
B Control Rods	**H** Turbine
C Reactor	**I** Cooling Water Condenser
D Steam Generator	**J** Cooling Tower
E Steam Line	**K** Fuel Rods
F Pump	**L** Transformers

© Artwork studio BKK/Shutterstock.com

Figure 3. Pressured water reactor nuclear power plant.

the nuclei of U-235 atoms are unstable. As they fission, neutrons are ejected and hit other uranium atoms, causing them to fission, until a chain reaction starts. At this point, fission becomes self-sustaining and produces enough heat needed to turn water into steam. Control rods contain the nuclear fuel and can be inserted or withdrawn to control the reaction.

Nuclear energy can also be released through a process called *fusion*. Fusion is when multiple nuclei join together to produce a heavier nucleus. Fusion is what powers the sun and stars. Iron and nickel are the standards by which other elements are measured. They have the strongest binding energies so are the most stable. Fusing two nuclei lighter than iron or nickel will release energy, because the combined mass of nuclei will be less than the sum of the masses of the individual ones. Examples include starlight and hydrogen bombs. Fusing two nuclei heavier than iron or nickel absorbs energy. An example is a supernova, which can pull in nearby space material until it explodes. Fusion power offers the prospect of an almost unlimited supply of energy but presents huge scientific and engineering challenges. It has been impossible to produce nuclear fusion so far because atoms must be heated to extreme temperatures or have enough force placed on them to overcome opposing electrostatic forces. The sun has the assistance of massive gravitational forces that are not found on Earth. Even so, researchers continue to work on this clean and plentiful power source. In fall 2015, researchers at the University of Gothenburg announced that they are working collaboratively with the University of Iceland to study a new type of small-scale nuclear fusion. Using a small laser-fired fusion reactor fueled by hydrogen, they were able to produce energy. They hope their new process may soon be used in small-scale power stations.

There are many benefits to using nuclear energy. It is relatively inexpensive over the long run, notwithstanding initial costs of building nuclear power plants. It is also reliable and more efficient than other sources. Nuclear energy facilities generate electricity every day with a 92% capacity factor, compared with natural gas (47.8%), coal (60.9%), and wind (33.9%). Most people would consider that the biggest disadvantage of using nuclear energy is the potential for a catastrophic accident. This is not just negative speculation. On March 28, 1979, the United States' most serious nuclear accident happened at Three Mile Island in Pennsylvania. A combination of equipment malfunctions and human error caused a partial meltdown. Nobody was injured, but the industry took a big public hit. On April 26, 1986, Ukraine's Chernobyl nuclear power plant exploded, spreading a radiation cloud over much of Europe. In this case, 200 people died. A ruptured reactor vessel and subsequent fire were found to be the cause. It is still not clear what the long-term health and environmental impact may be on the region. Most recently, on March 11, 2011, the Fukushima Daiichi nuclear power plant in Japan suffered an explosion after a massive earthquake. Large amounts of radiation were released into the atmosphere and residents who lived within 12 miles of the plant were evacuated. Events such as earthquakes and potential terrorist attacks are difficult to predict. However, nuclear facilities in the United States are held to the highest safety standards by the Nuclear Regulatory Commission. This independent monitoring agency was created by Congress in 1974 to protect both people and the environment. Inspectors oversee each reactor every day.

Technically, uranium is a nonrenewable source. However, reports show there is enough to power plants for 80 more years. The good news is that plans are already being made to replace uranium with thorium, and countries such as China have plans to start using it. Thorium is a greener alternative to uranium but challenges remain in making it as efficient. In the meantime, there are concerns regarding radiation and radioactive waste. According to the World Nuclear Association, safe practices have been used successful for many years to transport and house nuclear waste.

Renewable Energy Sources

Hydroelectric Power

Hydroelectric power is generated from flowing water. It is the largest, most inexpensive source of renewable energy in the United States and produces about 7% of the electricity. It has been used for thousands of years worldwide due to its easy accessibility. It continues to be used for this reason as well. In addition, it is a clean energy source that is continuously renewable. There are approximately 80,000 U.S. dams and less than 3% are powered. There is still great potential use for this energy source. New technologies continue to make the process more efficient for generating electricity and safer for the aquatic environment. For example, fish ladders can be built next to dams to allow them to swim upstream in a natural manner.

Hydropower is dependent on the water cycle. Water on the ground evaporates into the air, condenses into clouds, and returns to the earth as precipitation. Rain and snowmelt form into rivers and streams and flow into lakes and ocean. Kinetic energy is produced by naturally flowing water or water that is stored and released by dams. There are several ways that hydropower technology generates electricity.

1. *Impoundment*: also known as reservoir and most common type of hydropower plant. Water is stored behind a dam in a reservoir at a higher elevation and released when needed. It flows downhill using the natural force of gravity and spins a turbine, activating a generator that produces electricity.
2. *Diversion*: also called run-of-river, uses the natural flow of a river and channels it into a turbine and generator system. It may or may not require a dam.
3. *Pumped Storage*: water in a lower elevation reservoir is pumped uphill to a higher elevation one when energy demand is low. When energy demand is high, water is released downhill into a turbine and generator system.

To view the video, Energy 101: Hydropower, from the Office of Energy Efficiency and Renewable Energy on the basics of hydroelectric power, visit the following website: http://energy.gov/eere/energybasics/articles/hydropower-technology-basics.

All three hydroelectric systems use mechanical energy to generate electricity. When dams are built to form reservoirs, the stored water is a form of potential energy. When water is released, it travels downhill, which converts the potential energy into kinetic energy. The water spins a hydro turbine, which is a metal shaft that contains blades at one end and electromagnets at the other. As the turbine spins it goes up into the generator and converts mechanical energy into electricity using electromagnets. "The operation of a generator is based on the principles discovered by Faraday. He found that when a magnet is moved past a conductor, it causes electricity to flow" (Hydroelectric power: How it works, n.d., para. 5). Electricity is then transferred from the power plant through power lines to your house.

Hydroelectric power turbines are so efficient that they can convert up to 90% of the kinetic energy into electricity (comparable to about 50% in fossil fuel plants). One drawback of hydroelectric power, though, is that it is concentrated in the western part of the United States. More than 50% of hydropower capacity is found in the states of Washington, Oregon, and California. However, the capacity throughout the nation continues to grow in every region due to hydropower expansion. Most new projects are focused on opportunities using existing water infrastructure, such as nonpowered dams. The World Energy Council estimates that only about 30% of total worldwide capacity has been developed. However, rivers and dams are not the only source of hydropower. Marine and hydrokinetic sources are available as well.

Hydrokinetic power sources, such as tidal currents and wave systems, are currently being researched to evaluate their technical and economic viability. They are not ready for widespread use yet, but dozens of organizations worldwide are working on them. According the U.S. Department of Energy, there is enough energy in waves and tides along the U.S. coastline to meet a large portion of American energy needs if harnessed properly. For example, water can move buoys and flow through wind turbines to generate electricity. The challenge is extracting power from the moving water on a large scale.

To view a 3-minute video, Energy 101: Marine and Hydrokinetic Energy, visit the following U.S. Department of Energy website: https://www.youtube.com/watch?v=ir4XngHcohM.

Solar Power

No other energy source, renewable or not, has the potential of solar power. According to Dave Llorens of Solar Power Rocks, the sun produces in 1 hour enough energy to power 2,880 trillion light bulbs. He compares this to giving every man, woman, and child on the planet a light bulb that will burn for their entire lifetime. That is a lot of energy! And it is free. Another way to look at it is that every 50 minutes the sun provides enough energy to satisfy the annual power needs of the planet. Given its abundance and cost, why does not everyone already have a solar thingamajig that powers all their needs?

Solar energy, in its natural state, is inefficient for humans to use. It has been very challenging for the solar industry to develop effective technologies to capture, convert, and store the sun's energy. The best solar panels today can only covert about 35% of sunlight into energy, with 15% being about average. One key problem is that current solar cells require expensive, high-purity materials. Plus sunlight is often intermittent in many geographical areas, so energy needs to be stored. Large banks of batteries are one method of storage but can be expensive and require space. Even when energy is produced, there is no infrastructure equipped to deal with it. Current power grids cannot store the energy. Besides, large solar facilities need to be located where there is lots of sunshine, often in remote western desert areas that are distant from energy grid infrastructures. Given these constraints, solar power technologies remain expensive for many consumers even though the cost continues to decline. However, purchasers can reduce their electricity bills, potentially receive energy credits on their taxes, or possibly have the option to sell back extra energy they produce to the power companies. Realistically, though, until it becomes comparable to existing energy systems in terms of cost and ease-of-use, many consumers will not be interested. To view a 3-minute video on solar cells, visit http://sciencenetlinks.com/videos/solar-cells/.

Solar thermal energy

Solar energy challenges have not stopped researchers from moving forward with solar technologies. In fact, their ultimate goal is to make a solar-based infrastructure that not only rivals those of fossil fuels but may also ultimately replace them. Some are already in existence. California contains the longest operating solar thermal power facility in the world. The Solar Energy Generating System (SEGS) is located in the Mojave Desert and has operated since 1984. Generating electricity with solar energy is similar to generating it with fossil fuels. Instead of creating steam for the turbines with fossil fuels, the sun's rays are used instead. This type of solar system is called concentrating solar power, since it uses the sun's rays to create enough heat to generate electricity. At SEGS, parabolic troughs are used to focus the sun's ray on a receiver pipe (see Figure 4) which can tilt to follow the sun's path. The parabolic shape allows the sun's rays to focus between 30 and 100 times its normal intensity on the receiver pipe, which can reach temperatures higher than 750°F. Special fluid is heated as it circulates through the pipes. It then returns to heat exchangers at a central location to transfer its heat to water, producing high-pressure steam. This steam runs the turbine and generator to create electricity.

Parabolic troughs are not the only equipment used to run solar thermal power plants. Solar dishes and solar power towers are also used to concentrate solar energy

Figure 4. Solar electric power plant parabolic mirrors concentrating sunlight on a receiver pipe.

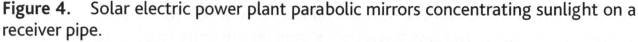

to a focal point. A solar dish can increase fluid temperature to higher than 1,380°F, although it does not produce steam. Instead, it converts heat to mechanical power by compressing the fluid when it is cold and then heating it. The fluid then expands and flows through a turbine. Solar dishes are preferred in remote locations. Solar power towers are promising technology for large-scale production. Using hundreds or thousands of flat sun-tracking mirrors called heliostats, the sun's energy can be concentrated up to 1,500 times directly to a single tower-mounted heat exchanger, or receiver. Energy loss is minimized since the heat exchanger receives the solar energy directly and it does not need to be transferred by a fluid as with the parabolic troughs.

Photovoltaic cells

Solar cells, also known as photovoltaic (PV) cells, directly convert sunlight into electricity at the atomic level. Some materials, including silicon, display a property known as the photoelectric effect. This is when solar energy, or other light sources, hit the material and cause it to absorb photons of light and eject electrons. When these free electrons are captured, they produce an electric current. A single PV cell does not produce much power. Multiple cells can be connected to produce a module (also known as panel), and multiple modules can be connected to produce an array. The larger the array, the more electricity produced. PV cells are found in many items, including calculators, refrigerators, traffic signs, and satellites. During the day, PV systems use sunlight to charge the batteries. Because of this self-reliance, they are particularly useful in remote areas where other power sources are scarce. For a fun slideshow that includes many uses of PV systems, visit the following website: http://www.slideshare.net/Sunworks/the-many-uses-of-pv-systems.

Solar panels can be put on your house or used in large solar farms. Either option produces electricity in the same way. Although solar farms put panels on motorized towers that can follow the path of the sun, the ones on your house are probably just on the south side of your roof (can you guess why?). The electricity generated in both scenarios is sold to the power companies. It is not all used directly by the homeowner. Any energy you do not use immediately goes into the grid. Solar panels generally do not produce the most power at the time you need it. And the size and cost of storage batteries are quite large. You still receive and pay for your electricity from the power company, but they buy your solar power from you.

There are many benefits of using PV solar cells for producing electricity. According to the Conserve Energy Future website, they are a clean energy source that gives off no contamination or air pollution. The sun is completely sustainable and renewable (at least for the next 6.5 billion years or so). Residential solar power systems are noiseless and easy to clean. PV systems are quite reliable and last 20–25 years with virtually no maintenance. They are more expensive to produce than other energy sources due to the cost of manufacturing PV systems and conversion efficiencies of some of the equipment. However, prices have been declining and may decline further as more people use solar energy. Most important, the environmental impacts are minor when compared to many other power sources. In 2015 solar power passed the 1% threshold globally. Italy, Germany, and Greece used PV systems to supply more than 7% of their electricity needs. China is the fastest growing market followed by Japan and the United States. However, Europe is still the largest user. Worldwide capacity is approximately 178 GW. Each GW is the equivalent of 33 coal-fired power stations.

Geothermal Energy

The United States is a world leader in using geothermal power. A large majority of that capacity (80%) is located in California. Geothermal power provides almost 7% of the state's electrical need. Geothermal energy is produced by heat from the Earth. Sources of geothermal heat include shallow hot water, hot springs, geysers, and deep molten rock (magma) at the center of the Earth. Very deep wells can be drilled down to underground reservoirs to bring steam and hot water to the surface. Geothermal energy is a clean, renewable, stable source of power and can reduce the United States' dependence on foreign oil. Geothermal energy only comprises 2% of renewable sources, but it has great potential.

According to the Union of Concerned Scientists, the surface of the Earth (to about 33,000 feet) holds about 50,000 times more energy than the oil and natural gas resources combined. Heat from the magma layer heat is continuously produced from decaying radioactive material. Highest underground temperatures are found in regions with active or young volcanoes. The Earth's crust is thin in these areas, and heat can easily get through. Some western states contain geographic hot spots as shown in Figure 5.

There are three different types of geothermal power plants to extract hydro-thermal fluids from the earth and convert them to electricity. The first is called a *dry steam plant*. They were the first ones used for geothermal power. These plants primarily collect steam and route it to a turbine. The energy from the steam causes the turbine blades to turn, engage the generator, and produce electricity. This is the same process you learned about for both hydroelectric and solar power sources. The process to produce electricity is the same. It is the source of energy needed to activate the turbine/generator that is different. *Flash steam power plants* are most commonly used today. High-temperature fluid (greater than 360°F) is pumped into a tank under high pressure. The tank is maintained at a lower pressure, so when the fluid enters it, the high-temperature fluid immediately vaporizes or "flashes." This vapor causes the turbine to turn, which drives the generator. In the *binary cycle power plant*, the fluids or steam never have direct contact with the turbine/generator units. Instead, more moderately heated geothermal fluid (below 400°F) travels through a heat exchanger and heats a second (or binary) liquid in a closed loop. The binary fluid has a lower boiling point and is heated to the point where it flashes and drives the turbine.

Geothermal energy production is convenient in that it can be used for supplying electricity directly to residential and commercial consumers. It is less expensive than fossil fuel sources by up to 80%. Hot water can be used to heat homes and build-ings, keep greenhouses warm enough to grown plants, and melt ice on sidewalks and roads. Geothermal energy is also environment friendly. In a study done by Bloomfield,

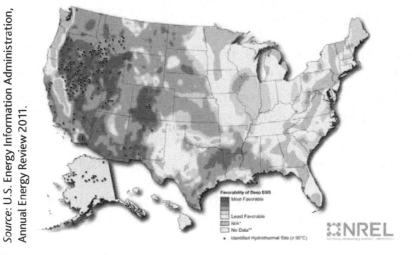

Source: U.S. Energy Information Administration, Annual Energy Review 2011.

Figure 5.   U.S. geothermal resources.

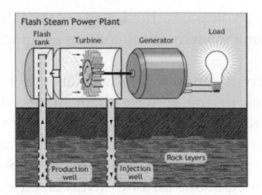

Figure 6.   A flash steam geothermal power plant. A dry steam plant is similar but without the flash tank. Instead, steam is routed directly into the turbine.

Moore and Neilson (2003), they found that using geothermal energy significantly reduced the amount of greenhouse gases that were released into the atmosphere. They also concluded that this was a viable way to meet the increasing power demands of the United States.

Wind Power

Wind power is one of the first sources of energy used by humans and has an extensive history. As early as 200 BC, the Chinese were using simple windmills to pump water. By the eleventh century, Middle Easterners were using windmills to grind grain. American colonists used windmills to cut wood at sawmills. It was a power source used extensively throughout the world until electric power supplanted it in many countries. While it was never entirely replaced, the oil shortages of the 1970s created a renewed interest in wind power. Since then, fluctuating oil prices and increased concern for the environment have propelled wind power research forward. Part of its appeal is that it is a free, renewable energy source.

> Unlike conventional power plants, wind plants emit no air pollutants or greenhouse gases. According to the U.S. Department of Energy, in 1990, California's wind power plants offset the emission of more than 2.5 billion pounds of carbon dioxide, and 15 million pounds of other pollutants that would have otherwise been produced. It would take a forest of 90 million to 175 million trees to provide the same air quality. (Wind Energy Development, para. 9)

Between 2011 and 2014 wind generation of electricity in the United States has increased by 50%. Currently, 40 states are operating utility-scale wind energy projects. According to the American Wind Energy Association, these projects serve the equivalent of 20 million American homes. The top three wind power capacity states are

© bon9/Shutterstock.com

Figure 7. A field of horizontal axis wind turbines generating green energy in Palmdale, California.

Texas (17,711 MW), Iowa (6,364 MW), and California (5,662 MW). By the end of 2015, there were 88,000 wind-related jobs in this country.

Wind power, itself, is kinetic energy. Wind turbines are used to convert this kinetic energy into mechanical power or electricity. There are three types of wind power. The first, *utility-scale wind*, includes turbines larger than 100 kW. They generate electricity for utility companies and send their power to the grid. *Distributed* or *small wind* includes turbines of 100 kW or smaller. These supply power directly to a home, farm, or business. *Offshore wind* includes turbines located in bodies of water. These have not been used yet in the United States. Wind power generators, or turbines, come in a variety of designs and power outputs. However, there are basic aspects of every model that work the same.

Turbines contain an anemometer that gauges the wind and sends the data to an internal controller. The anemometer works with a brake and controller to reduce the blade speed if the wind blows too hard (usually more than 55 miles per hour). Wind energy creates electricity by first blowing past the turbine blades, which causes them to rotate. Most turbines have three blades, but some contain two. The internal shaft begins to spin. It is connected to a gearbox, which can change rotation speed to ensure that the shaft is spinning fast enough to produce electricity (1,000–1,800 rotations per minute). Turbines are generally up to 325 feet tall and consist of a steel cylindrical tower. Wind generators also come in vertical axis turbine designs. They are not as common and use cups instead of blades to catch the wind and rotate the shaft. Increasingly, though, alternative energy companies are using a solar-wind hybrid design. Solar power can be used when the wind is not blowing; and wind power can be used when the sun is not shining. This gives consumers the best of both options.

There are some downsides to using wind power systems. Wind is intermittent and does not always blow when electricity is needed. Often wind power systems require higher initial investment than traditional fossil-fuel plants. There has been concern over the noise of the rotor blades and the fact that birds and bats having been killed by flying into them. However, according to the U.S. Department of the Interior Bureau of Land Management Utah, most of these problems have been addressed through new technologies and proper placement of wind farms.

Nanotechnology and Energy

You have been learning throughout this book the role that nanotechnology is playing in many fields and the potential it still holds as scientists learn more about it. Energy use is no exception. In fact, nanotechnology research is being done in a number of energy fronts. Scientists at Rice University have developed a solar-steam method that uses nanoparticles to convert solar energy directly into steam. The particles are first submerged in water and then exposed to sunlight. They heat up and vaporize immediately, causing steam. Steam is widely used for industrial purposes and producing electricity. While the technology has not yet reached the state of powering turbine/generator systems, it can be used for sanitizing tools and instruments as well as purifying water in developing countries.

Incandescent lightbulbs are being phased out worldwide due to their inefficiencies. Although many people preferred their warm glow over newer LED bulb models, more than 95% of their energy is wasted as heat. Now nanotechnology may allow them to make a comeback. Photonic crystal structures, which can manipulate beams of light, would surround the lightbulb's filament and reflect the waste heat back to be reabsorbed and reemitted as light. Nanotube sheets in thermocells could be wrapped around sources of heat, such as hot water or exhaust pipes, to also capture wasted heat and convert it to electricity. These thermocells are the size of button cell batteries. The temperature differential between a hot pipe and air produces an electrochemical reaction between the nanotube sheets which then continuously generates electricity.

Climate Change

What do you think about when you hear these words? Do you get concerned and think of melting glaciers, dust-bowl landscapes, and hotter summers? Or do you get irritated and think of pseudoscience, belligerent politicians, and the last cold winter?

You have been exposed to moral issues associated with some of the technologies found throughout this book. You have read the words of experts and debated the merits of their viewpoints; you may have felt strongly one way or the other about some of these topics. However, none have been debated so rigorously, for so long, by so many people as climate change. Figure 8 shows people marching through New York City to encourage world leaders to address climate change. It is the rare person who does not hold a strong opinion about this matter. Interestingly, whatever that opinion may be, chances are that no amount of scientific data, discussion, or new information will change it. If you hope to look to the experts for answers, you will probably be disappointed. They are as passionate about their own beliefs, if not more so, than the average citizen.

Several books were recently published on climate change. One was by Dr. Joseph Romm (2016), who holds a Ph.D. in physics from MIT and served as acting assistant secretary of energy in 1997. Within the first three pages of the book's preface, he states:

> Climate change is now an existential issue for humanity. Serious climate impacts have already been observed on every continent. Far more dangerous climate impacts are inevitable without much stronger action than the world is currently pursuing … The central purpose of the resulting United Nations Intergovernmental Panel on Climate Change (IPCC) was to provide the best science to policymakers. In the ensuing years, the science has gotten stronger, in large part because observations around the world confirmed the vast majority of the early predictions made by climate scientists. (pp. xiii–xv)

Figure 8.   Peoples Climate March through New York City in 2014.

The second book was published by the Institute of Public Affairs and edited by Alan Moran (2015). It contains 22 chapters written by professors and other experts on climate change. According to Moran in the six-page introduction, the book will do the following:

> Explodes the myth that 97 per cent of scientists regard human induced global warming as both likely and serious ... shows that any human effect on climate is trivial compared to natural variation ... reviews the farce of the 2009 Copenhagen conference ... notes the scandalous attribution of Nobel Prize status to all involved in the IPCC ... illustrates the trivial increase in global temperature that has occurred over the past century (with no increase in the past eighteen years) ... disinters the graveyards of failed forecasts by climate doomers ... [and] considers the warminists monumental failures are finally denting the faith in them by the commentariat and politicians. (pp. 1–6)

Clearly the debate surrounding climate change will not be answered any time soon. Instead, you will have to make your own informed decision on what you believe. The only caveat, though, is that you should keep an open mind, conduct your own research, and be open to considering new possibilities based on what you learn.

Greenhouse Effect

You need to first understand the greenhouse effect to understand the Earth's climate. An example of the greenhouse effect is found in Figure 9. During the day the sun's rays warm the earth, and the heat is released back into the air at night. But some of this heat is trapped in Earth's lower atmosphere by greenhouse gases, which include carbon dioxide, methane, nitrous oxide, fluorinated gases, and water vapor. Most greenhouse gases are stable and do not react to temperature or air pressure changes. However, water vapor is the main greenhouse gas and responds to atmospheric changes by evaporating or condensing into rain or snow. Whether cloud cover changes contribute to warming or cooling of the earth is still an unknown question. While clouds do hold in some of the heat, their bright white tops can reflect sunlight as well.

Greenhouse gases have a great influence on the Earth's temperature and keep it an average of between 57°F and 59°F. Carbon dioxide, methane, and nitrous oxide are emitted into the atmosphere through natural and man-made activities. It is the increase in these gases since the industrial revolution that is the major concern to many people. Burning fossil fuels is a major contributor. If the greenhouse effect gets too strong, the Earth gets warmer. Too much carbon dioxide also makes the ocean water more acidic.

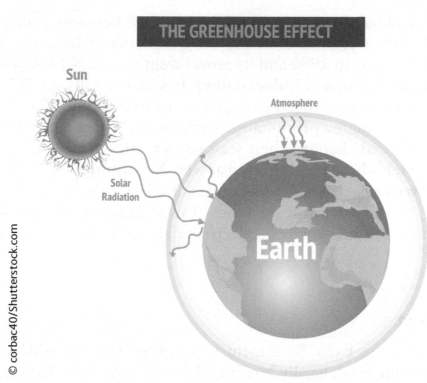

© corbac40/Shutterstock.com

Figure 9. Greenhouse effect.

Fluorinated gases are manmade and emitted into the atmosphere through manufacturing processes, aerosol propellants, solvents, and air conditioning systems (among others). Between 1990 and 2014, the fluorinated gases increased about 77%, primarily by hydrofluorocarbons that are used widely to replace ozone-depleting substances. Emissions of perfluorocarbons and sulfur hexafluoride have declined due to efforts by the aluminum and electricity industries.

One thing is clear: the Earth's climate is a very complex, interactive system that is influenced by numerous chemical, physical, and atmospheric changes over a vast geographic area. Understanding its complete model or making predictions about its future is a major challenge. However, there are numerous projects focused on answering important climate questions. It is important to gather as much valid data as we can and have a rational dialogue about it. After all, there is only one Earth, and we are all in this together. The following section is not comprehensive but includes developments in several areas.

Intergovernmental Panel on Climate Change

In 1988 the Intergovernmental Panel on Climate Change (IPCC) was created to assess the science related to climate change and inform policymakers about its impact and possible future risks. According to IPCC, it "embodies a unique opportunity to provide rigorous and balanced scientific information to decision-makers because of its scientific and intergovernmental nature" (IPCC Factsheet, n.d., p. 1). It published its first assessment report in 1990 and its most recent one in 2014 (Climate Change 2014: Impacts, Adaptation, and Vulnerability). In Climate Change 2014, the IPCC notes that climate change has been strongest on natural systems. Hydrological systems are affected by changing precipitation patterns or melting ice and snow. Marine species have moved and changed their seasonal activities as well as migration patterns. Some crop yields have been affected as well. Key future risks include disrupted livelihoods in low-lying coastal areas and small islands due to storm surges and coastal flooding caused by rising sea levels. This includes associated health problems, infrastructure breakdowns (e.g., electricity and food services), and ecosystem disruption.

Paris Agreement

In December 2015, the Paris Agreement was signed. This included 195 countries adopting the first universal, legally binding global climate deal. The agreement will start in 2020. According to the European Commission, the governments agreed, in part, to the following:

* A long-term goal of keeping the increase in global average temperature to *well below 2°C* above pre-industrial levels
* To aim to limit the increase to *1.5°C*, since this would significantly reduce risks and the impacts of climate change
* On the need for *global emissions to peak as soon as possible*, recognizing that this will take longer for developing countries
* To undertake *rapid reductions thereafter* in accordance with the best available science (European Commission, n.d. para. 3).

In addition, they will meet every 5 years, create a transparent accountability system, and encourage other countries to support the agreement.

An earlier international agreement, the Kyoto Protocol, was developed to reduce industrialized countries' emissions by 5.2% compared to 1990. It was linked to the United Nations Framework Convention on climate change and adopted in Japan in 1997. It entered into force in 2005. Japan, New Zealand, and Russia were early participants. Canada withdrew in 2012 after the first commitment period, and the United States never ratified it. The United Nations reported that this world's first emissions reduction treaty exceeded its goal and reduced emissions by over 20%. The second commitment period was agreed upon in 2012, and over 30 countries are committed.

U.S. National Aeronautics and Space Administration

U.S. National Aeronautics and Space Administration (NASA) currently has more than a dozen scientific satellites studying all aspects of the planet. This includes the land, oceans, and atmosphere. This satellite system can determine solar activity, sea levels, and changes in the air. Satellite photos can also provide time-series comparisons of geological structures such as ice sheets and snow packs. While NASA was created in 1958 to study space, it now has a large role in Earth science given its technological and scientific capabilities. With approximately 30 years of satellite data, the IPCC among others, turns to NASA when it needs information.

Climate scientists study the Earth as an integrated system and look at the following three categories. Notice how many different variables are included in each group. It is fairly easy to measure each variable independently. It is the interaction among them that challenge researchers:

1. *Forcings*: these are the initial drivers of climate and contain solar irradiance (including the sun's 11-year spot cycle); greenhouse gas emissions; and small borne particles such as dust, smoke, and soot.
2. *Climate feedbacks*: these can be positive (increases warming) or negative (reduces warming) and include cloud cover, precipitation levels, tree growth, and ice albedo or reflection (white ice reflects and dark ocean surfaces absorb heat).
3. *Climate tipping points*: these are events that cause the climate changes abruptly in areas such as ocean circulation, ice loss (and rising sea), and rapid release of methane (frozen deposits may thaw and release into atmosphere).

Initially scientists called the changes they were seeing as "global warming." But as years of data were analyzed, it became clear that while this may have been true in some areas, it was not true universally. It also became apparent how difficult it was to create valid climate models on a global scale. Eventually "global warming" was replaced with the term "climate change." Even so, NASA researchers found 2015 to be the warmest year on record since recordkeeping began in 1880. They created a 30-second graphic of the planet's temperatures from 1880 to 2015 that can be located at http://climate. nasa.gov/climate_resources/139/. They also found that the Antarctic has been accumulating enough snow and ice to outweigh increased losses from glacier melt. This is in contrast to some other studies, including one from the IPCC. This highlights the difficulties of creating consensus on something as complicated as climate change. However, there are numerous sites where you can do your own exploring. You can start at NASA's Vital Signs of the Planet site at http://climate.nasa.gov. Visit the global climate dashboard at http://climate.gov. Or explore the United States through the National Centers for Environmental Information. They are responsible for monitoring, assessing, and preserving climate and weather data. They partner with climate.

gov, weather.gov, drought.gov, and globalchange.gov, to bring you the most up-to-date maps and information. You can get started at http://www.ncdc.noaa.gov/.

The Science behind the Technology

Throughout this chapter you have learned about the science of energy and how it is used to create electricity. There are many forms, including PV cells, hydroelectric power, and nuclear fission. Regardless of the energy source, its purpose is to drive a turbine/generator system. This is what actually creates the electricity. Power sources are just the means to get there. But for all the talk about energy, what is it really? Simply put, energy is the ability to do work. If the energy is stored it is known as potential. If it is energy that is moving then it is known as kinetic. Energy can come in many forms such as chemical, electrical, thermal (heat), radiant (light), mechanical, and nuclear. Let us summarize each:

Chemical energy. Chemical energy results from a chemical reaction and is a type of potential energy that is stored within the bonds of atoms and molecules in a substance. It is released in a chemical reaction which often produces heat.

Electrical energy. Electrical energy is made available by the flow of an electric charge through a conductor. It is a form of kinetic energy since the electric charges are moving.

Thermal (heat) energy. Thermal energy is generated and measured by heat. It is a form of kinetic energy since it is moving.

Radiant (light) energy. Radiant energy travels in a wave motion (electromagnetic waves). It is a form of kinetic energy since the particles are moving as they carry light, heat, and radiation from one source to another.

Mechanical energy. Mechanical energy is the energy that is possessed by an object due to its position and motion. It is the energy acquired by the objects upon which work is done. This type of energy can be either kinetic or potential.

Nuclear energy. Nuclear energy is the energy in a nucleus of an atom (which holds neutrons and protons). There are two ways to create nuclear energy, fission and fusion. Fission occurs when atoms are split into smaller atoms, releasing energy. This is what is used by nuclear power plants to produce electricity. Fusion occurs when atoms are combined or fused to form a larger atom, releasing energy.

Measuring energy is not a straightforward process, because different sources use different measurements. One common measuring block is known as a British Thermal Unit (BTU). This is the amount of heat energy it takes to raise the temperature of one pound of water by 1°F when at sea level. You measure your air conditioner or heater with BTUs. Another common measuring block is known as a joule. A joule is equal to 1,000 BTUs. A watt is the energy consumption of one joule per second.

A kilowatt (kW), used to measure wind power, is equal to about 3.6 megajoules, which are 1,000,000 joules or 1,000 watts. The website rapidtables.com is a good source for easily converting one source to another.

Career Connections

There are numerous professions that cross many of the topics found in this chapter. Geologist is a primary one. According to geology.com, demand and salaries follow the price of commodities such as fuels, metals, and construction materials. Environmental concerns have also created more jobs in this sector as well. The average salary of geologists in the petroleum industry has been increasing steadily since 2000 and now average $83,000 a year. Geoscience positions require at least a bachelor's degree. Enrollment in these programs has been steady but is not expected to meet demand. According to Bureau of Labor Statistics, the job outlook for geoscientists is projected to grow 10% from 2014 to 2024 with a median pay of $89,700 a year. One holdback to meeting demand is that many students who enroll in these programs do not graduate. Part of the problem is that these programs often require "challenging" classes such as calculus, physics, and chemistry. You may even think about these subjects as difficult or unattainable. It is time to stop thinking that way. It closes down a whole area of study before you even get started. Most often science and math classes appear to be hard because they are new and different than what you may be used to doing in class. But if you focus on them, they are interesting, applicable, and the pathway to really cool, high-paying jobs.

According to the American Council on Renewable Energy, you can expect to see increasing job availability in renewable energy fields as well. Renewable energy now accounts for nearly 40% of all domestic power capacity. The array of technologies will need people to design and maintain them. Private sector investment in the U.S. clean energy sector surpassed $100 billion several years ago which stimulated economic development while supporting hundreds of thousands of jobs, which is expected to continue.

The field of nuclear science needs more qualified people both generally and in important subfields such as nuclear chemistry. A qualified workforce in these areas is necessary to remain competitive worldwide and contributes to a healthy economy. It can also contribute to a healthy salary. Nuclear physicists and astronomers earn an average salary of almost $111,000 a year. Many positions require graduate education, but if you are a student who is interested in these fields, complete a degrees in physics, chemistry, or related area. It is also very important to get involved in hands-on research during your undergraduate education. Talk directly to your professors and advisors as most universities have opportunities for student research. This is also critical to networking with professionals who can help you find jobs, graduate assistantships, and write letters of recommendation.

Modular Activities

Discussions

❖ Post an original discussion in the online discussion board on the following topic: Give an example of your own of how energy technology improves our lives as well as an example of how energy technology has negative effects (200 words minimum). Next, comment on the post(s) of a minimum of one other student in a thoughtful and academic way that enhances the conversation. See rubric for grading and assessment measures.

❖ Post an original discussion about why you think people are so passionate about climate change in comparison to other technological issues. What material do you use to inform your own decision? Do you think we will ever come to a consensus or at least a less emotional debate—or should people remain as passionate as they are? How do you think we should move forward on finding the answers in a way that most people would agree on? Next, comment on the post(s) of a minimum of one other student in a thoughtful and academic way that enhances the conversation.

❖ El Nino has been in the news a lot lately about its effect on the global climate. Research it and post a paragraph about what you found. First review the discussion to make sure you are adding original information and not repeating something. You can also make comments about El Nina as well. Next, comment on the post(s) of a minimum of one other student in a thoughtful and academic way that enhances the conversation. In the end, the class should have a good understanding of this phenomenon.

Tests

❖ Online graded quiz on overall chapter content, written in multiple choice format. When submitted, we recommend giving the correct answer along with the page number in which it is found for questions students did not answer correctly.

❖ Take the home energy IQ test found at http://energy.gov/articles/quiz-test-your-home-energy-iq and submit your final results to your instructor. Include a one-page self-assessment on what you learned from the test. For example, were you surprised by the findings, plan to make changes to your lifestyle, and so on. Teachers may also want to create a discussion about this test. Have students share some of their results and talk about what it means to them.

Research

❖ Hydraulic fracturing: Write a three-page report explaining what hydraulic fracturing is and how it works.

❖ Answer four (4) of the following questions for one specific energy source (e.g., solar, coal, and oil). You need to provide a reference of where you obtained the information. Remember to specify units of measurement used where applicable.

1. How much of this energy type does the United States produce?
2. How much of this energy type is produced globally?
3. Which country is the largest producer of this type of energy?
4. How much of this energy does the United States consume?
5. How much of this energy is consumed globally?
6. Give a brief description of this type of energy and how it is created

❖ The following site contains a case study titled Energy: The U.S. in Crisis? There are elements that can be done individually or in a group setting. http://sciencenetlinks .com/lessons/energy-the-us-in-crisis/

❖ Research one of the following types of energy and write a one-page summary of how it works: chemical, electrical, thermal (heat), radiant (light), mechanical, or nuclear. You must cite your sources. Images are appropriate but must not be the majority of your content. You will be assessed on the quality, depth, accuracy, and professionalism of your summary.

❖ The International Energy Agency has factsheets available from 2006 to 2015 at http://www.worldenergyoutlook.org/factsheets/. Select a factsheet from 2006 or 2007 and review the predictions that are made, since some should already have happened if their predictions are correct. Research the predictions to see if they did happen or explain what really happened or why. Or, select a factsheet from 2014 or 2015 and read through the predictions. Do additional research to determine why the predictions were made and if you believe they are accurate or not. Support your comments.

❖ One challenge of researching energy is that new data are posted often. Select one of the renewable energy sources and find the most up-to-date information on research, users, statistics, and benefits/disadvantages.

❖ The National Oceanic and Atmospheric website has a global climate dashboard located at the following site: https://www.climate.gov/#education/teaching Resources. It includes three dashboard options (climate change, climate variability, and climate projects) along with nine indicators (e.g., temperature, arctic sea ice, and snow). Review the dashboard and develop your own theory about the future of the Earth's climate. Integrate this information, along with other research, and write a paper about the state of the climate, future possibilities, and whether humans should be worried.

❖ Fluorinated gases are man-made greenhouse gases. Research what they are, who uses them, and their impact, if any, on the atmosphere and environment.

Design and/or Build Projects

❖ Create a solar cooker. It must heat to a minimum of two times the outdoor temperature in a 4-hour period (so, if it is 40° outside it must heat to 80° in a 4-hour period. This can be demonstrated using an oven thermometer and taking pictures of the before and after temperatures). Use the Web to find ideas on sites such as at http://solarcooking.org/plans/.

❖ Construct a model of a Trombe wall. You do not need to use glass (plastic wrap will do) or concrete/wood (cardboard will do). You are just trying to prototype the concept as accurately as possible. Web search to find samples of Trombe walls such as the following:

  ➤ http://www.nrel.gov/docs/fy04osti/36277.pdf or

  ➤ http://www.energysavers.gov/your_home/designing_remodeling/index.cfm/mytopic=10300 or

  ➤ http://www.ecowho.com/articles/17/What_is_a_Trombe_Wall_and_how_can_you_use_one?.html

  ➤ You will be assessed on the accuracy and quality of your model.

❖ Review the following web link and create a potential project in your own neighborhood using real structures. Create a plan using pictures, drawings, plant types, and examples of other street canyons in your city. http://sciencenetlinks.com/science-news/science-updates/urban-greening/.

❖ On the International Energy Agency, World Energy Outlook, there are two Excel databases: WEO 2015 Electricity access database and WEO 2015 Biomass database. Select one of these databases and review all worksheets. Create a presentation on what the data are telling you. Use pivot tables, charts, and graphs to analyze the data. Do additional research to support your ideas and add commentary to your project. Website is located at: http://www.worldenergyoutlook.org/resources/energydevelopment/energyaccessdatabase/.

Assessment Tasks

❖ Rising light pollution, caused by artificial lights which raise the night sky luminance, has been increasing globally. Review one of the following sources that discuss the issue and write a critical assessment of the site:

  ➤ http://www.bbc.com/future/story/20160617-what-rising-light-pollution-means-for-our-health

  ➤ http://advances.sciencemag.org/content/2/6/e1600377.full

  ➤ http://www.nature.nps.gov/night/light.cfm

  ➤ https://www.nps.gov/samo/learn/management/light-pollution.htm

Your critical assessment must include the following: (1) site you chose to review, (2) critical assessment of site content as it relates to educating readers about light pollution (200 words minimum), (3) critical assessment of site as it relates structure, appearance, and usability (200 words minimum).

Terms

Climate change—Changes in climate patterns due to increased greenhouse gases in the atmosphere, especially carbon dioxide.

Fission—The splitting of nuclei (such as uranium) into smaller parts to create energy.

Fusion—The combining of more than one nuclei to create or absorb energy.

Geothermal—Energy from the heat of the earth either at the surface or miles below.

Greenhouse effect—Process by which sun's energy is trapped in Earth's lower atmosphere.

Hydroelectric—Energy created by falling or flowing water.

Photovoltaic—Solar cells that convert sunlight directly into electricity.

References

Bloomfield, K. K., Moore, J. N., & Neilson, Jr., R. M. (2003). *Geothermal energy reduces greenhouse gasses*. Retrieved from https://geothermal.org/PDFs/Articles/greenhousegases.pdf

Hydroelectric power: How it works (n.d.). *The USGS water science school*. Retrieved from http://water.usgs.gov/edu/hyhowworks.html

Intergovernmental Panel on Climate Change (IPCC). *Climate change 2014: Impacts, adaptation, and vulnerability. Summary for policymakers*. Retrieved from http://www.ipcc.ch/pdf/assessment-report/ar5/wg2/ar5_wgII_spm_en.pdf

IPCC Factsheet (n.d.): What is the IPCC? Retrieved from http://www.ipcc.ch/news_and_events/docs/factsheets/FS_what_ipcc.pdf

Moran, A. (Ed.) (2015). Climate change: The facts. Woodsville, NH: Stockade Books.

Romm, J. (2016). *Climate change: What everyone needs to know*. New York, NY: Oxford University Press.

Further Reading

Alternative Energy. (n.d). *Solar energy*. Retrieved from http://www.altenergy.org/renewables/solar.html

American Council on Renewable Energy. (2014). *The outlook of renewable energy in America*. Retrieved from http://acore.org/files/pdfs/ACORE_Outlook_for_RE_2014.pdf

American Wind Energy Association. (n.d.). *Wind 101: The basics of wind energy*. Retrieved from http://www.awea.org/Resources/Content.aspx?ItemNumber=900

American Wind Energy Association. (n.d.). *Wind energy facts at a glance*. Retrieved from http://www.awea.org/Resources/Content.aspx?ItemNumber=5059

Bureau of Labor Statistics. (n.d). *Occupational outlook handbook*. Retrieved from http://www.bls.gov/ooh/a-z-index.htm#G

CBS News. (n.d). *World's worst nuclear accidents*. Retrieved from http://www.cbsnews.com/pictures/worlds-worst-nuclear-accidents/

Chandler, D. L. (2016, January 11). A nanophotonic comeback for incandescent bulbs? *MIT News Release*. Retrieved from http://news.mit.edu/2016/nanophotonic-incandescent-light-bulbs-0111

Charriau, P., & Desbrosses, N. (2016, June). *2015 Global energy trends*. Retrieved from http://www.enerdata.net/enerdatauk/press-and-publication/publications/peak-energy-demand-co2-emissions-2016-world-energy.php\

Energy in Depth. (n.d). *Just the facts*. Retrieved from http://energyindepth.org/just-the-facts/

Energy.gov. (2013, August 14). *Hydropower technology basics*. Retrieved from http://energy.gov/eere/energybasics/articles/hydropower-technology-basics

Energy.gov. (n.d). *Electricity generation*. Retrieved from http://energy.gov/eere/geothermal/electricity-generation

Energy.gov. (n.d). *Geothermal basics*. Retrieved from http://energy.gov/eere/geothermal/geothermal-basics

Energy.gov. (n.d). *Marine and hydrokinetic energy research & development*. Retrieved from http://energy.gov/eere/water/marine-and-hydrokinetic-energy-research-development

Energy.gov. (n.d). *Reaching for the horizon: The 2015 long range plan for nuclear science*. Retrieved from http://science.energy.gov/~/media/np/nsac/pdf/2015LRP/2015_LRPNS_091815.pdf

EPA. (n.d). *Overview of greenhouse gases*. Retrieved from https://www3.epa.gov/climatechange/ghgemissions/gases/fgases.html

European Commission. (n.d). *Paris agreement*. Retrieved from http://ec.europa.eu/clima/policies/international/negotiations/paris/index_en.htm

Geology.com. (n.d). *Geologist salaries and the economic slowdown*. Retrieved from http://geology.com/articles/geologist-salary.shtml

Holdren, J. P. (2015, September 10). The current state of energy technology. *White House*. Retrieved from https://www.whitehouse.gov/blog/2015/09/10/current-state-energy-technology

How a solar farm works. (n.d). *Convergence energy*. Retrieved from http://convergence-energy.com/uncategorized/how-a-solar-farm-works/

How stuff Works. (n.d). *How oil refining works*. Retrieved from http://science.howstuffworks.com/environmental/energy/oil-refining.htm

Hydroelectric power. (n.d.). *Alternative energy*. Retrieved from http://www.altenergy.org/renewables/hydroelectric.html

Hydropower. (n.d.). *U.S. Energy Information Administration*. Retrieved from http://www.eia.gov/kids/energy.cfm?page=hydropower_home-basics Energy Information Administration

Jodhka, S. S., & Prakash, A. (2011, December). The Indian middle class: Emerging cultures of politics and economics. *KAS International Reports*. Retrieved from http://www.kas.de/wf/doc/kas_29624-544-2-30.pdf

Knier, G. (n.d.). How to photovoltaics work? *NASA Science, Science News*. Retrieved from http://science.nasa.gov/science-news/science-at-nasa/2002/solarcells/

Llorens, D. (n.d.). How much energy does the sun produce? (and other fun facts). *Solar Power Rocks.* Retrieved from https://solarpowerrocks.com/solar-basics/3-reasons-the-sun/

Maehlum, M. A. (2013, May 3). Nuclear energy pros and cons. *Energy Informative.* Retrieved from http://energyinformative.org/nuclear-energy-pros-and-cons/

Markert, L. R., & Backer, P. R. (2010). *Contemporary technology: Innovations, issues, and perspectives* (5th ed). Tinley Park, IL: The Goodheart-Wilcox.

Martinez-Uria, R., O'Connor, P. W., & Johnson, M. M. (2015, April). 2014 hydropower market report. *U.S. Department of Energy.* Retrieved from http://www.energy.gov/sites/prod/files/2015/04/f22/2014%20Hydropower%20Market%20Report_20150424.pdf

Nanotechnology and Energy. (n.d.). *UnderstandingNano.com.* Retrieved from http://www.understandingnano.com/nanotechnology-energy.html

NASA. (n.d.). *Climate kids. What is the greenhouse effect?* Retrieved from http://climatekids.nasa.gov/greenhouse-effect/

NASA. (n.d.). *Mass gains of Antarctic ice sheet greater than losses.* Retrieved from http://www.nasa.gov/feature/goddard/nasa-study-mass-gains-of-antarctic-ice-sheet-greater-than-losses

NASA. (n.d.). *Taking a global perspective on earth's climate.* Retrieved from http://climate.nasa.gov/nasa_role/

NASA. (n.d.). *The study of earth as an integrated system.* Retrieved from http://climate.nasa.gov/nasa_role/science/

National Academy of Engineering. (n.d.). *Make solar energy economical.* Retrieved from http://www.engineeringchallenges.org/challenges/solar.aspx

National Energy Institute. (2015, July). *Fact sheet, quick facts: Nuclear energy in America.* Retrieved from http://www.nei.org/Master-Document-Folder/Backgrounders/Fact-Sheets/Quick-Facts-Nuclear-Energy-in-America

New Home Wind Power. (n.d.). *Solar wind power.* Retrieved from http://www.newhomewindpower.com/solar-wind-power.html

New Home Wind Power. (n.d.). *Wind power generators.* Retrieved from http://www.newhomewindpower.com/wind-power-generators.html

Nuclear Energy Institute. (n.d.). *How nuclear reactors work.* Retrieved from http://www.nei.org/Knowledge-Center/How-Nuclear-Reactors-Work

Nuclear Fission. (n.d.). *Hyper physics.* Retrieved from http://hyperphysics.phy-astr.gsu.edu/hbase/nucene/fission.html

Nuclear Fusion. *Hyper physics.* Retrieved from Hyper Physics at http://hyperphysics.phy-astr.gsu.edu/hbase/nucene/fusion.html

Planete Energies. (2015, February 4). *The two types of solar energy, photovoltaic and thermal.* Retrieved from http://www.planete-energies.com/en/medias/close/two-types-solar-energy-photovoltaic-and-thermal

Pure Energies. (n.d). *How solar panels work.* Retrieved from http://pureenergies.com/us/how-solar-works/how-solar-panels-work/

Rice unveils super-efficient solar-energy technology. (2012, November 19). *Rice University News & Media News Release.* Retrieved from http://news.rice.edu/2012/11/19/rice-unveils-super-efficient-solar-energy-technology/

Runyon, J. (2011, September 28). Solar shakeout continues: Stirling energy systems file for Chapter 7 Bankruptcy. *RenewableEnergyWorld.com.* Retrieved from http://www.rcnewableenergyworld.com/articles/2011/09/solar-shakeout-continues-stirling-energy-systems-files-for-chapter-7-bankruptcy.html

Science Daily. (n.d.). *Nuclear fusion*. Retrieved from https://www.sciencedaily.com/terms/nuclear_fusion.htm

Solar energy pros and cons. *Conserve energy future*. Retrieved from http://www.conserve-energy-future.com/pros-and-cons-of-solar-energy.php

Solar power passes 1% global threshold. (n.d.). *Clean Technica*. Retrieved from http://cleantechnica.com/2015/06/12/solar-power-passes-1-global-threshold/

Stronberg, J. B. (2016, June 24). *Natural gas: Bridge or barrier to a clean energy future?* Retrieved from http://www.renewableenergyworld.com/articles/2016/06/natural-gas-bridge-or-barrier-to-a-clean-energy-future.html

The week. (n.d.). *What is fracking and why is it so controversial?* Retrieved from http://www.theweek.co.uk/fracking/62121/what-is-fracking-and-why-is-it-so-controversial

Total energy consumption. (2015). *Enerdata global energy statistical yearbook 2016*. Retrieved from https://yearbook.enerdata.net/#energy-consumption-data.html

Union of Concerned Scientists. (n.d.). *How geothermal energy works*. Retrieved from http://www.ucsusa.org/clean_energy/our-energy-choices/renewable-energy/how-geothermal-energy-works.html#.V3AO36KmC74

United Nations. (n.d.). *Kyoto protocol 10th anniversary timely reminder climate agreements work*. Retrieved from http://newsroom.unfccc.int/unfccc-newsroom/kyoto-protocol-10th-anniversary-timely-reminder-climate-agreements-work/

United Nations. (n.d.). *Kyoto protocol*. Retrieved from http://unfccc.int/kyoto_protocol/items/2830.php

United States Department of Energy. (2004). *Geothermal technologies program*. Retrieved from http://www.nrel.gov/docs/fy04osti/36316.pdf

United States Department of the Interior, Bureau of Land Management, Utah. (n.d.). *Wind energy environmental impacts*. Retrieved from http://www.blm.gov/ut/st/en/prog/energy/wind_energy/wind_energy_environmental.print.html

United States Energy Information Administration. (2016, March 23). *Annual coal report*. Retrieved from http://www.eia.gov/coal/annual/

United States Energy Information Administration. (n.d.). *Coal explained*. Retrieved from http://www.eia.gov/energyexplained/index.cfm?page=coal_reserves

United States Energy Information Administration. (n.d.). *Oil: Crude and petroleum products explained*. Retrieved from http://www.eia.gov/energyexplained/index.cfm?page=oil_home#tab2

United States Energy Information Administration. (n.d.). *Solar explained, solar thermal power plants*. Retrieved from http://www.eia.gov/energyexplained/?page=solar_thermal_power_plants

United States Geological Survey. (n.d.) *Earthquakes induced by fluid injection FAQs*. Retrieved from https://www2.usgs.gov/faq/categories/9833/3428

United States Nuclear Regulatory Commission. (n.d.). *About NRC*. Retrieved from http://www.nrc.gov/about-nrc.html

University of Gothenburg. (2015, September 23). *Small-scale nuclear fusion may be a new energy source*. Retrieved from http://www.gu.se/english/about_the_university/news-calendar/News_detail//small-scale-nuclear-fusion-may-be-a-new-energy-source.cid1323710

University of Texas at Dallas. (2010, February 26). *Nanotube thermocells hold promise as energy source*. Retrieved from UTD at http://www.utdallas.edu/news/2010/2/26-1381_Nanotube-Thermocells-Hold-Promise-as-Energy-Source_article.html

What is a supernova? (n.d.). *Space.com*. Retrieved from Space.com at http://www.space.com/6638-supernova.html

What's your impact. (n.d.). *What are greenhouse gases?* Retrieved from http://whatsyourimpact.org/greenhouse-gases

Williams, B. (2014, May 19). *The 3 challenges solar energy needs to overcome to continue its growth.* Retrieved from http://solarenergy.net/News/3-challenges-solar-energy-needs-overcome-continue-growth/

Wind Energy Development Programmatic Environmental Impact Statement. (n.d.). *Wind Energy basics.* Retrieved from http://windeis.anl.gov/guide/basics/

Wind Energy Foundation. (n.d.). *History of wind energy.* Retrieved from http://windenergyfoundation.org/about-wind-energy/history/

World Coal Association. (n.d.). *Coal mining.* Retrieved from http://www.worldcoal.org/coal/coal-mining

World Energy Council. (2011). *Hydropower.* Retrieved from https://www.worldenergy.org/data/resources/resource/hydropower/

World Energy Outlook. (n.d.). *Energy access database.* Retrieved from http://www.worldenergyoutlook.org/resources/energydevelopment/energyaccessdatabase/

World Energy Outlook. (n.d.). *Factsheets 2006-2015.* Retrieved from http://www.worldenergyoutlook.org/factsheets/

World Nuclear Association. (2015, September). *Thorium.* Retrieved from http://www.world-nuclear.org/information-library/current-and-future-generation/thorium.aspx

World Nuclear Association. (2016, February). *Radioactive wastes—Myths and realities.* Retrieved from http://www.world-nuclear.org/information-library/nuclear-fuel-cycle/nuclear-wastes/radioactive-wastes-myths-and-realities.aspx

World Nuclear Association. (2016, January). *Nuclear power in the world today.* Retrieved from http://www.world-nuclear.org/information-library/current-and-future-generation/nuclear-power-in-the-world-today.aspx

World Nuclear Association. (2016, May). *Nuclear fusion power.* Retrieved from http://www.world-nuclear.org/information-library/current-and-future-generation/nuclear-fusion-power.aspx

Yan, S. (2015, October 14). *China has a bigger middle class than America.* Retrieved from http://money.cnn.com/2015/10/14/news/economy/china-middle-class-growing/

What's your impact. (n.d.). *What are greenhouse gases?* Retrieved from http://whatsyourimpact.org/greenhouse-gases

Williams, B. (2014, May 19). *The 3 challenges solar energy needs to overcome to continue its growth.* Retrieved from http://solarenergy.net/News/3-challenges-solar-energy-needs-overcome-continue-growth/

Wind Energy Development Programmatic Environmental Impact Statement. (n.d.). *Wind Energy basics.* Retrieved from http://windeis.anl.gov/guide/basics/

Wind Energy Foundation. (n.d.). *History of wind energy.* Retrieved from http://windenergy foundation.org/about-wind-energy/history/

World Coal Association. (n.d.). *Coal mining.* Retrieved from http://www.worldcoal.org/coal/coal-mining

World Energy Council. (2011). *Hydropower.* Retrieved from https://www.worldenergy.org/data/resources/resource/hydropower/

World Energy Outlook. (n.d.). *Energy access database.* Retrieved from http://www.worldenergy outlook.org/resources/energydevelopment/energyaccessdatabase/

World Energy Outlook. (n.d.). *Factsheets 2006-2015.* Retrieved from http://www.worldenergy outlook.org/factsheets/

World Nuclear Association. (2015, September). *Thorium.* Retrieved from http://www.world-nuclear.org/information-library/current-and-future-generation/thorium.aspx

World Nuclear Association. (2016, February). *Radioactive wastes—Myths and realities.* Retrieved from http://www.world-nuclear.org/information-library/nuclear-fuel-cycle/nuclear-wastes/radioactive-wastes-myths-and-realities.aspx

World Nuclear Association. (2016, January). *Nuclear power in the world today.* Retrieved from http://www.world-nuclear.org/information-library/current-and-future-generation/nuclear-power-in-the-world-today.aspx

World Nuclear Association. (2016, May). *Nuclear fusion power.* Retrieved from http://www.world-nuclear.org/information-library/current-and-future-generation/nuclear-fusion-power.aspx

Yan, S. (2015, October 14). *China has a bigger middle class than America.* Retrieved from http://money.cnn.com/2015/10/14/news/economy/china-middle-class-growing/